T0262187

Environmental Pollution and Control

Environmental Pollution and Control

Edited by **Alfred Muller**

New York

Published by Callisto Reference,
106 Park Avenue, Suite 200,
New York, NY 10016, USA
www.callistoreference.com

Environmental Pollution and Control
Edited by Alfred Muller

© 2016 Callisto Reference

International Standard Book Number: 978-1-63239-627-3 (Hardback)

This book contains information obtained from authentic and highly regarded sources. Copyright for all individual chapters remain with the respective authors as indicated. All chapters are published with permission under the Creative Commons Attribution License or equivalent. A wide variety of references are listed. Permission and sources are indicated; for detailed attributions, please refer to the permissions page and list of contributors. Reasonable efforts have been made to publish reliable data and information, but the authors, editors and publisher cannot assume any responsibility for the validity of all materials or the consequences of their use.

The publisher's policy is to use permanent paper from mills that operate a sustainable forestry policy. Furthermore, the publisher ensures that the text paper and cover boards used have met acceptable environmental accreditation standards.

Trademark Notice: Registered trademark of products or corporate names are used only for explanation and identification without intent to infringe.

Printed in the United States of America.

Contents

Preface

Environmental pollution is a growing concern of environmentalists, scientists and ecologists around the globe. It refers to the contamination of natural resources due to harmful chemicals, paint, plastic and other pollutants. The effects of environmental pollution are becoming very severe and therefore, its immediate control is a matter of utmost importance. This book covers in detail some existent theories and innovative concepts revolving around environmental pollution control. It is a compilation of topics, ranging from the basic to the most complex advancements in this field. It is appropriate for students seeking detailed information in this area as well as for experts. Thus, it will serve as a valuable source of reference for readers and contribute to the progress of this field.

The information shared in this book is based on empirical researches made by veterans in this field of study. The elaborative information provided in this book will help the readers further their scope of knowledge leading to advancements in this field.

Finally, I would like to thank my fellow researchers who gave constructive feedback and my family members who supported me at every step of my research.

Editor

Aerobic treatment of selective serotonin reuptake inhibitors in landfill leachate

Ove Bergersen[1*], Kine Østnes Hanssen[2] and Terje Vasskog[2,3]

Abstract

Background: Pharmaceuticals used in human medical care are not completely eliminated in the human body and can enter the municipal sewage sludge system and leachate water from landfill both as the parent compound and as their biologically active metabolites. The selective serotonin reuptake inhibitors (SSRIs) have a large potential for unwanted effects on nontarget organisms in the environment. Leachates from active or old closed landfills are often treated with continuous stirring and simple aeration in a pond/lagoon before infiltration into the environment. The aim of this work was to simulate the reduction of five SSRIs (citalopram, fluoxetine, paroxetine, sertraline and fluvoxamine) and three of their metabolites (desmethylcitalopram, didesmethylcitalopram and norfluoxetine) during aerobic treatment of leachate from landfills. This landfill leachate-simulation experiment was performed to see what happens with the pharmaceuticals during aerated treatment and continuous stirring of landfill leachate for 120 h. It is important to establish whether different pollutants such as pharmaceuticals can be removed (oxidized or otherwise degraded) or not before infiltration into the environment.

Results: All the SSRIs had a significant concentration reduction during the aeration treatment process. Total SSRI concentrations were reduced significantly during aerobic treatment, and the individual SSRIs were reduced by 89% to 100% after 120 h. Among the high-concentration samples, fluoxetine (10 mg L^{-1}) was the least degraded with 93% concentration reduction. Among the low-concentration samples, paroxetine was the least degraded with 89% concentration reduction. Fluvoxamine and citalopram were most effectively eliminated and were completely removed from both the high- and low-concentration samples. The samples were also investigated for the metabolites desmethylcitalopram, didesmethylcitalopram and norfluoxetine but only norfluoxetine in the high-concentration fluoxetine sample was detected.

Conclusions: Our results suggest that aeration is an effective method for eliminating pharmaceuticals such as SSRIs from landfill leachate water. Comparing the results of all SSRIs with different treatment methods, paroxetine and fluvoxamine seem to be the easiest compounds to eliminate independent of method, while fluoxetine and sertraline seem to be the most stable.

Keywords: Aeration; Biological treatment; Environment; Landfill leachate; Pharmaceuticals; SSRI

Background

For several years, the occurrence of pharmaceuticals and personal care products (PPCPs) in the aquatic environment has been recognized as one of the emerging issues in environmental risk assessment [1-4]. Pharmaceuticals used in human medical care are not completely eliminated in the human body and can enter the municipal sewage sludge system both as the parent compound and as their biologically active metabolites. These compounds have a large potential for unwanted effects on nontarget organisms in the environment. Several investigators have examined the removal of pharmaceutical compounds during the passage through municipal sewage treatment systems [5-8]. Effluent concentrations and elimination rates for different compounds vary significantly. Various pharmaceutical compounds have been detected in concentrations up to $\mu g \ L^{-1}$ in sewage effluents, downstream of sewage treatment plants and in surface and groundwater [2,9,10]. In addition to sewage treatment plants, landfill sites where unused drugs and different

* Correspondence: ove.bergersen@bioforsk.no
[1]Norwegian Institute for Agricultural and Environmental Research (Bioforsk), Soil and Environment Division, Fredrik A Dahls vei 20, N-1432 Ås, Norway
Full list of author information is available at the end of the article

personal care products have been disposed may release these compounds into the environment. There are several old landfill sites that were established when the disposal of such compounds through garbage was tolerated, and drugs have been detected in leachates from such municipal landfills [11]. Leachates from landfill sites are also known to penetrate down to the groundwater causing pollution [12].

Six selective serotonin reuptake inhibitors (SSRIs) are on the Norwegian market today, fluoxetine, fluvoxamine, paroxetine, sertraline, citalopram and escitalopram (the pure S-enantiomer of citalopram). SSRIs are typically used as antidepressants and are used to treat conditions such as depression, anxiety disorders and some personality disorders. The total consumption of SSRIs has slowly increased in Norway the last years, although when adjusted to population growth, the number of SSRI users has been relatively stable at just below 4% of the population. Citalopram/escitalopram is most widely distributed and constitutes approximately 591 kg of the total amount, while sertraline with its higher defined daily dosage is sold in the largest amounts contributing with approximately 621 kg [13]. SSRIs have been found in wastewater in Norway [8,14], and it is reasonable to believe that not all pharmaceuticals are used or delivered back to the pharmacies but might end up on landfill sites through private garbage disposal. Contaminants from landfill sites might reach the environment through leachate water run-off, and water soluble chemicals can reach the aquatic ecosystems or the groundwater. SSRIs have been shown to have a number of unwanted effects on aquatic organisms, such as the behavioural effect fluoxetine has on fish [15] and the induction of spawning and parturition in bivalves [16,17]. Fluoxetine is probably the most studied SSRI when it comes to effects on nontarget organisms, but effects from the other SSRIs have been found as well.

Depletion of SSRIs in sewage sludge has previously been found during an aerobic composting process [18]. Vasskog [18] showed that the depletion rate was highest for fluoxetine (1.23 mg (kg ash/day)$^{-1}$) and paroxetine (1.31 mg (kg ash/day)$^{-1}$) and lowest for citalopram (0.88 mg (kg ash/day)$^{-1}$). In addition, three out of four known SSRI metabolites were detected in all compost samples, and two of them showed a significant increase in concentrations during the composting period.

PPCPs have also been investigated under anaerobic digestion of sewage sludge [19-22], but SSRIs were not part of these investigations. Carballa et al. [19] observed high removal efficiencies for antibiotics, natural estrogens, musk and naproxen, while for example carbamazepine showed no elimination under anaerobic treatment of sewage sludge. Vieno et al. [20] also observed no elimination of carbamazepine in their anaerobic treatment of raw sewage, while fluoroquinolones were eliminated by >80%.

Beta-blockers were found to be eliminated less than 65%. Degradation was found for acetylsalicylic acid, while a mixture of degradation and abiotic removal mechanisms on other pharmaceuticals was observed during an anaerobic degradability study by Musson et al. [21]. The fate of hormones and pharmaceuticals during combined anaerobic treatment and nitrogen removal in black water has been investigated by de Graaff et al. [23]. They found that only a few compounds were partly removed, and anaerobic treatment was effective only to remove the majority of paracetamol. Falås et al. [24] have demonstrated that there are large variations in removal of different pharmaceuticals within the same pharmaceutical class as well in activated sludge. In this experiment, the nonsteroidal anti-inflammatory drugs (NSAIDs) ibuprofen and naproxen showed the highest removal rates, while diclofenac, mefenamic acid and clofibric acid showed little or no removal. Bergersen et al. [25] have shown that anaerobic treatment of sewage sludge gives a reduction in SSRI concentrations from 32% (fluoxetine) to 98% (citalopram) in a 24-day experiment, again demonstrating large variations within the same pharmaceutical class.

Eggen et al. [26] have described that PPCPs from old landfill leachate with treatment based on aeration and sedimentation may represent a significant source of concern for new and emerging pollutants in groundwater.

To our best knowledge, no studies have been conducted on SSRIs in treatment of aerated landfill leachate. When pharmaceuticals enter the environment through landfill leachate systems, they may either be found in the water phase or bound to particles. Landfill leachate aeration ponds could treat the water via aeration and particulate matter sedimentation.

Based on earlier results showing that the removal of SSRIs from sewage sludge is different during aerobic and anaerobic treatment, it was interesting to compare this with what would happen in the aeration process that landfill leachate goes through.

The aim of this work was to simulate the possible elimination of five SSRIs (citalopram, fluoxetine, paroxetine, sertraline and fluvoxamine) and three of their metabolites (desmethylcitalopram, didesmethylcitalopram and norfluoxetine) during aerobic treatment of leachate from landfills. Today, landfill leachates are treated with aeration in a pond/lagoon before infiltration into the environment. This landfill leachate-simulation experiment was performed to see what happens with the pharmaceuticals during aerated treatment and continuous stirring of landfill leachate for 120 h. This experiment mimics the common process used for treatment of landfill leachate ponds and aims to predict the efficiency of pharmaceutical removal in aerated leachate ponds. It is important to establish whether different pollutants such as pharmaceuticals can be removed or not before infiltration into the environment.

Results and discussion

Aerobic treatment and oxygen consumption

Due to the use of the pharmaceutical compounds as a carbon source combined with the carbon source in the leachate, we followed the changing concentrations of each SSRI in flasks with continuous stirring. Simultaneously, we followed the changing consumption of oxygen within the flasks headspace through 120 h (Figure 1). In landfill leachate, prescription and nonprescription pharmaceutical concentrations generally range from 100 to 10,000 ng L^{-1} [27]. The concentrations used in this experiment are somewhat higher than what is found before to enhance the effects seen on the experimental outcomes but are still regarded as environmentally relevant. Each type of SSRI with different concentrations could also give different oxygen consumption when present in landfill leachate, which again will be valuable information about the aerobic treatment of the landfill leachate.

Table 1 presents the ThOD for each SSRI, the total OC after 48 and 120 h and the measured OC minus the control sample (oxygen consumption of leachate) without SSRI. All the calculated ThODs show lower values

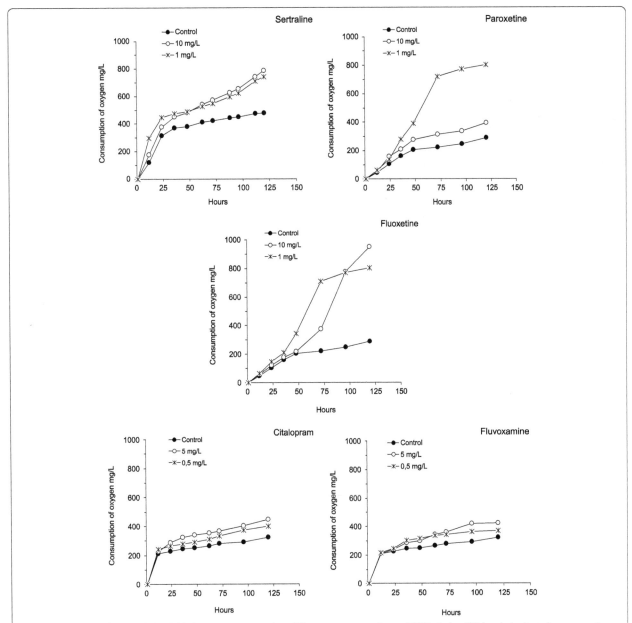

Figure 1 Total OC after 120 h in Voith Sapromat exposed to different concentrations of SSRIs in landfill leachate. Sertraline, paroxetine and fluoxetine 10 and 1 mg L^{-1}; fluvoxamine and citalopram 5 and 0.5 mg L^{-1}. Measured background respiration as control from leachate without SSRIs is included.

Table 1 Theoretical and measured concentration of each SSRI spiked in leachate water before an aerobic treatment

Source	Theoretical SSRI mg L⁻¹	Measured SSRI mg L⁻¹	(ThOD) SSRI mgO2 L⁻¹	Measured 48 h Tot. OC mgO2 L⁻¹	Measured 120 h Tot. OC mgO2 L⁻¹	Measured 48 h OC* mgO2 L⁻¹	Measured 120 h OC* mgO2 L⁻¹
Fluvoxamine	5	6	11.0	300	424	49	99
	0.5	0.61	1.1	317	370	66	55
Citalopram	5	3.2	13.1	340	445	89	120
	0.5	0.39	1.3	290	401	39	76
Sertraline	10	10	30.2	482	787	101	311
	1	1.33	3.0	488	740	107	264
Fluoxetine	10	11	22.8	340	950	135	661
	1	0.84	2.3	290	802	85	513
Paroxetine	10	14	22.4	278	395	73	106
	1	1.31	2.2	393	803	188	514

Measured oxygen consumption (Tot. OC) and OC* (Equation 1) after 48 and 120 h with each separate SSRI is compared with theoretical oxygen demand (ThOD) of SSRI tablets. The results of measured total OC are given as mean values ($n = 2$). *Equation 1.

than the measured OC after Equation 1 for each SSRI. All flasks spiked with the different SSRI tablets show dissimilar oxygen consumption, although considerably higher than the ThOD for each SSRI. This increased oxygen consumption in the flasks might be due to (i) co-metabolism between each SSRIs plus tablets excipients and low decomposable carbon sources in the leachate, which was not accessible alone in the control flask; (ii) microbial co-metabolism between SSRIs and the carbon source from the excipients in the tablets alone. Each tablet mixture used contained approximately 20% SSRIs and 80% excipients. Each SSRI tablet contains a large number of additives, e.g. carbon sources such as cellulose, starch and glycerol.

The high induced oxygen consumption observed in the flasks with paroxetine or fluoxetine in leachate substrate is most likely due to the excipients (Figure 1 and Table 1). Both these SSRI tablets have a higher amount of excipients than the other three SSRIs. The increased amount of excipient gave considerably higher oxygen consumption compared to the control samples and the other flasks spiked with fluvoxamine, citalopram or sertraline. Figure 1 and Table 1 also show that each SSRI gave an increased oxygen consumption compared to background consumption in the control flasks after a minimum of 50 h independent of SSRI concentration. After 120 h fluvoxamine shows less oxygen consumption in the flask with 0.5 mg L⁻¹, while the other SSRIs show increased oxygen consumption (Table 1). The diluted leachate solution without SSRIs had an initial pH of 7.1, and when spiking the samples with SSRIs, the pH increased in all samples due to the basic nature of the SSRIs (Table 2). The pH at start in

Table 2 SSRI concentrations and pH at the beginning and after 120 h of the experiment

SSRI	Measured at start SSRI mg L⁻¹	STDEV mg L⁻¹	Rel. STDEV %	pH	Measured at end SSRI mg L⁻¹	STDEV mg L⁻¹	Rel. STDEV %	pH	Reduction %
Fluvoxamine	6.0	1.104	18	8.7	0.006	0.004	68	8.4	100
	0.6	0.123	20	8.6	0.002	0.002	87	8.3	100
Citalopram	3.2	0.102	3	9.1	0.005	0.001	16	8.5	100
	0.4	0.026	7	9.0	0.002	0.001	11	8.6	100
Sertraline	10.0	1.227	12	8.6	0.355	0.046	13	7.9	96
	1.3	0.171	13	8.5	0.038	0.017	45	7.7	97
Fluoxetine	11.0	0.873	8	9.2	0.805	0.063	8	8.1	93
	0.8	0.156	18	9.0	0.003	0.001	14	8.0	100
Norfluoxetine	n.d.				0.006	0.001	11		
	n.d.			n.d	n.d.				
Paroxetine	14.0	0.163	1	9.8	0.105	0.011	11	8.0	99
	1.3	0.143	11	9.6	0.141	0.003	2	7.9	89

n.d., not detected.

the samples varied between 8.5 (sertraline, low concentration) and 9.8 (paroxetine, high concentration). During the 120-h experiment, the pH was found to decrease in all samples, and the exact values can be seen in Table 2. This decrease in pH together with high oxygen consumption could support an explanation that the SSRIs spiked in the leachate have gone through an oxidation or are otherwise reduced in concentration.

Earlier SSRIs have been investigated under aerobic and anaerobic treatment processes of sewage waste to see if the SSRIs deplete or accumulate. Vasskog et al. [18] showed that all the SSRIs had a significant decrease in concentration during aerobic treatment in a composting process. Bergersen et al. [25] showed that paroxetine and citalopram were almost completely reduced during anaerobic treatment of sewage sludge while the lower reduction of fluoxetine, fluvoxamine and sertraline indicate that these three compounds have a higher potential for accumulation than citalopram and paroxetine under anaerobic conditions. A review describing removal or sorption of pharmaceuticals and other personal care products in water treatment such as natural wetland systems concludes that these systems have a high potential for removal of such chemicals [28]. One SSRI (fluoxetine) has been found to be removed from wetland systems by sorption into duckweed [29]. To our best knowledge, no studies have been conducted on SSRI depletion in landfill leachate ponds treated with aeration.

Analysis and reduction of selective serotonin reuptake inhibitors

The five selective serotonin reuptake inhibitors citalopram, sertraline, paroxetine, fluvoxamine and fluoxetine were analysed at the beginning and at the end of a 120-h experiment in a Voith Sapromat respirometer to investigate their concentration reduction during the process. The concentrations were calculated as mg L^{-1} and compared with the total theoretical amount of each SSRI if no reduction had occurred in the samples. Table 1 shows the theoretical concentration, measured concentration and recovery of each SSRI before treatment. After treatment for 120 h, all the SSRIs were measured at lower concentrations at the end of the experiment. The results shown in Table 2 give a detailed view of the concentration of each SSRI and the depletion throughout the experiment. All compounds showed a reduction in concentration of 89% to 100%, independent of high- or low-start concentration. The recovery was measured at the beginning of the experiment and was generally high (between 64% and 140%). The high recoveries of the SSRIs also imply that there is low or no binding to particles and humic acids in this experimental setup. There was no possibility to measure the recovery at the end of the experiment, but it is expected to be equal to that at the beginning. The highest recoveries (above 100%) might be explained partly by the standard deviation but maybe more likely by the fact that the added SSRI is taken from crushed tablets. The weighed amount of the tablet might not have been homogenous and the sample withdrawn not representative. However, the main outcome of the experiment would be the same, showing the concentration reduction of the SSRIs during the experiment. There are small differences between the recovery of the high and low concentrations at the beginning of the experiment, which also indicate that the recovery should be more or less equal at even lower concentrations. Therefore, it is most likely that the low concentrations found at the end of the 120-h treatment were due to depletion of the compounds and not low recovery.

Likewise, their metabolites desmethylcitalopram, didesmethylcitalopram and norfluoxetine were analysed from the treated samples to determine whether they were formed during the experiment. Only one metabolite was found in the samples, and norfluoxetine was detected in a very low concentration (0.006 mg L^{-1}) in the sample with an initially high fluoxetine concentration (Table 2).

The experiment is set up to simulate the conditions in a landfill leachate pond, and it is reasonable to believe that the pharmaceuticals will have a similar degradation profile in a full-size treatment system treated with aeration for days or weeks.

The results in this experiment differ from earlier experiments, where the same SSRIs were measured during an aerobic composting and anaerobic sewage sludge treatment [8,25]. By comparing the results from the aerobic and anaerobic experiment with this experiment, both similarities and differences are found.

Table 3 illustrates a comparison of three different treatment processes that show reduction efficiency and reduction rate for all five SSRIs. First: All SSRIs show considerable higher reduction rate in an aqueous aerated media (Table 3). Second: In the aerobic composting process [18], fluvoxamine and paroxetine had the highest reduction rates, while in the anaerobic experiment

Table 3 Comparison of the reduction rates of SSRIs with different treatment methods

SSRI	Aerobic sludge composting [18]	Anaerobic sludge treatment [25]	Aeration treatment landfill leachate
	Reduction rate (% per 21 days)	Reduction rate (% per 24 days)	Reduction rate (% per 5 days)
Fluvoxamine	88	53	100
Citalopram	26[a]	85	100
Sertraline	46[a]	38	96
Fluoxetine	35[a]	32	93[a]
Paroxetine	46	98	90

[a]Metabolites detected.

[25], paroxetine still had a high reduction rate while fluoxetine had a very low reduction rate. The metabolite norfluoxetine was detected in both aerobic experiments while it was absent after anaerobic treatment. Third: Citalopram was the most stable SSRI in the aerobic composting experiment, while in the anaerobic sewage sludge and in this aerated treatment experiment, it was quickly reduced in concentration. Based on this experiment, we can only assume that the differences in stability in these three experiments might be due to the different bacteria present in aerobic and anaerobic systems, but it could also originate from different chemical reactions in the three systems since we cannot determine whether the concentration reduction is due to chemical or biological degradation.

We are limited in the extent to which it is possible to compare three functionally different systems, especially when we consider the variability in pharmaceutical stability under different conditions. Table 3 clearly shows that the availability factor for degradation of each SSRI is higher in aqueous medium compared to solid-phase medium like the composting system. SSRI degradation in a solid-phase medium also gave detection of three different metabolites from five different SSRIs after the biological treatment. Independent of anaerobic or aeration treatment, SSRIs in aqueous solution gave considerably higher reduction rates.

Conclusions

Selective serotonin reuptake inhibitors can be effectively reduced in concentration by aerobic treatment of landfill leachates. The concentration reduction was close to complete for all compounds in both concentrations during the experimental time of 120 h. Only one of the known human metabolites (norfluoxetine) was found in the aeration samples, which suggests other degradation products than those formed by the human metabolism. Comparing the results of all SSRIs with different treatment methods, paroxetine and fluvoxamine seem to be the easiest compounds to eliminate independent of method, while fluoxetine and sertraline seem to be the most stable.

Our results suggest that aeration is an effective method for minimising the impact of pharmaceuticals such as SSRIs from landfill leachates water. However, further studies are required to determine what kind of degradation products are formed, and based on such results, preventive actions in the form of more sophisticated water treatment systems can be made.

Methods
Aerobic treatment experiment
Leachate from an active landfill (3 L) was diluted with water (10 L), and the composition of the leachate is given in Table 4. Table 4 shows the content of the leachate with carbon source as total organic carbon (TOC), organic acids and Total-N that will enhance microbial activity. The strategy of the experiment was to measure the microbial respiration in the leachate solution spiked with SSRI tablets and compare it to control flasks with diluted leachate water without addition of tablets.

Pharmaceutical pollution from landfill leachates will be as unused tablets or from tablet packages. Therefore, the stock solution was made up of crushed tablets of each SSRI that was dissolved in 30 mL methanol (MeOH, Merck, Darmstadt, Germany) and then diluted with 270 mL distilled water, stirred and heated for 2 h at 35°C and stored at 4°C until used. Diluted leachate in 500-mL flasks was spiked with a stock solution of the different SSRIs. The theoretical starting concentration was 5 and 0.5 mg L^{-1} for fluvoxamine and citalopram and 10 and 1 mg L^{-1} for sertraline, fluoxetine and paroxetine.

The flasks with leachate water were incubated in a water bath in a Voith Sapromat respirometer (Voith, Heldenheim/Brenz, Gemany) [30]. The Voith Sapromat continuously measure consumption of oxygen in solutions and the developed CO_2 gas in the flask is trapped with NaOH (Merck). The experiment was performed in duplicates, and the samples were incubated for 120 h at 20°C in darkness. The samples were stirred continuously in closed flasks with a magnet. Total oxygen consumption (Tot. OC) was calculated from the whole tablet mixture with all excipients, with the background respiration from the leachate after 48 and 120 h, while the

Table 4 Chemical analysis of leachate water after dilution used in the Voith Sapromat respirometer experiment

Parameter	Units	Raw material	Diluted leachate
TOC	mg L^{-1}	773	178
Tot-N	mg L^{-1}	322	74
Tot-S	mg L^{-1}	21	4.8
SO4	mg L^{-1}	0.3	0.1
Fe	mg L^{-1}	16	3.7
Formic acid	mg L^{-1}	31	7.2
Acetic acid	mg L^{-1}	5.6	1.3
Propionic acid	mg L^{-1}	1.1	0.25
Butanoic acid	mg L^{-1}	1.5	0.3
Pentanoic acid	mg L^{-1}	3.2	0.7
Hexanoic acid	mg L^{-1}	2.4	0.6
Heptanoic acid	mg L^{-1}	1.3	0.3
DL- lactat	mg L^{-1}	5.5	1.3
BOD 7	mg L^{-1}	430	
Redox potential	mV	−148	
pH		7.1	6.9

oxygen consumption (OC)* (Equation 1) was calculated from the whole tablet mixture with all excipients, with the background respiration from the leachate (Equation 1) subtracted. Theoretical oxygen demand (ThOD) for each SSRI was calculated using the Buswell equation [31]. The SSRI tablets contain a large number of additives, e.g. carbon sources such as cellulose, starch and glycerol. The tablet mixture used in our experiment contains approximately 20% SSRIs and 80% excipients.

$$Measured\ OC_j = Measured(OC\ h_j\text{-}OC\ h, control) \quad (1)$$
$$(mg\ oxygen\ L^{-1})$$
$$OC\ hours_j = Measured\ oxygen\ consumption\ of\ each$$
$$SSRI\ (mg\ oxygenL^{-1})$$
$$OC\ hours,\ control = Measured\ oxygen\ consumption$$
$$leachate\ (mg\ oxygenL^{-1})$$

Chemicals and solvents

Standards of the SSRIs citalopram (1-[3-(dimethylamino) propyl]-1-(4-fluorophenyl)-1,3-dihydro- 5-isobenzofuran carbonitrile), sertraline ((1S-cis)-4-(3,4-dichlorophenyl)-1,2,3,4-tetrahydro-N-methyl-1-naphtylamine), paroxetine (trans-(–)-3-[(1,3-benzo-dioxol-5-yloxy) methyl]benzene propanamine), fluvoxamine (5-methoxy-1-[4-(trifluoro methyl)-phenyl]-1-pentanone-O-(2-aminoethyl)oxime) and fluoxetine ((±)-N-methyl-γ-[4(trifluoromethyl)phenoxy] benzenepropanamine), and the metabolites desmethyl citalopram (1-(4-fluorophenyl)-1,3-dihydro-1-[3-(methy lamino)propyl]-5-isobenzofurancarbonitrile), didesmet hylcitalopram (1-(3-aminopropyl)-1-(4-fluorophenyl)-1,3-dihydro-5-isobenzofurancarbonitrile) and norfluoxe-tine (γ-[4-(trifluoromethyl)phenoxy]-benzenepropanamine),

and the deuterated standards D3-sertraline, D4-paroxetine, D5-fluoxetine and D6-citalopram were purchased from Toronto Research Chemicals (Toronto, ON, Canada). The mass spectrometric identification criteria, recoveries, LOD and LOQ and pKa-values for the compounds are given in Table 5. pKa values are calculated by Advanced Chemistry Development (ACD/Labs) Software V8.14 for Solaris, obtained from SciFinder Scholar 2010.

The isotope labelled compounds were used as internal standards (IS) for quantification. All standards were delivered with purity above 98%. Escitalopram, which is the pure S-enantiomer of citalopram, is not possible to distinguish from the R-enantiomer with our analytical techniques, and these enantiomers are measured as the same compound.

The standards were dissolved in HPLC-grade MeOH from Merck in concentrations of 100 or 200 μg mL^{-1}, stored at –18°C in the dark and used as stock solutions. From this stock solution, the daily work solutions were made by dilution with purified water.

Solvents for HPLC were HPLC-grade acetonitrile from Merck, purified water obtained from a MilliQ purification unit from Millipore (Bedford, MA, USA) and pro-analysis-grade formic acid from Merck. The same water and formic acid were used for liquid-phase microextraction (LPME), while dihexyl ether (purum ≥ 97%) from FLUKA (Buchs, Switzerland) was used as organic solvent to fill the pores of the fibre.

Chemical analysis

The samples were mixed with deionized water (ratio 1:2 by volume) and the pH was measured 30 min after mixing with a Ross electrode (Orion Instruments, Baton Rouge, USA). The redox potential in the leachate was

Table 5 Identification criteria, extraction recoveries, LOD, LOQ, pKa and mass spectrometric detection parameters for the different SSRIs

Compound	pKa	[M + H] + (measured)	Fragment (measured)	Cone voltage	Collision energy	Retention time (min)	LOD (pg L^{-1})	LOQ (pg L^{-1})	Extraction recovery (%)
Citalopram	9.59	325.5	108.6	35	25	3.3	20	60	64/78[a]
Desmethylcitalopram	9.41	311.1	108.5	28	25	3.1	230	760	-
Didesmethylcitalopram	10.14	297.1	108.5	29	20	2.9	200	780	-
Citalopram-D6 (IS)	-	331.2	108.6	41	27	3.3	-	-	-
Paroxetine	10.05	330.2	69.6	40	22	4.1	50	205	140/131[a]
Paroxetine-D4 (IS)	-	334.2	73.7	40	29	4.1	-	-	-
Fluvoxamine	9.39	319.1	70.6	24	17	4.8	130	380	120/122[a]
Fluoxetine	10.05	310.1	43.8	22	15	6.0	150	490	110/84[a]
Norfluoxetine	9.05	296.1	133.7	15	5	5.5	160	540	-
Fluoxetine-D5 (IS)	-	315.2	43.8	20	12	6.0	-	-	-
Sertraline	9.47	306.1	158.6	18	25	6.4	160	520	100/133[a]
Sertraline-D3 (IS)	-	309.1	274.8	17	12	6.4	-	-	-

[a]Extraction recoveries at high/low concentrations. LOD, limit of detection; LOQ, limit of quantitation.

analysed by an ATI ORION (Orion Instruments) and conductivity by a Phillips PW 9527 digital meter (Philips, Eindhoven, Netherlands).

Inorganic parameters of the leachate illustrated in Table 4 were analysed by an accredited laboratory (Eurofins AS) using Norwegian standards regarding total nitrogen (Tot-N), TOC, biochemical oxygen demand (BOD7), total sulphur (Tot-S), SO_4 and iron [32]. The leachate water was also characterized with respect to content of short-chained organic acids. The organic acids in the leachate water used in the experiment were alkaline extracted by adding 0.1 M NaOH (Merck) to the samples. Concentrated HCl (pro analysis, Merck) was then added to the water phase and undissociated acids were then extracted with diethyl ether (>98%, Merck) and further derivatized by t-butyldimethylsilyl (>99% Sigma Aldrich, St. Louis, USA) and measured on a gas chromatograph-mass spectrometer (GC-MS) with electron impact ionization (EI) after a method described by Schooley et al. [33].

All samples taken during the experiment were stored in a freezer at −18°C before chemical extraction and analysis. Three replicate liquid-phase microextraction analyses of the SSRIs and metabolites were performed.

An already developed extraction method known as hollow fibre-supported liquid-phase microextraction (HF-LPME or just LPME) was employed to extract the SSRIs and their metabolites from the samples. The method development is described elsewhere [34], and only a brief description will be given here. Since the SSRI concentration in the leachate water samples was relatively high, and the LPME method has proved to be very sensitive for the SSRIs [8], the samples were diluted to a concentration in the high end of the calibration curve prior to extraction (200 ng L^{-1}). The dilution was done with deionized water to a total volume of 1.1 L. The isotope labelled internal standards was added in a concentration of 50 ng L^{-1}, and the pH was adjusted to 12 with 5 M NaOH (Merck) before the samples were extracted by LPME. Three replicates were analysed from each of the selected days.

A 28-cm-long plasmaphan polypropylene hollow fibre (Membrana, Wuppertal, Germany) was used for the extraction, and the pores of the fibre were filled with dihexyl ether by dipping the hollow fibre into the solvent for 10 s. The excess ether on the fibre surface was removed by dipping it in an ultrasonic bath for 2 to 3 s. Then the fibre lumen was filled with 20-μL acceptor phase (purified water/formic acid at pH 2), and the ends of the fibre were sealed with a thin metal wire. The same wire was used to keep the fibre hanging in the middle of the sample bottle. The samples were stirred for 2 h at 800 RPM. After extraction, the extracts were transferred to Waters (Milford, MA, USA) total recovery HPLC-vials and analysed using HPLC-MS/MS. The separation

was conducted on a Waters Acquity UPLC BEH C_{18} column (1.0 × 150 mm with 1.7-μm particles), and the mobile phases consisted of A: H_2O and 0.1% formic acid and B: 90% acetonitrile, 9.9% H_2O and 0.1% formic acid. The gradient started with a 1-min isocratic elution with 70% A and 30% B; from 1 to 7 min, there was a linear change in composition to 55% A and 45% B and finally a linear change from 7 to 12 min to 50% A and 50% B. The flow rate was set to 50 μL min^{-1}. The mass spectrometric detection was done on a triple quadrupole instrument (Waters Quattro Premier XE) in multiple reactions monitoring mode with positive electrospray ionization for optimal selectivity and sensitivity. The protonated molecular ion $[M + H]^+$ was used as a precursor ion for all compounds, and the most intense product ion was selected for quantification (see Table 2).

Calibration curves

Calibration curves were made in the concentration range 1 to 250 ng L^{-1}. The standard samples were set up and extracted in the same way as the leachate water samples. Standard samples were extracted and analysed throughout the experiment period to ensure that the calibration curves were constant over time. Four internal standards were used for quantification. Sertraline was quantified with D_3-sertraline as IS; paroxetine with D_4-paroxetine; fluvoxamine, fluoxetine and norfluoxetine with D_3-fluoxetine; and citalopram, desmethylcitalopram and didesmethylcitalopram with D_6-citalopram as IS. The linearity (R^2) of the method ranged between 0.9997 (paroxetine) and 0.9894 (fluvoxamine) and was considered as good for all compounds in this concentration range. The linearity of fluvoxamine was a little lower than for the rest of the compounds, most likely because no isotope labelled IS was available for this compound.

The concentration measurements were done on the basis of the peak area of the analyte divided by the peak area of the IS, and all concentrations were analysed in triplicate for each concentration.

Quality control

The identification criteria of the SSRIs are given in Table 5 and consist of correct retention time and the correct ion transition. Ideally, two or more product ions should be analysed in the mass spectrometric analysis, but for four of the five SSRIs, the second product ion is of too low intensity to be seen at the relevant concentrations; hence, only the most intense ion was used.

The methods limit of detection and limit of quantification was found by extracting and analysing a dilution series of standards in leachate water, and the limits are given in Table 5. Three parallels of each concentration were analysed.

All leachate water samples were extracted and analysed in triplicate. Between every third sample (nine injections), a blank sample of purified water was analysed to check for carry-over signals. No carry-over was detected.

Competing interests
The authors declare that they have no competing interests.

Authors' contributions
The authors OB and TV are responsible for the concept and design of the study. OB has also performed the aerobic treatment of the landfill leachate including the measurements of TOC, Tot-N, Tot-S, SO$_4$ and Fe, as well as writing of the manuscript. TV also developed the extraction and LC-MS methods used in the experiment, as well as result interpretation and writing of the manuscript. KØH performed the extractions of the samples, set up standard curves and calculated concentrations, as well as participated in discussions of methods and results. All authors read and approved the final manuscript.

Acknowledgements
Financial support for this project was provided by Bioforsk and the Norwegian Research Council through the 'Pharmatreat' project (project number 171738/S30).

Author details
[1]Norwegian Institute for Agricultural and Environmental Research (Bioforsk), Soil and Environment Division, Fredrik A Dahls vei 20, N-1432 Ås, Norway. [2]Department of Pharmacy, Faculty of Health Sciences, University of Tromsø, N-9037 Tromsø, Norway. [3]Norut (Northern Research Institute), N-9294 Tromsø, Norway.

References
1. Daughton CG, Ternes TA. Pharmaceuticals and personal care products in the environment: agents of subtle change. Environ Health Perspect. 1999;107:907–38.
2. Heberer T. Occurrence, fate, and removal of pharmaceutical residues in the aquatic environment: a review of recent research data. Toxicol Lett. 2002;131:5–17.
3. Sanderson H, Johnson DJ, Reitsma T, Brain RA, Wilson CJ, Solomon KR. Ranking and prioritization of environmental risks of pharmaceuticals in surface waters. Regul Toxicol Pharmacol. 2004;39:158–83.
4. Fent K, Weston AA, Caminada D. Ecotoxicology of human pharmaceuticals. Aquat Toxicol. 2006;76:122–59.
5. Ternes TA. Occurrence of drugs in German sewage treatment plants and rivers. Water Res. 1998;32:3345–260.
6. Simonich SL, Federle TW, Eckhoff WS, Rottiers A, Webb S, Sabaliunas D, et al. Removal of fragrance materials during U.S. and European wastewater treatment. Environ Sci Technol. 2002;36:2839–47.
7. Carballa M, Omil F, Lema JM, Llompart M, Garcia-Jares C, Rodriguez I, et al. Behavior of pharmaceuticals, cosmetics and hormones in a sewage treatment plant. Water Res. 2004;38:2918–26.
8. Vasskog T, Anderssen T, Pedersen-Bjergaard S, Kallenborn R, Jensen E. Occurrence of selective serotonin reuptake inhibitors in sewage and receiving waters at Spitsbergen and in Norway. J Chrom A. 2008;1185:194–205.
9. Jux U, Baginski RM, Arnold H-G, Krönke M, Seng PN. Detection of pharmaceutical contaminations of river, pond, and tap water from Cologne (Germany) and surroundings. Int J Hyg Environ Health. 2002;205:393–8.
10. Kolpin DW, Furlong ET, Meyer MT, Thurman EM, Zaugg SD, Barber LB, et al. Pharmaceuticals, hormones, and other organic wastewater contaminants in U.S. streams, 1999–2000: a national reconnaissance. Environ Sci Technol. 2002;36:1202–11.
11. Holm JV, Rügge K, Bjerg PL, Christensen TH. Occurrence and distribution of pharmaceutical organic compounds in the groundwater downgradient of a landfill (Grindsted, Denmark). Environ Sci Technol. 1995;29:1415–20.
12. Lapworth DJ, Baran N, Stuart ME, Ward RS. Emerging organic contaminants in groundwater: a review of sources, fate and occurrence. Environ Pollut. 2012;163:287–303.
13. The Norwegian Institute of Public Health. Report; Drug consumption statistics 2012:1 - drug consumption in Norway 2007–2011. ISBN-electronic: 978-82-8082-491-2.
14. Vasskog T, Berger U, Samuelsen P-J, Kallenborn R, Jensen E. Selective serotonin reuptake inhibitors in sewage influents and effluents from Tromsø, Norway. J Chrom A. 2006;1115:187–95.
15. Pittman JT, Ichikawa KM. iPhone® applications as versatile video tracking tools to analyse behaviour in zebrafish (Danio rerio). Pharmacol Biochem Behav. 2013;106:137–42.
16. Fong PP, Molnar N. Norfluoxetine induces spawning and parturition in estuarine and freshwater bivalves. Bull Environ Contamin Toxicol. 2008;81:535–8.
17. Bringolf RB, Heltsley RM, Newton TJ, Eads CB, Fraley SJ, Shea D, et al. Environmental occurrence and reproductive effects of the pharmaceutical fluoxetine in native freshwater mussels. Environ Toxicol Chem. 2010;29:1311–8.
18. Vasskog T, Bergersen O, Anderssen T, Jensen E, Eggen T. Depletion of selective serotonin reuptake inhibitors during sewage sludge composting. Waste Managem. 2009;29:2808–15.
19. Carballa M, Omil F, Ternes T, Lema JM. Fate of pharmaceutical and personal care products (PPCPs) during anaerobic digestion of sewage sludge. Water Res. 2007;41:2139–50.
20. Vieno N, Tuhkanen T, Kronberg L. Elimination of pharmaceuticals in sewage treatment plants in Finland. Water Res. 2007;41:1001–12.
21. Musson SE, Campo P, Tolaymat T, Suidan M, Townsend TG. Assessment of the anaerobic degradation of six active pharmaceutical ingredients. Sci Total Environ. 2010;408(9):2068–74.
22. Paterakis N, Chiu TY, Koh YKK, Lester JN, McAdam EJ, Scrimshaw MD, et al. The effectiveness of anaerobic digestion in removing estrogens and nonylphenol ethoxylates. J Hazard Mater. 2012;199–200:88–95.
23. de Graaff MS, Vieno NM, Kujawa-Roelveld K, Zeeman G, Temmink H, Buisman CJN. Fate of hormones and pharmaceuticals during combined anaerobic treatment and nitrogen removal by partial nitritation-anammox in vacuum collected black water. Water Res. 2011;45:375–83.
24. Falås P, Baillon-Dhumez A, Andersen HR, Ledin A, la Cour JJ. Suspended biofilm carrier and activated sludge removal of acidic pharmaceuticals. Water Res. 2012;46:1167–75.
25. Bergersen O, Hanssen KO, Vasskog T. Anaerobic treatment of sewage sludge containing selective serotonin reuptake inhibitors. Biores Technol. 2012;117:325–32.
26. Eggen T, Moeder M, Arukwe A. Municipal landfill leachate: a significant source for new and emerging pollutions. Sci Total Environ. 2010;408:5147–57.
27. Demas A, Kolpin D. Technical announcement: pharmaceuticals and other chemicals common in landfill waste. US Department of the Interior: US Geological Survey; 8/11/2014.
28. Haarstad K, Bavor HJ, Mæhlum T. Organic and metallic pollutants in water treatment and natural wetlands: a review. Water Sci Technol. 2012;65:176–99.
29. Reinholdt D, Vishwanathan S, Park JJ, Oh D, Saunders FM. Assessment of plant-driven removal of emerging organic pollutants by duckweed. Chemosphere. 2010;80:687–92.
30. Bouchez M, Blanchet D, Besnainou B, Leveau J, Vendercasteele JP. Kinetic studies of biodegradation of insoluble compounds by continuous determination of oxygen consumption. J Appl Microbiol. 1997;82:310–6.
31. Tchobanoglous G, Theisen H, Vigil S. Integrated solids waste management. New York: McGraw-Hill; 1993.
32. Norwegian standards (NS-EN ISO/IEC 17025): Tot N (modified NS 4743), TOC (AJ 31 /NS-EN 13137), BOD7 (AV110/NS 4758) Tot S (AVJ and AVI/ ICP-AES) SO$_4$ (AV10/ IC) and Fe by (AVJ and AVI/ ICP-AES). Metal analyses from the filter material by (AVI/ ICP-AES). http://www.standard.no/imaker.exe?id=4162. (accessed 2008).
33. Schooley DL, Kubiak FM, Evans JW. Capillary gas chromatographic analysis of volatile and non-volatile organic acids from biological samples as the t-butyldimetylsilyl derivatives. J Chrom Sci. 1985;23(9):385–90.
34. Ho TS, Vasskog T, Anderssen T, Jensen E, Rasmussen KE, Pedersen-Bjergaard S. 25,000-fold pre-concentration in a single step with liquid-phase microextraction. Anal Chim Acta. 2007;592:1–8.

2

A multi-disciplinarily designed mesocosm to address the complex flow-sediment-ecology tripartite relationship on the microscale

Holger Schmidt[1], Moritz Thom[1], Kerstin Matthies[2], Sebastian Behrens[3], Ursula Obst[2], Silke Wieprecht[1] and Sabine Ulrike Gerbersdorf[1*]

Abstract

Background: The stabilization of fine sediments via biofilms ('biostabilization') has various economic and ecological implications but is presently unaddressed within lotic waters. To investigate natural biofilm growth and functionality in freshwater sediments under controlled boundary conditions, a unique mesocosm was constructed that combines established know-how from engineering and natural sciences and consists of six straight flumes. To test the comparability of biofilm growth within one flume and between the flumes, extracellular polymeric substances (EPSs), microbial biomass and microbial community composition were closely monitored over time and space as well as in relation to biofilm adhesiveness (proxy for biostabilization).

Results: Most importantly, biofilm development and biostabilization capacity revealed no significant differences within flume regions or between the flumes and the biofilms significantly stabilized the substratum as compared to abiotic controls. However, interesting temporal successions in biofilm growth phases became visible in shifting abundance and diversity of bacteria and microalgae resulting in varying EPS secretion and biostabilization.

Conclusions: These findings demonstrated the importance of biostabilization for fine sediment dynamics in freshwaters. Secondly, this unique setup allows comparable biofilm growth under controlled environmental conditions, an important requisite for future research on the ecological significance and impact of biostabilization for ecosystem functioning at varying environmental scenarios.

Keywords: Biofilm; Biostabilization; Sediment stability; Adhesion; Extracellular polymeric substances (EPSs); MagPI; DGGE; Flume; Diatoms

Background

During biofilm growth, microbes secrete extracellular polymeric substances (EPSs) gluing cells to the sediment and sediment particles to each other. Hereby, especially fine sediments (0.02 to 0.2 mm) are granted a higher resistance against erosive processes [1,2]. This constitutes a significant ecosystem service with broad economic and ecological implications. Understanding the dynamics of fine sediments is essential for maintaining waterways, dams and harbours, e.g. biostabilized fine sediments can significantly complicate reservoir flushing. Additionally,

a high quantity of macro and micropollutants can bind to fine sediment grains (and its organic matrix) to be jointly deposited in river regions with lower flow velocities [3,4]. The immobilization of pollutants within sediments largely depends on the substratum stability. Thus, biofilms can delay or prevent the re-suspension of these legacies into the water body as well as their bioavailability, while microbial bioremediation might even degrade those substances.

Knowledge about biostabilization of fine sediments is steadily increasing, and new techniques to measure adhesiveness of biofilms at high resolution have been established [5-7]. Moreover, the influence of various environmental factors on the formation of biofilms has been investigated, mostly by experimental procedures (e.g. [8,9]). In the

---
* Correspondence: sabine.gerbersdorf@iws.uni-stuttgart.de
[1]Institute for Modelling Hydraulic and Environmental Systems, University Stuttgart, Pfaffenwaldring 61, 70569 Stuttgart, Germany
Full list of author information is available at the end of the article

laboratory, parameters can be adjusted and reproduced relatively easy while insights into interspecies relations and larger-scaled ecosystem developments are limited, especially since most studies concentrate on monospecies biofilm (e.g. [10,11]). Other projects focused on field observations that illustrate diverse ecological phenomena [12,13]. Nevertheless, the environmental conditions are barely controllable and it is difficult to unravel specific links between species abundances, community composition, physiology and ecosystem functionality.

The presented project maximizes the advantages of laboratory (controlled reproducible settings) and field investigations (natural relevance) while combining engineering and biological expertise. Several studies investigated biofilm formation in microcosms [14-16]. Here, a continuous flow system is used consisting of six straight flumes run with natural river water. Detailed information about the construction, the hydraulic regimes and flow velocities of the straight flumes are given in Thom et al. [17]. The new flow channel system is unique because of the following: (1) the flume dimensions guarantee fully developed turbulence, uniform water flow and constant discharge as important requirements in hydraulic research; (2) the inoculation and development of biofilm from natural water on natural-like substratum minimizes behavioural artefacts of the microorganisms as a response to a more artificial physical environment and (3) microbial growth and development can be linked to biofilm functionality, here biostabilization as one important ecosystem service. Thereby, well-established methods from engineering science meet protocols in microbial and chemical analysis as well as molecular approaches to gain insights into the process of biostabilization with its various complex interactions. However, the main focus of the present paper is the evaluation of the newly designed straight flume setup with respect to the following question: is biofilm growth and development within and between the individual flumes comparable under controlled boundary conditions despite the well-known heterogeneity of natural waters? This is an essential prerequisite for further research into the phenomenon biostabilization at different habitats and environmental scenarios in order to reliably relate the manipulated boundary conditions to the observed effects on biofilm functionality. A 4-week experiment was conducted in which biofilm growth was evaluated intra-flume and inter-flume wise. The analytical focus of this study was on biochemical and molecular biological parameters of the developing biofilm and on biofilm adhesiveness as a proxy for substratum stability.

Results
Water chemistry
Nutrient concentrations of the water samples were constant over the experimental time and except for nitrate, at the

detection limit (according to LAWA [18]): phosphate < 0.2 mg L^{-1}, ammonium and nitrate approx. 0.04 ± 0.03 and 2.9 ± 0.1 mg L^{-1}, respectively, and sulphate with 48.1 ± 0.4 mg L^{-1}. Concentrations of fluoride and chloride were below 0.2 and around 58.7 ± 0.7 mg L^{-1}, respectively.

Inter- and intra-flume comparison
Comparison of the data on biochemical analysis, microbial biomass and surface adhesiveness showed no significant difference neither between the different regions within one flume nor between the different flumes (Tables 1, 2 and 3).

Development of the biofilm during the experiment
EPS matrix
Generally, contents of colloidal EPS carbohydrates and proteins exhibited an overall increase throughout the experiment (Figure 1), e.g. mean carbohydrate contents increased about fourfold from 11.6 ± 3.5 µg gDW^{-1} (day 2) to 42.8 ± 13.4 µg gDW^{-1} (day 26) while mean protein values increased about fivefold from 1.0 ± 0.8 µg gDW^{-1} to 5.2 ± 3.0 µg gDW^{-1}. However, the increase between two subsequent sampling points was only significant between days 11 and 14, for EPS carbohydrates (KWT; $n = 120$; $p = 0.0162$) as well as for EPS proteins (KWT; $n = 120$; $p < 0.0001$). Nevertheless, the mean values measured from day 14 onwards were significantly higher than those determined until day 7 (for both, EPS carbohydrates and proteins: KWT; $n = 120$; $p < 0.0001$). Overall, the contents of EPS carbohydrates and proteins showed a strong positive correlation during the experiment (Spearman; rs = 0.70; $n = 120$; $p < 0.0001$).

Chlorophyll a content and bacterial cell counts
Chlorophyll a contents and bacterial cell counts (BCC) increased during the experiment (Figure 2). For instance, mean chlorophyll a contents increased from 0.1 ± 0.1 µg gDW^{-1} (day 2) to 3.7 ± 2.1 µg gDW^{-1} (day 26). Meanwhile, mean BCC increased tenfold from $4.6 \pm 1.1 \times 10^6$ gDW^{-1} to $4.3 \pm 1.4 \times 10^7$ gDW^{-1}. Thus, highly significant differences were detected for both parameters between earlier biofilm stages (until day 7) and matured biofilms (from day 22 onwards) (KWT; $p < 0.0001$; $n = 120$ for chlorophyll a; $n = 40$ for BCC).

However, while BCC showed significant increases from day 5 to day 7 and from day 7 to day 11 (KWT; $n = 40$; $p = 0.0471$ and $p = 0.0074$, respectively), the values dropped significantly from day 11 to day 14 (KWT; $n = 40$; $p = 0.0009$). In contrast, chlorophyll a increased significantly from day 11 to day 14 (KWT; $n = 120$; $p = 0.0210$), with an even more pronounced microalgal growth between day 14 to day 18 (KWT; $n = 120$; $p = 0.0053$). Over the total experimental time, BCC and chlorophyll a as a proxy for algal biomass were positively

Table 1 Intra-flume comparison

Flume region	Carbohydrates (µg gDW^{-1})	Proteins (µg gDW^{-1})	Chlorophyll a (µg gDW^{-1})	Bacterial cells (10^7 gDW^{-1})	Surface adhesiveness (mA)
Front	27.9 ± 12.5	2.4 ± 1.6	1.3 ± 1.6	1.5 ± 0.8	618 ± 99
Middle	27.3 ± 14.7	2.9 ± 2.2	1.4 ± 1.7	1.6 ± 0.9	603 ± 115
Back	27.0 ± 13.3	2.9 ± 2.1	1.4 ± 1.8	1.7 ± 1.0	599 ± 104

Mean values of EPS (carbohydrates and protein) and chlorophyll a contents ($n = 144$), bacterial cell counts ($n = 24$) and surface adhesiveness ($n = 162$) over all measuring dates of the experiment within all six flumes (with STDev).

related (Spearman; rs = 0.69; $n = 40$; $p < 0.0001$). Still, between days 11 and 14, the relation was negative although not significant (Spearman; rs = −0.56; $n = 10$; $p = 0.089$). In addition, chlorophyll a values were positively related to EPS carbohydrates (Spearman; rs = 0.75; $n = 120$; $p < 0.0001$) as well as EPS proteins (Spearman; rs = 0.60; $n = 120$; $p < 0.0001$) during the entire experiment.

Microalgae community

Intra-flume comparisons of the diatom community displayed a high proportional similarity index (PSI) from the beginning onwards that even increased over the experiment: PSI 0.71 (day 7), 0.76 ± 0.09 (day 14) and 0.86 ± 0.03 (day 22). The PSI of the inter-flume comparison was similar (e.g. 0.72 ± 0.05 at day 22).

While most algal species were determined sporadically in single flumes (listed in Table 4); three diatoms occurred ubiquitously in all flumes: *Nitzschia fonticola*, *Nitzschia paleacea* and *Surirella brebissonii var. kuetzingii*. Apparently, *N. paleacea* was the dominant species throughout all flumes while the other two diatoms showed varying abundances in the different flumes (*N. fonticola* dominant in flume 2 and *S. brebissonii* dominant in flume 4). Besides the generally prevailing diatoms, the green algae *Scenedesmus sensu lato* was detected within the biofilms from day 7 onwards.

Bacterial community

The biofilms' bacterial range-weighted richness (RWR) showed no significant inter-flume variations over time (one-way ANOVA; $n = 20$; $p = 0.9740$), but indicated two different stages during biofilm growth. During the first 2 weeks, the initially high (mean) RWR of the biofilms (day 5 41.4 ± 9.8) decreased to a medium (day 7 19.6 ±

6.0) and virtually low level (day 14 10.7 ± 4.5). In contrast, a strong increase could be detected in RWR over the last 2 weeks (day 22 51.4 ± 14.1).

Generally, high values of dynamics were observed (mean rate of weekly change 31.5 ± 8.0%), indicating severe changes within the bacterial community. Over time, mean dynamics (rate of change) within the bacterial community increased from day 5 to day 7 (from 25.0 ± 10.7 to 30.8 ± 13.7%), followed by a decrease until day 11 (to 25.4 ± 10.5%, Figure 3). Until day 14, mean dynamics reached a new maximum (42.9 ± 7.5%) and decreased again subsequently (to 25.0 ± 8.3% until day 26). The inter-flume comparison of the dynamics development pattern showed no significant differences (one-way ANOVA; $n = 35$; $p = 0.506$).

Over the experiment, the functional organization of the biofilms' bacterial community increased steadily (Figure 4) and showed no significant inter-flume difference (one-way ANOVA; $n = 20$; $p = 0.3910$): on day 5, 20.0% of the bacterial DGGE bands corresponded to a mean proportion of 52.9 ± 7.2% of the cumulative band abundance indicating a medium degree of functional organization. This value slightly increased until day 7 (59.8 ± 4.7%) and subsequently reached a high level at 73.8 ± 10.8% on day 14. Until day 22 and for the rest of the experiment, this increase was mitigated, but still detectable (77.2 ± 5.2% on day 22).

Biostabilization

The developing biofilms established a surface adhesiveness which was up to four times higher than the abiotic sediment (Figure 5). This increase from 232 ± 7 mA at the start of the experiment to 652 ± 90 mA at day 5 was highly significant (KWT; $n = 135$; $p < 0.000$, days 1 to 2;

Table 2 Inter-flume comparison

Flume	Carbohydrates (µg gDW^{-1})	Proteins (µg gDW^{-1})	Chlorophyll a (µg gDW^{-1})	Bacterial cells (10^7 gDW^{-1})	Surface adhesiveness (mA)
1	23.1 ± 9.9	2.0 ± 1.7	1.1 ± 1.0	2.2 ± 1.7	630 ± 107
2	26.5 ± 11.2	2.7 ± 1.2	1.0 ± 1.0	1.9 ± 1.7	600 ± 96
3	25.5 ± 8.2	3.1 ± 1.6	1.4 ± 1.5	1.9 ± 1.0	558 ± 91
4	31.0 ± 16.5	2.5 ± 2.0	1.6 ± 2.1	2.0 ± 1.4	623 ± 132
5	32.7 ± 18.0	3.4 ± 3.0	2.1 ± 2.5	1.5 ± 0.9	623 ± 88

Mean values of EPS (carbohydrates and protein) and chlorophyll a contents ($n = 120$), bacterial cell counts ($n = 40$) and surface adhesiveness ($n = 135$) over all measuring dates of the experiment within all five flumes (with STDev).

Table 3 Results of the Kruskall-Wallis tests

Comparison	Carbohydrates	Proteins	Chlorophyll *a*	Bacterial cells	Surface adhesiveness
Intra-flume	$p = 0.8203$ ($n = 144$[a])	$p = 0.5865$ ($n = 144$)	$p = 0.9492$ ($n = 144$)	$p = 0.8540$ ($n = 24$[b])	$p = 0.7670$ ($n = 162$[c])
Inter-flume	$p = 0.3364$ ($n = 120$[d])	$p = 0.1223$ ($n = 120$)	$p = 0.5432$ ($n = 120$)	$p = 0.9522$ ($n = 40$[e])	$p = 0.0631$ ($n = 135$[f])

Intra-flume and inter-flume comparisons of the measured data. [a]3 regions × 6 flumes × 8 sampling points; [b]3 regions × 1 flume × 8 sampling points; [c]3 regions × 6 flumes × 9 sampling points (incl. blanks); [d]3 regions × 5 flumes × 8 sampling points; [e]1 region (middle) × 5 flumes × 8 sampling points; [f]3 regions × 5 flumes × 9 sampling points.

$p = 0.0131$, days 2 to 5). After a decline between days 5 and 7, new maximum values up to 956 mA were measured on day 11, with a mean value of 684 ± 119 mA. Subsequently, adhesiveness slightly decreased and stagnated at a level of ca. 607 ± 89 mA for the rest of the experiment.

Discussion

Biofilm growth within the new sophisticated mesocosm

The main purpose of the present study was to test this new design of straight flumes in terms of nature-like biofilm settlement and cultivation. Despite the absolutely identical setup of the six flumes, biofilm development could still differ due to smallest, possibly undetected variations in, e.g. the flow field. Moreover, biofilm growth is *per se* characterized by spatial heterogeneity. Thus, for the new flume design, it remained to be shown whether deviations in biofilm growth increase strongly along the test sections of one flume or between the individual flumes. This is very important since the comparability of biofilm growth and composition in the different straight flumes as well as within different flume regions is an essential prerequisite for further studies on the impact of environmental parameters on biofilm ecology and functionality.

The high spatial heterogeneity in biofilms and their complex mutual interactions with the environment have been described manifold (e.g. [19-21]). Biofilms have

even been described as microbial landscapes that not only are shaped in their spatial configuration by multiple physico-chemical factors but also alter their environment in considerable dimensions due to their growth [22]. Thus, not surprisingly, biofilms settling in the flumes exhibited high small-scale heterogeneity (on a single cartridge), but these pattern were clearly similar in all flume regions and flumes. This reflects the exact same settings of temperature, illumination and, most importantly, hydrodynamics in all flumes to provide the same settling and growth conditions for biofilms. Most common, an integral approach (Reynolds) is applied to account for the flow conditions within a flume as a precisely controlled hydrodynamic regime is either not necessary or not practicable. However, an exact determination of near-bed turbulences and bed shear stress directly at the sites of interest is essential for this study because various boundary effects such as eddy developments along the walls of a flume may influence hydrodynamics and near-bed turbulence significantly. Consequently, this might impact the erosive forces acting on biofilms as well as their nutrient replenishment to affect biofilm morphology and activity. Hence, in this study, the turbulence distribution was determined via high-resolution laser Doppler anemometry (LDA) before this experiment and later on checked with acoustic Doppler velocimetry (ADV) [17]. In addition, the long inlet flow section as well as the design of

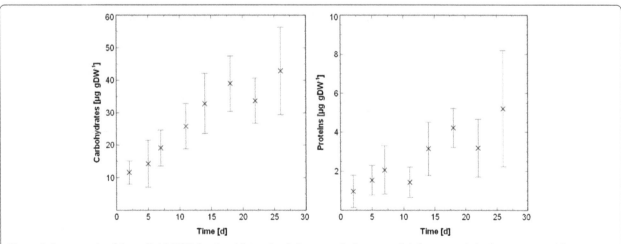

Figure 1 Compounds of the colloidal EPS fraction (determined via spectral photometry). *Left*: mean carbohydrate contents; *right*: mean protein contents (for both $n = 15$; with STDev).

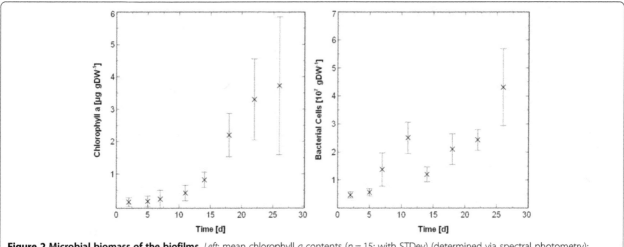

Figure 2 Microbial biomass of the biofilms. *Left:* mean chlorophyll *a* contents (*n* = 15; with STDev) (determined via spectral photometry); *Right:* mean bacterial cell counts (*n* = 5; with STDev) (determined via epi fluorescence microscopy).

the flumes (with a sufficient distance of the biofilm cartridges to the flume walls) ensures a homogenous turbulence distribution over the biofilm growth section. Another essential fact to keep in mind is that bed shear stress is an exponential function of the flow velocity emphasizing the importance of constant discharges within the flumes. This is why high-resolution discharge measurements were performed in all flumes throughout the experiment. These data gave evidence on the constant and similar discharges in all six flumes and thus on the identical bed shear stress levels over all biofilm growth sections in the presented experiment. This is one important prerequisite for the high

reproducibility of biofilm cultivation that could be shown for the new experimental setup in this study.

Biomass and EPS

Several studies investigating intertidal biofilms suggested a complex non-linear relationship between microbial biomass and EPS production [23-25]. The occasionally fast shifting nutritional and physiological states of the biofilm microbes might be of major importance for EPS production and secretion. This was underpinned by Underwood *et al.* [26] who described that diatom EPS production was, among others, dependent on the nutritional condition of the cells. The natural riverine water used in the flumes contained moderate nutrient concentrations comparable to oligotrophic habitats that remained stable over the experiment. Besemer *et al.* [27] and Artigas *et al.* [12] reported microbial biomass similar to the values determined in the present experiment under comparable conditions of nutrient availability (e.g. bacterial cell densities up to 1.13×10^8 cells cm^{-2} or chlorophyll *a* concentrations of 0.03 to 6.28 µg cm^{-2}). In contrast, biofilms in a eutrophic river (Neckar) were characterized by high microbial biomass and EPS production [28], e.g. an up to 16-fold higher chlorophyll *a* content than in this study emphasizing the effects of different water qualities upon biofilm development, corresponding metabolic rates and possibly, functionality. Low supply of nutrients may have restricted EPS production during biofilm development in the present experiment; still, the biostabilization effect (as discussed later) was impressive.

Table 4 Algae species of the mature biofilms (22nd day)

	Flume no.				
	1	2	3	4	5
Cyclotella menighiana	-	3.8	-	3.0	-
Fragilaria construens f. venter	-	-	3.0	-	-
Myamaea atomus var. permitis	-	-	3.6	-	-
Navicula capitatoradiata	-	3.5	-	-	-
Navicula reichardtiana	-	2.0	2.1	-	-
Nitzschia abbreviata	-	2.6	-	-	-
Nitzschia acicularis	2.1	-	-	-	-
Nitzschia fonticola	11.2	21.9	8.9	4.5	6.1
Nitzschia fonticola - romana form	2.9	10.5	5.6	-	3.4
Nitzschia palea var. debilis	6.6	-	-	-	-
Nitzschia palea var. palea	4.1	2.0	-	2.7	-
Nitzschia paleacea	48.5	37.0	61.4	54.5	69.2
Stephanodiscus sp.	3.7	-	-	-	-
Surirella brebissonii var. kuetzingii	2.9	3.5	3.0	26.9	9.8
Others	17.8	13.1	12.5	8.4	11.6

Proportional abundances of algal species [%]; species with a relative abundance of less than 2.0% were added up and summarized as 'others'.

Microbial community

As Marzorati *et al.* [29] stated, RWR, dynamics and functional organization calculated from the DGGE fingerprints should be seen as qualitative indicators, not as absolute measures due to the known drawbacks of molecular

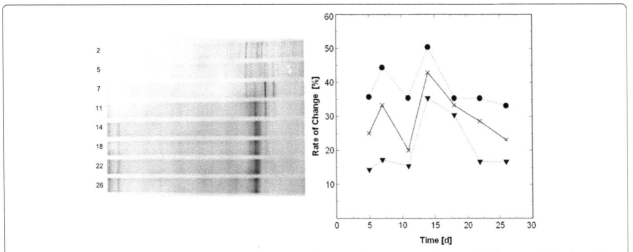

Figure 3 Dynamics within the bacterial community. *Left*: DGGE band patterns of 16S rDNA gene diversity of biofilms exemplified shown for flume 1 (stained with ethidium bromide; *inverted picture*; *numbers* represent days of growth); *Right*: moving window analysis of the bacterial community's dynamics in the same flume (*crosses*) based on densitometry similarity matrices. Additionally shown are the maximal (*round dots*) and minimal (*triangular*) change rates over the residual flumes.

fingerprinting techniques. Nevertheless, they are an important tool to describe, compare and interpret different DGGE band patterns in order to obtain a higher level of information about ecological processes in biofilms. Generally, observing the microbial community of biofilms in this study revealed two distinct stages during development. The first phase could be described as an initial colonization of the abiotic substrate by planktonic bacteria. The findings of Beier *et al.* [30] or Crump and Hobbie [31], e.g. 40 bacterial TGGE bands in comparable riverine

water, support the initially high bacterial RWR in this experiment. During the subsequent establishment of a bacterial biofilm within the first 11 days, competition and specialization of the bacterial community became visible by decreasing RWR and increasing functional organization. Manz *et al.* [32] and Araya *et al.* [33] described similar shifts within the bacterial community composition during the formation of lotic biofilms. In the period between days 11 and 14, severe changes within the bacterial community could be observed. Along with significantly decreasing cell

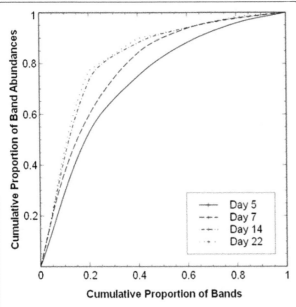

Figure 4 Functional organization of bacterial community expressed as Parento-Lorenz curves obtained via densitometry and normalization of DGGE peak patterns.

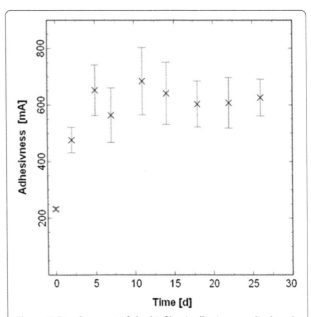

Figure 5 Development of the biofilms' adhesiveness displayed as mean determined current (*n* = 15; with STDev) during MagPI measurement.

counts and RWR, increasing dynamics and functional organization of the bacteria might be due to bacterial biofilm detachment, the initial algal colonization or settlement of grazing ciliates as described by Wey et al. [34] - or a combination of these processes.

Besemer et al. [27] highlighted the role of diatoms as key players in river biofilms as they shape the bacterial community in a combination of physical and biological processes. The importance of diatoms in this experiment became obvious since parallel to the development of the diatom community (from day 14 on), a steady increase of vertical and horizontal heterogeneity of the biofilms could be visually observed. As described by Besemer et al. [16], this spatial small-scale heterogeneity was a major driver affecting the development of the microbial community. Thus, diatoms apparently influenced the bacterial community in different ways. On one hand, their photosynthetic activity and EPS production may have supported the growth of associated bacteria as indicated by high increase rates of the total cell counts after day 14 of the experiment, a phenomenon already described by Espeland et al. [35]. On the other hand, the pronounced three-dimensional structures of their colonies may have allowed the formation of various physical and biochemical transition/gradient zones, e.g. O_2 gradients as described by Glud et al. [36] or Fenchel and Glud [37]. Due to this formation of many microenvironments, accumulation of different metabolic products of the diatoms as well as their associated bacteria in the biofilm matrix and several metabolic cascades, various niches can develop for a broad range of bacteria with different physiologies [38]. In this context, specialized bacteria may have found optimal conditions in the matured biofilms as reported by Manz et al. [32]. Summarizing, the diatom development apparently led to dominance of a few bacterial species but may also have led to the establishment of various niches where diverse bacterial species might occur in low abundances, indicated by steadily decreasing dynamics within the bacterial community and parallel increasing bacterial RWR.

Biostabilization

Due to its high economic and ecological importance, sediment dynamics in lotic systems plays a major role in hydraulic engineering. Different modelling approaches approximated the highly complex sediment-water system. Briefly, the Shields model [39] is commonly applied to determine the stability of sediments versus erosional forces by defining characteristic diameters and the density of the sediment grains. However, this approach is not applicable for sediments with strong biological influence. Righetti et al. [40] described the first model based on the Shields equation introducing adhesiveness/adhesion as a new parameter in order to incorporate the influence

of biofilms on sediment stability. Finally due to the development of the MagPI device that has been applied successfully in the marine habitat [6,7,41], the determination of biological-induced adhesiveness could be achieved at high temporal and spatial resolution. While most studies focused on brackish/intertidal areas (e.g. [23,42,43]), up to now, biostabilization of lotic fine sediments is virtually unaddressed. Spears et al. [44] suggested the major importance of biostabilization in marine/brackish habitats supporting the current doctrine that high quantities of strong ionic bounds significantly strengthen the EPS matrix. Nevertheless, despite observing significantly lower EPS values as well as microbial biomass (as compared to brackish/marine biofilms), a significant biostabilization effect was detected. This emphasized the importance of biofilm-induced stabilization of fine sediments in lotic waters (and contradicted the current doctrine).

Furthermore, this study gave insights in the development of various geochemical and (micro-) biological parameters during biofilm growth affecting this important ecosystem function. In this context, the content of colloidal EPS compounds could be seen as an approximate marker for biostabilization capacity. Further investigation of the quality of these polymers could lead to the identification of single carbohydrate or protein moieties with high gluing and stabilizing capability. While biofilm stability appeared to be related to total cell counts in nascent biofilm stages, the development of the bacterial community composition turned out to be even more crucial. Thereby, short and long-term shifts within the bacterial and algal community occurred simultaneously to significantly change biofilm stability. This emphasizes the importance of the molecular biology tools used in this study to address biofilm composition and diversity. Identification of bacterial key players in biostabilization and their interactions with algae could be the next step to further elucidate the contribution of microbiology to sediment stability. Future studies should also consider the role of the protozoa community in shaping microbial biofilm community and functionality, here biostabilization.

Conclusions

The stabilizing effect of biofilms upon lotic fine sediment is currently unaddressed despite its broad range of economic and ecological implications. To investigate the complex interactions between the biofilm and its environment, a sophisticated and unique setup was designed combining biological and engineering expertise.

The evaluation of biofilm growth in these new flumes is presented in this paper to demonstrate that comparable biofilms (intra- and inter-flume wise) could be cultivated while exposed to the same abiotic environment. Furthermore, the biofilm cultivation under strictly controlled boundary conditions gave evidence on the importance of biostabilization (known to be substantial in intertidal mudflats) within lotic fine sediments. In doing so, first insights into various ecological processes which shape the microbial community and impact the overall biofilm functionality could be gained; for instance, the change of a bacteria-dominated nascent biofilm to a diatom-dominated matured development stage resulted in biofilms which stabilized the underlying sediment significantly.

Summarizing, regarding their biological and biochemical features, representative biofilms can be grown in this novel system - an essential prerequisite for further research into natural biofilm colonization and development. In particular, the mutual interactions of various environmental parameters impacting biofilms can now be addressed and reliably related to each other. With the link to biofilm functionality, the significance of biostabilization can thus be investigated for different niches in freshwater habitats.

Methods

Experimental setup and sampling

Biofilms grew in six flumes, each with an individual, separate water circuit (see Thom et al. [45] and Figure 6) under constant natural-like environmental conditions. Briefly, the flumes (length × width × height, 3.00 m × 0.15 m × 0.15 m) were designed to allow a homogeneous flow field and constant shear stress across the biofilm cultivation section (length 1.00 m). This section contained 12 substratum cartridges (length × width × height, 0.08 m × 0.06 m × 0.02 m) that could be transferred outside the flume for further measurements. Cartridges were illuminated by two parallel fluorescent tubes (Osram Biolux; 480 to 665 nm) and homogenous irradiation was confirmed by measurements of light intensity and wavelength irradiance of the photosynthetic active radiation (PAR) spectrum using a high-resolution spectroradiometer (SR-9910, Macam Photometric Ltd., Livingston, Scotland) as described by

Gerbersdorf and Schubert [46]. Discharge was set at 0.80 ± 0.10 L s^{-1} by adjusting the by-pass (flow velocity 0.07 ± 0.01 m s^{-1} within each flume) and continuous measured with an installed mini-flow meter (Bürkert 8030, Ingelfingen, Germany). Fluvial water was retrieved from the middle reach of the River Enz (Baden-Württemberg, Germany). While adjusted to constant temperature ($15°C \pm 0.3°C$) by a cooling water circuit, 200 L were circulated in each flume by a circulatory pump (BADU Eco Touch, Speck Pumpen, Neunkirchen am Sand, Germany) for 4 weeks (1 August 2012 to 28 August 2012). Thus, indigenous microorganisms within the river water settled the cartridges filled with inert glass beads (diameter 100 to 200 μm) eventually forming a biofilm. Sampling of each flume was performed on the days 2, 5, 7, 11, 14, 18, 22 and 26. One litre of water was withdrawn for subsequent nutrient analysis. The 12 substratum cartridges of each flume were classified in three regions 'front', 'middle' and 'back' of four consecutive cartridges each. At each sampling, ten samples of 0.5 cm^3 were randomly withdrawn from the sediment of each flume regions by a cut-off syringe and pooled. Subsamples (0.5 cm^3) were transferred into Eppendorf tubes for further analyses. Due to pump failure and flow stagnation in the sixth flume during the first week, this flume was excluded from inter-flume comparison.

Water chemistry, EPS and bacterial cell counts

Water samples were analysed according to DIN EN ISO 10304 and using a quick test (Hach Lange GmbH, Berlin, Germany) based on DIN 38406-E5-1: the concentrations of fluoride, chloride, nitrate, ammonia, sulphate and phosphate ions were quantified. The colloidal (water-extractable) EPS fractions of the biofilms were extracted according to Gerbersdorf et al. [47]. Afterwards, carbohydrates and protein contents were determined by phenol assay and modified Lowry procedure,

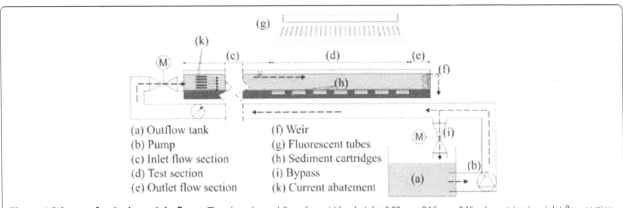

(a) Outflow tank
(b) Pump
(c) Inlet flow section
(d) Test section
(e) Outlet flow section
(f) Weir
(g) Fluorescent tubes
(h) Sediment cartridges
(i) Bypass
(k) Current abatement

Figure 6 Scheme of a single straight flume. The glass channel (length × width × height, 3.00 m × 0.15 m × 0.15 m) contained an inlet flow section (c) (2.00 m) and a biofilm cultivation section (d) (1.00 m).

respectively [48-50]. Chlorophyll a/pheophytin were measured by ethanol extraction before and after acidification, respectively (DIN 38 412/16). Bacterial cell counts were performed in the three regions of flume 4 (intra-flume comparison) as well as in the middle regions of the flumes 1, 2, 3 and 5 (inter-flume comparison) on the eight sampling points of the experiment. Samples were fixed with 4% paraformaldehyde (final concentration) and shaken horizontally for 15 min to be treated afterwards (5 pulses à 5 s at 20% intensity) with a Sonopuls UW 3100 ultrasonic probe (Bandelin electronic, Berlin, Germany). After a settlement period for 1 min, subsamples of 99 µL were taken from the supernatant. Suspended cells were stained with 1 µL SYTO 13 (500 µM) (Life Technologies, Carlsbad, CA, USA) for 15 min and counted at 488 nm excitation using an Axioscop fluorescence microscope (Carl Zeiss, Oberkochen, Germany).

Algal community composition

To investigate inter-flume heterogeneity, samples fixed with Lugol solution (2%) were analysed from the middle region of the flumes 1 to 5 at day 22. For the intra-flume comparison, samples from all three regions of flume 3 were investigated at days 7, 14, and 22. All samples were separated in two subsamples; one was used for direct microscopic cell counts by Axioscope A1 microscope (Carl Zeiss, Oberkochen, Germany). The other was further treated to remove organic matter and embedded in order to determine diatom species by their frustules [51-53]. The data functioned as basis for further calculations on the PSIs of the algal communities [54].

Bacterial community composition

DNA was isolated using the Nucleospin Kit for Soil (Macherey and Nagel, Düren, Germany) according to the manufacturer's information. A PCR assay was conducted using the universal bacterial 16S rRNA gene primers 27f (5'-AGA GTT TGA TCM TGG CTC AG-3') and 517r (5'-ATT ACC GCG GCT GCT GG-3') [55,56]. The forward primer had an attached GC-clamp (5'-CGC CCG CCG CGC CCC GCG CCC GTC CCG CCG CCG CCC CCG CCC C-3') for subsequent DGGE (Denaturing Gradient Gel Electrophoresis). Each reaction (25 µL) contained 15 to 20 ng DNA, 0.125 µL dNTPs (200 µM), 0.25 µL of each primer (40 µM), 0.13 µL Taq DNA polymerase (HotStart™ Polymerase, 5 U/µL, QIAGEN, Venlo, Netherlands), 2.5 µL × 10 PCR buffer (QIAGEN, Venlo, Netherlands) provided by the enzyme manufacturer and 16.38 µL sterile PCR water (Merck Millipore, Darmstadt, Germany). Amplification was performed with a GeneAmp PCR system 9700 (Applied Biosystems, Carlsbad, CA, USA) as follows: 30 s at 94°C, 35 cycles of 30 s at 94°C, 30 s at 55°C and 60 s at 72°C, followed by a final extension of 7 min at 72°C. The PCR products were loaded onto 1% agarose gels in × 1 TAE buffer (pH 8.0), ethidium-bromide-stained and studied under UV illumination (600 nm).

DGGE was performed with a Bio Rad DCode (Bio-Rad Laboratories, Hercules, CA, USA) system as described by Muyzer et al. [57]: PCR products of similar size were loaded on a 1.5-mm-thick, vertical gel containing 7.5% (w/v) polyacrylamide (37.5:1 acrylamide/bisacrylamide) with a linear denaturing gradient of urea and formamide (40 to 70%). Electrophoresis was performed in a × 1 TAE buffer (pH 8.5) for 17 h at 70 V and 56°C. Gels were stained with SYBR Gold (Invitrogen, Karlsruhe, Germany) for 15 min and analysed using a Lumi-Imager F1 Working Station (Roche Diagnostics, Mannheim, Germany). An image of each gel was taken with a CCD camera system (The Imager, Appligene, Illkirch, France), and the software Lumi Analyst 3.1. DGGE banding patterns were analysed with the program GelCompar II (6.0) and processed with ImageJ (148 a) and DiGit (Kramer 2013, personal communication). RWR, community dynamics and functional organization were calculated according to Marzorati et al. [29].

Surface adhesiveness

Biofilm adhesiveness - a measure for biostabilization capacity - was determined with a modified magnetic particle induction (MagPI) system (Figure 6), e.g. using an improved electromagnetic inductor containing a highly magnetisable µ metal core with low remanence. Briefly, ferromagnetic particles were spread on the biofilm surface and the electromagnet was set up vertically in a defined distance of 0.4 cm to the surface. The strength of the magnetic field needed to retrieve the particles from the surface is equivalent to its adhesiveness. The magnetic field can be calculated from the current (amperage) via calibration of the used electromagnet.

Statistics

Statistics were conducted using the software Analyze-it225 (1.0.5.0.): Shapiro-Wilks tests (confidence interval 95%) were performed to check normal distribution of the data sets. In case of normally distributed data, a one-way ANOVA (confidence interval 95%; chi-square approximation) was conducted; otherwise, a Kruskal-Wallis test (KWT) (chi-square approximation; Bernoulli correction for ties) was performed to compare the different groups. To test correlation of different parameters, Spearman's rank correlation coefficients were calculated.

Abbreviations

ANOVA: analysis of variance; BCC: bacterial cell counts; ca.: circa; °C: degree Celsius; DIN EN ISO: Deutsche Industrie Norm Europäische Norm International Organization for Standardization; DGGE: denaturing gradient gel electrophoresis; DNA: deoxyribonucleic acid; DW: dry weight; h: hour; KWT: Kruskal-Wallis test; mA: milliampere; mT: millitesla; min: minute; µM: micromolar; µL: microliter; ng: nanogram; PAR: photosynthetic active radiation; PCR: polymerase chain reaction; rDNA: ribosomal DNA; resp.: respectively; RWR: range-weighted richness; S: Svedberg; s: seconds; STDev: standard deviation; TAE: tris acetic acid EDTA; UV: ultraviolet; V: volt; w/v: weight to volume.

Competing interests

The authors declare that they have no competing interests.

Authors' contributions

HS compiled and analysed the biochemical, biological and molecular biology data. MT performed the adhesiveness measurements. KM, UO and SB supported the molecular biology analysis and data computation. SW and SUG headed the investigation. HS and SUG developed the experimental design and wrote the manuscript. The final manuscript was read and approved by all authors.

Acknowledgements

The authors are grateful to the DFG (Deutsche Forschungsgemeinschaft) for financing the project 'Ecosystem Engineering: Sediment entrainment and flocculation mediated by microbial produced extracellular polymeric substances (EPS)' (GZ: GE 1932/3-1). We thank for the excellent support of the molecular analyses at the Institute of Functional Interfaces, Karlsruhe Institute of Technology (KIT) (Prof. U. Obst) and at the Centre for Applied Geosciences (ZAG), University Tübingen (Prof. A. Kappler). The authors much appreciate the cooperation with Dr. M. Schweikert, Institute of Zoology, University Stuttgart to get access to the Microscopic Facility.

Author details

[1]Institute for Modelling Hydraulic and Environmental Systems, University Stuttgart, Pfaffenwaldring 61, 70569 Stuttgart, Germany. [2]Institute of Functional Interfaces, Karlsruhe Institute of Technology (KIT), Hermann-von-Helmholtz-Platz 1, 76344 Eggenstein-Leopoldshafen, Germany. [3]Geomicrobiology/ Microbial Ecology, Centre for Applied Geosciences (ZAG), Eberhard-Karls-University Tübingen, Sigwartstrasse 10, 72076 Tübingen, Germany.

References

1. Underwood GJC, Paterson DM. Seasonal changes in diatom biomass, sediment stability and biogenic stabilization in the Severn Estuary. J Mar Biol Assoc UK. 1993;73:871–87.
2. Tolhurst J, Gust G, Paterson D. The influence of an extracellular polymeric substance (EPS) on cohesive sediment stability. In: Winterwerp JC, Kranenburg C, editors. Fine Sediment Dynamics in the Marine Environment. Amsterdam: Proceedings in Marine Science 5; 2002. p. 409–25.
3. Karickhoff SW, Brown DS, Scott TA. Sorption of hydrophobic pollutants on natural sediments. Water Res. 1979;13:241–8.
4. Audry S, Schafer J, Blanc G, Jouanneau JM. Fifty-year sedimentary record of heavy metal pollution (Cd, Zn, Cu, Pb) in the Lot River reservoirs (France). Environ Pollut. 2004;132:413–26.
5. Cuadrado DG, Carmona NB, Bournod C. Biostabilization of sediments by microbial mats in a temperate siliciclastic tidal flat, Bahia Blanca estuary (Argentina). Sediment Geol. 2011;237:95–101.
6. Larson F, Lubarsky H, Gerbersdorf SU, Paterson DM. Surface adhesion measurements in aquatic biofilms using magnetic particle induction: MagPI. Limnol Oceanogr Methods. 2009;7:490–7.
7. Lubarsky H. The impact of microbial extracellular polymeric substances on sediment stability. Doctoral dissertation, University of St Andrews. 2011.
8. Chavant P, Martinie B, Meylheuc T, Bellon-Fontaine MN, Hebraud M. Listeria monocytogenes LO28: surface physicochemical properties and ability to form biofilms at different temperatures and growth phases. Appl Environ Microbiol. 2002;68:728–37.
9. Shrout JD, Chopp DL, Just CL, Hentzer M, Givskov M, Parsek MR. The impact of quorum sensing and swarming motility on Pseudomonas aeruginosa biofilm formation is nutritionally conditional. Mol Microbiol. 2006;62:1264–77.
10. Rogers J, Dowsett AB, Dennis PJ, Lee JV, Keevil CW. Influence of temperature and plumbing material selection on biofilm formation and growth of Legionella pneumophila in a model potable water system containing complex microbial flora. Appl Environ Microbiol. 1994;60:1585–92.
11. Lemon KP, Higgins DE, Kolter R. Flagellar motility is critical for Listeria monocytogenes biofilm formation. J Bacteriol. 2007;189:4418–24.
12. Artigas J, Fund K, Kirchen S, Morin S, Obst U, Romani AM, et al. Patterns of biofilm formation in two streams from different bioclimatic regions: analysis of microbial community structure and metabolism. Hydrobiologia. 2012;695:83–96.
13. McKew BA, Taylor JD, McGenity TJ, Underwood GJC. Resistance and resilience of benthic biofilm communities from a temperate saltmarsh to desiccation and rewetting. ISME J. 2011;5:30–41.
14. Battin TJ, Kaplan LA, Newbold JD, Hansen CME. Contributions of microbial biofilms to ecosystem processes in stream mesocosms. Nature. 2003;426:439–42.
15. Singer G, Besemer K, Hoedl I, Chlup A, Hochedlinger G, Stadler P, et al. Microcosm design and evaluation to study stream microbial biofilms. Limnol Oceanogr Methods. 2006;4:436–47.
16. Besemer K, Hoedl I, Singer G, Battin TJ. Architectural differentiation reflects bacterial community structure in stream biofilms. ISME J. 2009;3:1318–24.
17. Thom M, Schmidt H, Wieprecht S, Gerbersdorf S. Dimensionierung von Fließrinnen zum Aufwuchs von Biofilm unter definierten hydraulischen Randbedingungen. DGL-Jahrestagung (24. - 28. September 2012, Koblenz); 2012.
18. Länderarbeitsgemeinschaft Wasser. Beurteilung der Wasserbeschaffenheit von Fließgewässern in der Bundesrepublik Deutschland - Chemische Gewässergüteklassifikation. Berlin: LAWA; 1998.
19. Donlan RM. Biofilms: microbial life on surfaces. Emerg Infect Dis. 2002;8:881–90.
20. Wimpenny J, Manz W, Szewzyk U. Heterogeneity in biofilms. FEMS Microbiol Rev. 2000;24:661–71.
21. Bernard CS, Giraud C, Spagnolo J, de Bentzmann S. Biofilms: the secret story of microbial communities. In: Locht C, Simonet M, editors. Bacterial pathogenesis: molecular and cellular mechanisms. Caister Academic Press; 2012. p. 129–68.
22. Battin TJ, Sloan WT, Kjelleberg S, Daims H, Head IM, Curtis TP, et al. Microbial landscapes: new paths to biofilm research. Nat Rev Microbiol. 2007;5:76–81.
23. Yallop ML, Paterson DM, Wellsbury P. Interrelationships between rates of microbial production, exopolymer production, microbial biomass, and sediment stability in biofilms of intertidal sediments. Microb Ecol. 2000;39:116–27.
24. Smith DJ, Underwood GJC. The production of extracellular carbohydrates by estuarine benthic diatoms: the effects of growth phase and light and dark treatment. J Phycol. 2000;36:321–33.
25. Hanlon ARM, Bellinger B, Haynes K, Xiao G, Hofmann TA, Gretz MR, et al. Dynamics of extracellular polymeric substance (EPS) production and loss in an estuarine, diatom-dominated, microalgal biofilm over a tidal emersion-immersion period. Limnol Oceanogr. 2006;51:79–93.
26. Underwood GJC, Boulcott M, Raines CA, Waldron K. Environmental effects on exopolymer production by marine benthic diatoms: dynamics, changes in composition, and pathways of production. J Phycol. 2004;40:293–304.
27. Besemer K, Singer G, Limberger R, Chlup A-K, Hochedlinger G, Hoedl I, et al. Biophysical controls on community succession in stream biofilms. Appl Environ Microbiol. 2007;73:4966–74.
28. Gerbersdorf SU, Jancke T, Westrich B, Paterson DM. Microbial stabilization of riverine sediments by extracellular polymeric substances. Geobiology. 2008;6:57–69.
29. Marzorati M, Wittebolle L, Boon N, Daffonchio D, Verstraete W. How to get more out of molecular fingerprints: practical tools for microbial ecology. Environ Microbiol. 2008;10:1571–81.
30. Beier S, Witzel K-P, Marxsen J. Bacterial community composition in central European running waters examined by temperature gradient gel electrophoresis and sequence analysis of 16S rRNA genes. Appl Environ Microbiol. 2008;74:188–99.
31. Crump BC, Hobbie JE. Synchrony and seasonality in bacterioplankton communities of two temperate rivers. Limnol Oceanogr. 2005;50:1718–29.
32. Manz W, Wendt-Potthoff K, Neu TR, Szewzyk U, Lawrence JR. Phylogenetic composition, spatial structure, and dynamics of lotic bacterial biofilms investigated by fluorescent in situ hybridization and confocal laser scanning microscopy. Microb Ecol. 1999;37:225–37.
33. Araya R, Yamaguchi N, Tani K, Nasu M. Change in the bacterial community of natural river biofilm during biodegradation of aniline-derived compounds determined by denaturing gradient gel electrophoresis. J Health Sci. 2003;49:379–85.
34. Wey JK, Juergens K, Weitere M. Seasonal and successional influences on bacterial community composition exceed that of protozoan grazing in river biofilms. Appl Environ Microbiol. 2012;78:2013–24.
35. Espeland EM, Francoeur SN, Wetzel RG. Influence of algal photosynthesis on biofilm bacterial production and associated glucosidase and xylosidase activities. Microb Ecol. 2001;42:524–30.
36. Glud RN, Kuhl M, Kohls O, Ramsing NB. Heterogeneity of oxygen production and consumption in a photosynthetic microbial mat as studied by planar optodes. J Phycol. 1999;35:270–9.

37. Fenchel T, Glud RN. Benthic primary production and O-2-CO2 dynamics in a shallow-water sediment: spatial and temporal heterogeneity. Ophelia. 2000;53:159–71.

38. Stewart PS, Franklin MJ. Physiological heterogeneity in biofilms. Nat Rev Microbiol. 2008;6:199–210.

39. Shields A. Application of similarity principles and turbulence research to bed-load movement. Pasadena, CA: California Institute of Technology; 1936. p. 43–58.

40. Righetti M, Lucarelli C. May the Shields theory be extended to cohesive and adhesive benthic sediments? J Geophys Res. 2007;112:C05039. doi:10.1029/2006JC003669.

41. Anderson A, Spears B, Lubarsky H, Davidson I, Gerbersdorf S, Paterson D. Magnetic particle induction and its importance in biofilm research. In: Fazel-Rezai R, editor. Biomedical Engineering - Frontiers and Challenges. Rijeka: InTech; 2011. p. 189–216.

42. Austen I, Andersen TJ, Edelvang K. The influence of benthic diatoms and invertebrates on the erodibility of an intertidal a mudflat, the Danish Wadden Sea. Estuar Coast Shelf Sci. 1999;49:99–111.

43. Gerbersdorf SU, Bittner R, Lubarsky H, Manz W, Paterson DM. Microbial assemblages as ecosystem engineers of sediment stability. J Soils Sediments. 2009;9:640–52.

44. Spears BM, Saunders JE, Davidson I, Paterson DM. Microalgal sediment biostabilisation along a salinity gradient in the Eden Estuary, Scotland: unravelling a paradox. Mar Freshw Res. 2008;59:313–21.

45. Thom M, Schmidt H, Wieprecht S, Gerbersdorf S. Physikalische Modellversuche zur Untersuchung des Einflusses von Biofilm auf die Sohlenstabilität. Wasserwirtschaft. 2012;6:32–6.

46. Gerbersdorf SU, Schubert H. Vertical migration of phytoplankton in coastal waters with different UVR transparency. Environ Sci Eur. 2011;23:36.

47. Gerbersdorf SU, Manz W, Paterson DM. The engineering potential of natural benthic bacterial assemblages in terms of the erosion resistance of sediments. FEMS Microbiol Ecol. 2008;66:282–94.

48. Dubois M, Gilles KA, Hamilton JK, Rebers PA, Smith F. Colorimetric method for determination of sugars and related substances. Anal Chem. 1956;28:350–6.

49. Raunkjaer K, Hvitvedjacobsen T, Nielsen PH. Measurement of pools of protein, carbohydrate and lipid in domestic wastewater. Water Res. 1994;28:251–62.

50. Frolund B, Palmgren R, Keiding K, Nielsen PH. Extraction of extracellular polymers from activated sludge using a cation exchange resin. Water Res. 1996;30:1749–58.

51. Battarbee R. Diatom analysis. Handbook of Holocene Palaeoecology and Palaeohydology. New York: John Wiley & Sons Ltd; 1986.

52. Hofmann G, Werum M, Lange-Bertalot H. Diatomeen im Süßwasser-Benthos von Mitteleuropa. A.R.G. Gantner Verlag K.G: Ruggell; 2011.

53. Krammer K, Lange-Bertalot H. Bacillariophyceae. In Ettl H, Gerloff J, Heyning H, Mollenhauer D, editors. Süßwasserflora von Mitteleuropa Band 1–4. Jena: Gustav Fischer Verlag; 1986–1991

54. Rosef O, Kapperud G, Lauwers S, Gondrosen B. Serotyping of Campylobacter jejuni, Campylobacter coli, and Campylobacter laridis from domestic and wild animals. Appl Environ Microbiol. 1985;49:1507–10.

55. Lane D. 16S/23S rRNA sequencing. In: Stackebrandt E, Goodfellow M, editors. Nucleic acid sequencing techniques in bacterial systematics. New York: Wiley; 1990. p. 115–48.

56. Emtiazi F, Schwartz T, Marten SM, Krolla-Sidenstein P, Obst U. Investigation of natural biofilms formed during the production of drinking water from surface water embankment filtration. Water Res. 2004;38:1197–206.

57. Muyzer G, Dewaal EC, Uitterlinden AG. Profiling of complex microbial populations by denaturing gradient gel electrophoresis analysis of polymerase chain reaction-amplified genes coding for 16S rRNA. Appl Environ Microbiol. 1993;59:695–700.

3

Estrogenic activity in drainage water: a field study on a Swiss cattle pasture

Andreas Schoenborn[1]*, Petra Kunz[2] and Margie Koster[3]

Abstract

Background: Dairy cow manure applied to pastures is a significant potential source of estrogenic contamination in nearby streams. One possible pathway is through infiltration via preferential flow to drainage pipes, particularly after heavy rainfall events. In a period of 73 days in the spring of 2010, a drainage catchment in a cattle pasture in the Swiss lowlands was closely monitored.

Manure was applied three times during the study, and part of the catchment was also subjected to grazing. During five field campaigns, water samples from two sampling locations were taken for 4–24 h in consecutive sampling intervals. 17β-estradiol equivalents (EEQ) were determined with the yeast estrogen screen (YES) and the ER-CALUX assay. Some water chemistry parameters, pH, conductivity, oxygen content and soil moisture tension were also monitored.

Results: Washout of estrogenic activity was highest during or right after heavy rainfall events, shortly after manure spreading, when peak values of >10 ng/l EEQ were found in several samples. However, in two field campaigns, high EEQ values were also found 14 and 28 days, after the last manure application, in one case during a dry weather period. This indicates that estrogenic compounds are more stable in natural soils than what is expected from data gathered in lab studies.

Conclusions: Streams in agricultural areas with a high proportion of drained land may be subject to numerous peaks of EEQ during the course of the year. This may have a negative effect on aquatic organisms, namely fish embryos, living in these streams.

Keywords: Drainage water; Manure; Dairy cattle; estrogenic activity; EEQ; YES; ER-CALUX; Aquatic organisms

Background

Treated wastewater is a potential source of estrogenic activity of anthropogenic origin in natural water bodies. estrogenic activity in wastewater has been linked to sexual changes in fish [1] and is suspected to be "a major causal factor in the evolution of intersexuality" in roach [2]. Following these findings, estrogens in treated wastewater were closely examined in a number of countries in the last 10–15 years, e.g. in Britain [2], The Netherlands [3], Denmark [4] and Switzerland [5, 6].

The role of agriculture as source of estrogenic activity for natural water bodies has received much less attention. A "normalised cow" excretes two orders of magnitude more and a "normalised pig" one order of magnitude more steroid estrogens than a "normalised human" [7]. A conservative estimate for Switzerland shows that the total annual estrogen load released onto the environment from livestock exceeds that excreted by humans by at least a factor 5 (Table 1). Johnson et al. calculated the same factor 5 for the UK. In a review on sex hormones originating from livestock, Lange et al. [8] concluded that "discussion on environmental endocrine disrupters has to be extended by this important aspect" even though they did not find causal links in literature to "any known severe adverse effect on wildlife or human endocrine system".

A number of studies demonstrate the presence of farm-animal-derived steroid hormones in manure and wastewater from dairy farms [9–11]. Manure is usually spread on soil surfaces or (more recently) is injected. With the exception of karst areas [12] it is still largely unclear to what extent these hormones can reach nearby

* Correspondence: andreas.schoenborn@zhaw.ch
[1]Institute of Natural Resource Sciences, Zurich University of Applied Sciences, P.O. Box CH-8200, Waedenswil, Switzerland
Full list of author information is available at the end of the article

Table 1 Estimated annual oestrogen load in Switzerland (2007) from excreta of humans and three common livestock animals

Species	Excretion of oestrogens (mg/individual/year)	Number of individuals	Annual oestrogen load (kg/year)	Share (%)
Cows	110[c]	708,340[a]	77.9	43
Pigs	43[c]	1,573,090[a]	67.6	37
Sheep	8.4[c]	443,584[a]	3.7	2
Humans	4.38[d]	7,593,500[b]	33.3	18

[a]From Bundesamt für Statistik [36]
[b]From Bundesamt für Statistik [47]
[c]After [25] (cycling females)
[d]Calculated from estimated average excretion of 12 µg/person/day (sum of E1, E2 and E3, pregnant women excluded), after Table two of [48]

water bodies. Sorption/desorption studies showed a rapid degradation and high sorption of estrogens in soils (several studies, as cited in [13]). In one of the few field studies (on grassland soils treated with cattle and sheep manure), Lucas and Jones [14] showed that estrone (E1) and 17ß-estradiol (E2) are "not persistent in agricultural soils" and calculated a half-life from 5–25 days for these two estrogens.

However, individual soil conditions can modify the persistence of estrogens. The presence of sheep urine enhances and prolongs the amount of estrogen leaching from soil [15]. The association with manure-borne dissolved organic carbon (DOC) reduces the bioavailability of estrogens and increases their persistence [16]. Anaerobic conditions slow down the degradation of some estrogens [17]. This coincides with the observation that E2 "was widespread, persisted much longer, and was more mobile than previously determined" in soils of a pig farm in North Dakota, United States of America [18].

One pathway from soil to water is through preferential flow channels and drainage pipes (also called "tile drains" or "mole drains"). For the monitoring of this pathway, the sampling procedure seems to be crucial. Based on grab samples taken in creeks and from "tile drains" on four different dates, [19] found no enhanced estrogen concentrations in surface water collected upstream and downstream of a large confinement dairy operation in the mid-western United States. In contrast to that, a Danish one-year study on two field sites with "structured, loamy soil" [13] found E1 and E2 in the drainage pipes within 14–30 days after application of pig manure slurry as well as continued leaching in high concentrations after 3 months. They sampled drainage water flow-proportionally for approximately 1 day, following the onset of "typical" storm events. Matthiessen et al. [20] used "POCIS" passive samplers to monitor 10 streams in England and Wales, from November 2005 to January 2006. Their study sites lay upstream and downstream of intensive livestock farms and were chosen due to a high predicted steroid load. estrogenic activities were

higher in 50 % and steroid concentrations were higher in 60 % of the downstream sites. However, estrogenic activity could not solely be attributed to E1 and E2.

The aim of the field study presented here was to assess the role of dairy cow manure as a source of estrogenic activity in drainage water of a cattle pasture in the Swiss lowlands. Assuming a short half-life of estrogens in soil and a good sorption capacity of the local soil, we hypothesised that peaks of estrogenic activity in drainage water should be highest during or right after heavy rainfall events, ideally shortly after manure spreading. The experimental setting was designed to catch these peaks.

Results and discussion
General conditions at the Guettingen field site
Characterisation of the soil
Three soil horizons were characterised regarding their organic carbon content, volume of macropores and porosity (Table 2). Grain size analyses indicated loam to sandy loam with a low saturated hydraulic conductivity K_{sat} of 1.52×10^{-6} (equivalent to a pK_{sat} of 6.2) (C. Boesiger, ZHAW Bachelor's thesis 2010, unpublished). The infiltration experiment revealed traces of blue stain as deep as 80 cm underneath the surface, demonstrating the importance of preferential flow paths in this soil (C. Boesiger, ZHAW Bachelor's thesis 2010, unpublished).

Manure application and grazing periods
Pastures "West" and "East" (Fig. 1) were fertilised with manure on March 24, 2010, followed by a grazing period (Fig. 2). Shortly before the first field campaign (FC1), on April 21 and 28, two doses of manure were applied to pasture "West", while pasture "East" was grazed until May 12, 2010 (the density of the animals on the pastures was not monitored). During the following cool and wet period, the grass grew slowly and harvesting grass was not possible to avoid soil compaction. Grass was cut after a few warm days on June 9, 2010. On June 15, 2010, shortly before FC4, manure was spread on pastures "West" and "East".

Precipitation and outflow
Precipitation in April, May and June 2010 was 25.6 mm (FC1), 128.3 mm (FC2 and FC3) and 111.8 mm (FC4 and FC5), respectively. April 2010 was drier and warmer, while May and the beginning of June were wetter and cooler than the long-term average of 1976–1990 (source:

Table 2 Soil characteristics at the Guettingen field site

Soil horizon[a]	Depth (cm)	Organic carbon content (%)	Macropore volume (%)	Porosity (%)
Ah	0–17	3.26 ± 0.16	11	48
Bg	17–37	1.497 ± 0.07	10	38
BC	>37	0.269 ± 0.01	11	34

[a]Classification according to Swiss soil taxonomy [49]

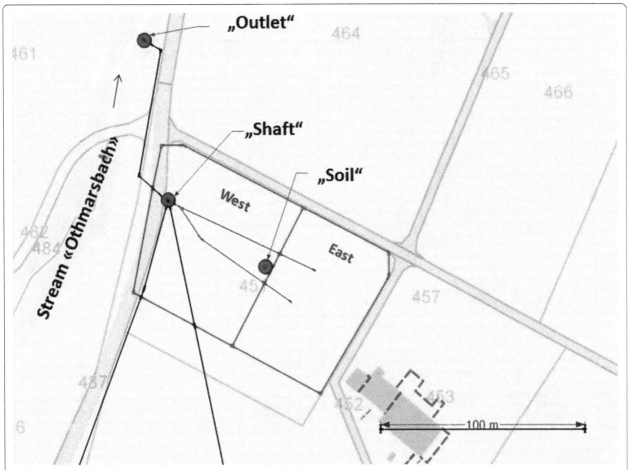

Fig. 1 Sketch of sampling location in Guettingen, Canton of Thurgau, Switzerland, showing the three sampling sites "Shaft", "Outlet" and "Soil", the drainage pipes (*green lines*) collecting at "Shaft" and the pastures "West" and "East". Source of GIS map: Gemeinde Guettingen

MeteoSwiss). FC4 (June 17–18) coincided with heavy rainfall (45.1 mm = 40.3 % of the total precipitation in June 2010).

The regular drainage outflow at "Shaft" lay between 0.05 and 0.1 l/s and exceeded this value only after rainfall events that were not absorbed by the soil. The wet weather conditions from end of April until mid-May led to rapid and transient peak outflow for a few hours following rainfall. After May 13, the general outflow increased moderately, most likely due to the increased groundwater level. Only prolonged rainfall led to persistently higher outflow at "Shaft". Figure 2 gives an overview on precipitation, outflow at "Shaft", pasture management, as well as the dates of the FC manure application and grass cutting.

Chemical and physical characteristics of drainage water
Water quality of the drainage water at "Shaft" (Table 3) reflected the agricultural influence, the geology of the catchment and the weather situation. Average nutrient concentrations (PO_4-P, NH_4-N) at times exceeded the limits set for treated wastewater by Swiss regulations. PO_4-P median concentration was 4.6 times higher and NH_4-N median concentration was 10 times higher than the mean level measured in 2009 during regular monitoring in the stream Hornbach, to which the Othmarsbach contributes. NO_3-N median concentration lay within the range of values found in literature for drains of grassland and pastures [21]. Chloride median concentration lay four to eight times above the natural background concentration of 2–4 mg/l in Switzerland [22].

Under dry weather (base flow) conditions, both hardness level and electrical conductivity were high, and oxygen saturation was mostly close to 100 %, indicating low dissolved organic matter. This shows that base flow consisted of drained groundwater from uphill. Under peak flow conditions, conductivity, in coincidence with rainfall, dropped sometimes sharply within minutes. Sudden rapid drops of oxygen saturation were also observed and coincided with visible leaching of organic material from manure into the drainage pipe. Thus, peak flow consisted of rainwater and sometimes of leached manure.

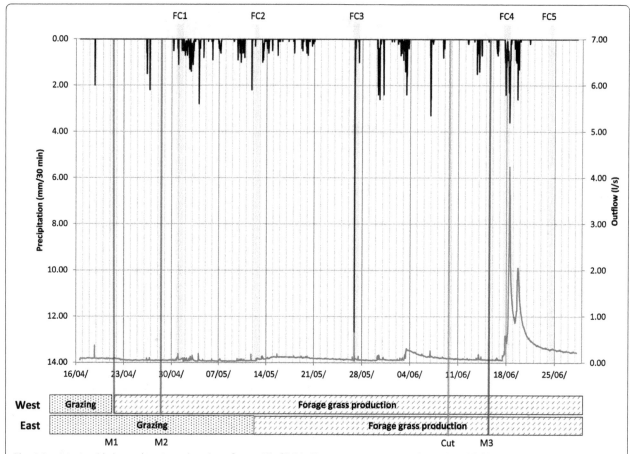

Fig. 2 Precipitation (*dark grey, hanging columns*), outflow at "Shaft" (*black*), pasture management (*bottom*) and field campaigns (FC, *light grey bars*) at the Guettingen field site. Manure applications (M1, M2, M3) are indicated by *vertical dashed lines*. A *vertical dotted line* indicates grass cutting

Soil moisture tension

Rainfall and soil moisture tensions (SMT) at 12.5 and 25 cm depth are summarised in Fig. 3. The sensors at 50 cm depth did not work reliably and these data were therefore excluded.

The Ah-horizon (12.5 cm) reached a SMT of >900 hPa (very dry) three times within the field period. After rainfall, soil water tension dropped (sometimes sharply) depending on the amount of rain. The largest drop of SMT was during FC3 (from 510 to 105 hPa within 24 h). In the rainy period between May 1 and May 22, SMT of the Ah-horizon was mostly below 100 hPa (saturated).

In the Bg-horizon (25 cm), the SMT was usually lower than in the Ah-horizon. One of the three exceptions was

Table 3 Results of the water chemistry and physical parameters at "Shaft": measurements between April 16 and June 28, 2010

Parameter	Median	Min	Max	Values below range[c]	Values above range[c]	Hornbach 2009
PO₄-P (mg/l), $n = 21$	0.23	0.06	0.91	3	8	0.05
NO₃-N (mg/l), $n = 21$	8.28	5.93	10.6	0	0	5.15
NH₄-N (mg/l), $n = 21$	0.2	0.02	0.35	9	1	0.02
Chloride (mg/l), $n = 19$	15.9	7.42	23.5	0	0	14.65
Hardness (°dH), $n = 3$	18.5	14.5	18.8	0	11	n.a.
Electrical conductivity (µS/cm), $n = 16$	692[b]	595[b]	786[b]	–	–	611
Oxygen (mg/l), $n = 4498$[a]	9.10[a]	6.12[a]	10.21[a]	–	–	10.55
pH, $n = 16$[b]	7.76	7.01	8.13	0	0	8.2

n.a. no data available

[a]Measurement with Troll 9500

[b]Measurement with Hach HQ40

[c]Measurement ranges: PO₄-P 0.05–1.5 mg/l, NO₃-N 0.23–13.5 mg/l, NH₄-N 0.015–2.0 mg/l, Cl 1–1000 mg/l, hardness 1–20 °dH, electrical conductivity 0.01µS/cm–200 ms/cm

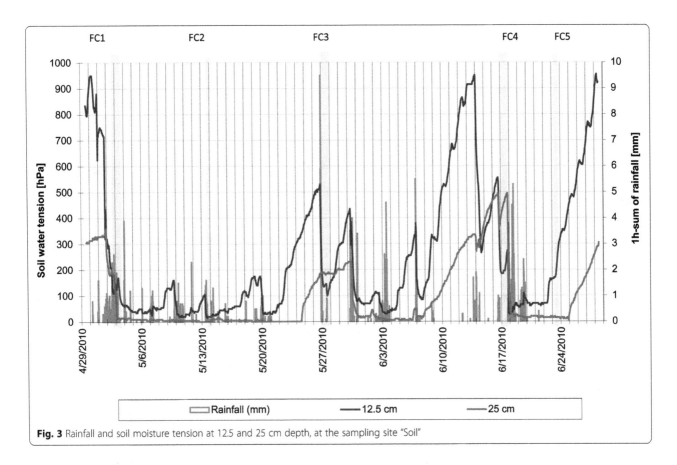

Fig. 3 Rainfall and soil moisture tension at 12.5 and 25 cm depth, at the sampling site "Soil"

during FC3 (May 27–28), when the SMT of the Ah-horizon dropped below the SMT in the Bg-horizon, due to the short but intense rainfall event of FC3. In the rainy period between May 3 and May 24, the Bg-horizon was completely saturated, with an SMT <15 hPa from May 3 to May 24.

estrogenic activity of manure and drainage water samples

estrogenic activity and bioassays

Receptor-based estrogen assays examine the sum of all estrogenic activity in a sample by measuring the response of a cell system exposed to a sample. The denotation "estrogen" summarises different natural and synthetic estrogens (e.g. estrone (E1), 17ß-estradiol (E2) and 17α-ethinylestradiol (EE2)) as well as their numerous conjugates (e.g. glucuronides, sulphates, disulphates, e.g. estrone-3-sulphate (E1-3S)). estrogen conjugates are relevant in this context because their receptor-binding potency is much lower than that of estrogens. However, they can be transformed to estrogens by deconjugation. Their potential estrogenic activity cannot be detected with bioassays prior to deconjugation. Apart from estrogens, the estrogen receptor can also be activated by non-steroidal substances that imitate estrogens and bind to the estrogen receptor (xenoestrogens).

In an attempt to standardise the effects of the different estrogens in receptor-based assays, their relative estrogenic potency (REP) has been defined in relation to the effect of E2 in the respective assay (e.g. [23]). The REP is bioassay specific and can vary for one bioassay between different laboratories [11].

Substances stimulating the estrogen receptor are called agonists, while substances with an inhibiting effect are called antagonists. Various natural and synthetic substances are known to have antagonistic effects in receptor-based estrogen assays [24]. In environmental samples, such as manure, soil or drainage water, agonistic and antagonistic effects may be modulated by matrix effects, caused by adsorption to particles or by chemical binding to colloidal organic substances [25].

estrogenic activity in manure samples

The estrogenic activity of the manure applied at the Guettingen field site varied from 201 to 2675 $\mu g/m^3$ EEQ (ER-CALUX) and from 955 to 7888 $\mu g/m^3$ EEQ (yeast estrogen screen (YES)) (Table 4). The highest EEQ (YES)-value was more than 20 times higher than the lowest EEQ (ER-CALUX)-value. The EEQ in all manure samples was 2.9 to 14.2 times higher if measured with the YES than with the ER-CALUX. In the

Table 4 Composition, date of application, applied load and EEQ of three manure mixtures (M1–M3) applied at the Guettingen field site during this study

No.	Date of manure application	Manure composition/substrate extracted	Load (m³/ha)[a]	EEQ (ER-CALUX) (µg/m³)	EEQ (YES) (µg/m³)
M1	April 21, 2010	50 % cattle/50 % chicken manure	30	201	955
M2	April 28, 2010	100 % cattle manure	20	2675	7888
M3	June 15, 2010	66 % cattle manure/33 % water	30	480	6816
K1	–	Ultra pure water	–	<LOD	<LOQ
K2	–	Ultra pure water spiked 2 µg/kg E2	–	1078	611
K3	–	M2, spiked with 2 µg/kg E2	–	2909	9856

K1 to K3 are procedural controls. The EEQ was calculated assuming a manure density of 990 kg/m³
[a]According to W. Vogt (personal communication)

control K2, EEQ (ER-CALUX) and EEQ (YES) differed only by a factor of 0.6.

In comparison to values reported in literature, our measurements are in the lower range. Dyer et al. (2001, cited by Hanselmann et al. [26]) measured 3300 ng/kg (wet weight) of E2 in liquid dairy manure, which is equivalent to 3333 µg/m³. Based on earlier work of Raman (2004, as cited by [7]), Johnson et al. [7] calculated a 17ß-estradiol equivalent of 31 µg/kg for typical dairy cow manure, corresponding to 31,313 µg EEQ per m³ of manure.

The composition of manure depends on different factors: storage time influences the concentration of estrogens and the distribution between the different estrogens [27]. Age structure of the cattle herd, the ratio of pregnant cows, livestock husbandry conditions (grazing vs. confinement) and the use of feed additives may influence the amount of estrogens in manure. Finally, farmers mix manure with water or other types of manure (Table 4), according to their experience, needs, manure type and availability. In this study, EEQ values of manure samples were systematically higher in the YES than in the ER-CALUX. This finding will be discussed in the "Oestrogenic activity in drainage pipe water" section.

estrogenic activity in drainage pipe water

The conditions during the five field campaigns are summarised in Table 5. estrogenic activities (EEQ) found at the drainages "Shaft" and "Outlet" are summarised in Fig. 4 and Fig. 5, respectively. The LOD of ER-CALUX and YES are reported in the "Analysis of estrogenic activity" section.

In field campaign FC1 (2 days after manure application on pasture "East"), both ER-CALUX and YES recorded a slight increase in EEQ at "Shaft", close to the LOD of both assays. At "Outlet", where the drainage waters of the whole system merge, the EEQ values were below the level of quantification (ER-CALUX) and below the level of detection (YES). Outflow at "Shaft" was only moderately increased by the 4.6 mm of rain. Soil moisture tension in the lower part of the soil remained high and did not drop, as it would be expected during an infiltration event (Fig. 3). This indicates that the upper soil layer was able to absorb the rainwater and no significant washout of manure constituents took place.

Field campaign FC2 (14 days after manure application on pasture "East") was conducted on a sunny day between 11:30 and 15:30. A rain event had been forecasted for this

Table 5 Frame conditions and oestrogenic activity of the five field campaigns

ID	Duration of FC, date, time	Days since manure application	Rain during FC (mm)	Soil moisture tension at 12.5 cm (hPa) Start	End	Max. EEQ (ER-CALUX) ± MU (ng/l EEQ)[a] Shaft	Outlet	Max. EEQ (YES) ± SD (ng/l EEQ)[b] Shaft	Outlet
FC1	24 h April 30, 16:00 to May 1, 16:00	2 days (East)	4.6	820	720	0.054 ± 0.01 (LOD 0.0008)	<LOQ (0.023) (LOD 0.0008)	0.14 ± 0.02	n.d.
FC2	4.5 h May 12, 11:00 to May 12, 15:30	14 days (East)	0	40	50	4.7 ± 1.22 (LOD 0.0008)	0.15 ± 0.04 (LOD 0.0008)	7.98 ± 2.65	0.69 ± 0.21
FC3	24 h May 26, 16:00 to May 27, 16:00	28 days (East)	18.8	510	105	0.74 ± 0.19 (LOD 0.0008)	10 ± 2.60 (LOD 0.0008)	2.9 ± 0.22	11.07 ± 2.31
FC4	24 h June 17, 16:00 to Jun 18, 16:00	2 days (East + West)	45.1	260	70	9 ± 2.34 (LOD 0.0008)	0.34 ± 0.09 (LOD 0.0008)	14.08 ± 1.05	0.65 ± 0.02
FC5	4 h June 24, 11:20 to June 24, 15:20	9 days (East + West)	0	355	380	0.049 ± 0.01 (LOD 0.0008)	0.048 (LOD 0.0008)	n.d.	n.d.

MU measurement uncertainty, 26 %, SD standard deviation, n.d. not detectable
[a]Data as reported by BDS
[b]Data as reported by Ecotox Centre

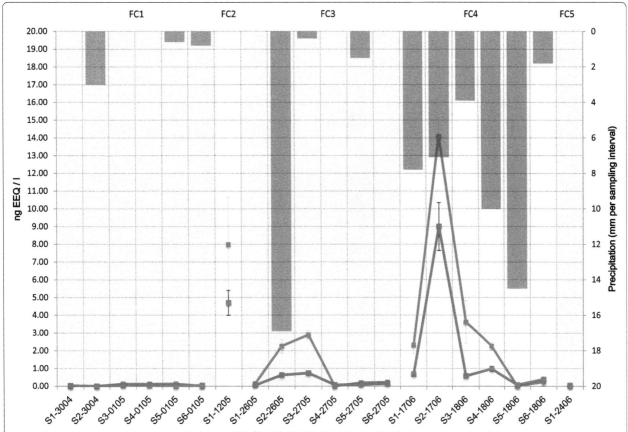

Fig. 4 Oestrogenic activity in water samples taken at "Shaft": EEQ ER-CALUX (*red line*), YES (*blue line*), precipitation (*hanging columns, right y-axis*), and field campaigns (FC). Error of ER-CALUX and standard deviation of YES are indicated by error bars. Sampling interval was 4 h

afternoon, and when it did not occur, we intended to use the FC as a dry weather reference. In the 12 days before FC2, a total of 67.9 mm of rain was recorded. The last rain fell 17.5 h before FC2. Outflow at "Shaft" was moderately stable during FC2, and the soil moisture tension of the upmost soil layer indicates near water-saturation of the soil (Fig. 3). Regarding the stable hydraulic conditions, EEQ values found at "Shaft" were unexpectedly high. The estrogenic activity may originate from cowpats and cow urine, since the grazing period had just ended a day before. Another possible explanation is that estrogens from manure application bound to soil particles were released due to the water-saturation of the soil, which is usually combined with anaerobic conditions.

Field campaign FC3 was conducted 28 days after the last manure application on pasture "East". The largest share of the rain fell within 30 min around 20:00. Outflow at "Shaft" increased within 30 min after the onset of the rainfall and decreased rapidly within an hour after its end. The soil moisture tension of the top soil dropped sharply during FC3, from rather dry to almost saturated (Fig. 3). EEQ values at "Shaft" were low before the rainfall, showed an increase between 20:00 and 4:00 to a maximum of 0.74 ± 0.19 ng/l (ER-CALUX) and 2.9 ±

0.22 ng/l (YES), and dropped to very low values right after that (Fig. 4). The origin of these estrogenic compounds must be located in the direct catchment of "Shaft". If manure or cow excreta were the origin, the estrogenic compounds must have been stable in the soil for at least 14 days—the time since the end of the grazing period. Surprisingly, at "Outlet", EEQ values increased to 11.07 ± 2.31 ng/l (YES) and 10 ± 2.60 ng/l (ER-CALUX) between 0:00 and 8:00 (Fig. 5). An explanation for these high EEQ peaks at "Outlet" is manure application by farmers in the upper parts of the catchment of "Outlet" on the afternoon just before FC3, followed by direct washout of manure from these pastures by rainfall.

Field campaign FC4 was conducted 2 days after manure application on the catchment of "Shaft", following the grass harvest. Abundant rain fell during FC4, as part of a cold front. Drainage pipe outflow increased from the usual low values around 0.1 l/s to more than 4 l/s during FC4 and only declined slowly afterwards. Soil moisture tension of the top soil dropped sharply during FC4, from moderately dry to almost saturated (Fig. 3). All samples taken during FC4 were brownish in colour. The sample taken from 20:00 to 24:00 at "Shaft" strongly smelled like manure, and the others a little less

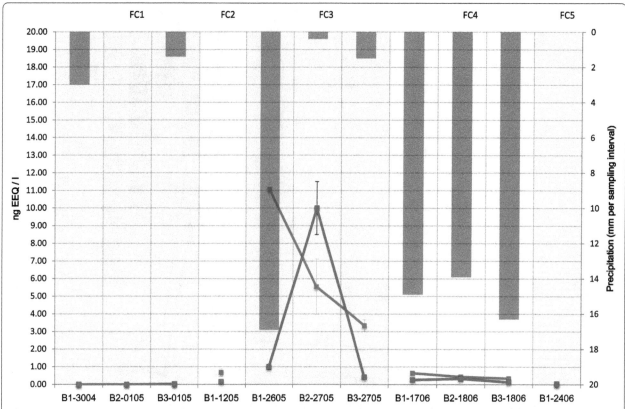

Fig. 5 Oestrogenic activity in water samples taken at "Outlet": EEQ ER-CALUX (*red line*), EEQ YES (*blue line*), precipitation (*hanging columns, right y-axis*) and field campaigns (FC) at the Guettingen field site. Error of ER-CALUX and standard deviation of YES are indicated by *error bars*. Sampling interval was 8 h

but still perceivably. The EEQ at "Shaft" reached a maximum of 9 ± 2.34 ng/l (ER-CALUX) and 14.08 ± 1.05 ng/l (YES) in this sampling period and rapidly dropped to lower values afterwards (Fig. 4). The high EEQ peaks after the onset of the rain can be explained by direct washout of manure-borne estrogenic compounds by rainfall. At "Outlet", outflow was torrential during FC4, leading to a dilution of the estrogenic load, resulting in EEQ values of 0.34 ± 0.09 ng/l (ER-CALUX) and 0.65 ± 0.02 ng/l (YES) (Fig. 5).

Field campaign FC5 was conducted 9 days after manure application on a sunny day, in a phase of declining outflow following FC4, and more than 4 days after the last rainfall. Outflow at "Shaft" was 0.3 l/s during all of FC5. The EEQ (ER-CALUX) was close to the level of detection. With the YES, no EEQ was detectable.

The three procedural blanks (K1-0705, K3-0705, K1-0306) showed no detectable EEQ in the YES. In the ER-CALUX, the procedural blank was below the level of quantification (LOQ, <0.016 ng/l) in two of the three cases, and 0.035 ± 0.009 ng/l in one case. The five procedural blanks spiked with 2 ng/l E2 had a mean EEQ of 0.81 ± 0.30 ng/l (ER-CALUX) and 1.94 ± 0.53 ng/l (YES). The procedural blank spiked with 10 ng/l E2 had

an EEQ of 5.5 ± 1.43 ng/l (ER-CALUX) and 20.1 ± 0.7 ng/l (YES) (see Table 6).

Throughout this study, the EEQ (ER-CALUX) values in water and manure samples were consistently lower than those measured with the YES. Several studies have already shown that different ER-bioassays lead to different EEQ values when analysing the same environmental sample. Reasons for these differences are known and most likely due to the differences of the assays. In the specific case of agricultural estrogens, one explanation is the difference in sensitivity of the two assays towards estrone. estrone is 10 times less potent in the ER-CALUX (REP of 0.02, E2 has an REP of 1) than in the YES (REP of 0.265). This may already explain the continuously lower EEQ values measured in the ER-CALUX, as large parts of the estradiol present in manure is oxidised to estrone within hours [28, 13]. However, the lower EEQ values of the ER-CALUX could have also been caused by an unknown antagonistic effect. The scope of this study does not allow a final conclusion.

The field campaigns showed that, when heavy rainfall occurred after manure spreading, EEQ values in drainage pipe water at "Shaft" and "Outlet" reached peak values higher than 10 ng/l during short periods of 4–8 h.

Table 6 Water samples, procedural controls: expected and measured EEQ

Sample	Procedural control processed with SPE	SPE processed vol. (l) (conc. factor)	Expected EEQ (ng/l)	EEQ (ER-CALUX) ± MU (ng/l)[a]	EEQ (YES) ± SD (ng/l)[b]
K1-0705	HPLC-grade water	0.991(1713 x)	0	<LOQ (0.016) (LOD 0.008)	n.d.
K2-0705	HPLC-grade water, spiked to 2 ng/l with E2	0.967(1672 x)	2	0.75 ± 0.195 (LOD 0.008)	1.72 ± 0.78
K3-0705	9:1 HPLC-grade water : acetone	0.350(601 x)	0	<LOQ (0.079) (LOD 0.008)	n.d.
K1-0306	HPLC-grade water	0.977(1656 x)	0	0.035 ± 0.009 (LOD 0.008)	n.d.
K2-0306	HPLC-grade water, spiked to 2 ng/l with E2	0.843(1425 x)	2	0.45 ± 0.117 (LOD 0.008)	1.81 ± 0.15
K1-0107	HPLC-grade water, spiked to 2 ng/l with E2	0.838(1425 x)	2	1.3 ± 0.338 (LOD 0.008)	1.91 ± 0.05
K1-0809	HPLC-grade water, spiked to 10 ng/l with E2	0.992(1993 x)	10	5.5 ± 1.43 (LOD 0.013)	20.1 ± 0.7
K2-0809	HPLC-grade water, spiked to 2 ng/l with E2	0.996(1996 x)	2	0.77 ± 0.2 (LOD 0.013)	2.83 ± 0.18
K3-0809	HPLC-grade water, spiked to 2 ng/l with E2	0.995(2003 x)	2	0.8 ± 0.208 (LOD 0.013)	1.42 ± 0.1

MU measurement uncertainty, 26 %, *SD* standard deviation, *n.d.* not detectable
[a]Data as reported by BDS
[b]Data as reported by Ecotox Centre

Different timing or different practices of neighbouring farmers concerning manure application can lead to a series of EEQ peaks in the receiving stream: during FC3, at "Outlet", EEQ values >10 ng/l were detected, while EEQ at "Shaft" remained much lower. This was due to manure application in the upper part of the catchment of "Outlet". In contrast, during FC4, the EEQ was much higher at "Shaft" than at "Outlet". Furthermore, in two of our five field campaigns, estrogenic activity was found in the drainage pipe water at "Shaft" without an apparent link to manure application directly before the field campaign (FC3), or even rainfall during the field campaign (FC2). This coincides with the results of Kjaer et al. [13]. They found that a washout of estrogens through drainage pipes can still occur months after manure application, and related it to soil conditions. Gall et al. [29] observed that "significant export (of hormones) was found during the spring prior to the addition of animal wastes". This suggests that "soil may act as a long-term reservoir for E2 in the environment" [30].

Contribution of phytoestrogens to total estrogenic activity

In Fig. 6, EEQ (ER-CALUX) values of eight selected samples are compared to the calculated estrogenicity (calEEQ) of six phytoestrogens measured in this study. In six out of eight samples, these phytoestrogens explain less than 13 % of the EEQ (ER-CALUX). Under peak outflow conditions, as they were found at "Outlet" in B2-2705 and at "Shaft" in S2-1706, the calEEQs of these phytoestrogens explain less than 3 % of the total EEQ (ER-CALUX). The unexpectedly high EEQ value in sample S1-1205, which was sampled under low-outflow and dry weather conditions, thus cannot be attributed to phytoestrogens.

Conclusions

The results of this study can be summarised as follows:

- EEQ values in manure vary greatly.
- Under base flow conditions, the EEQ values in drainage water are either below the level of detection (LOD) or in the lower range of all reported measurements.
- Manure-borne estrogenic activity in drainage pipe waters can temporarily reach EEQ values higher than 10 ng/l for 4–8 h.
- The highest peak values were found during heavy rainfall events, 1–2 days after manure spreading.
- Two neighbouring drainage catchments can show different patterns regarding EEQ peak values in drainage water.
- High EEQ peaks in drainage water can also occur weeks after manure application. In two cases, we found high EEQ values at "Shaft" 14 days (FC2) and 28 days (FC3) after the last manure application. This supports the hypothesis of Schuh et al. [18] that "soil may act as a long-term reservoir for E2 in the environment".
- Washout in these cases seems to be linked with soil moisture tension.
- Grab water samples from drainages at base flow conditions are therefore not useful for assessing the EEQ load from a drainage catchment.
- Peak EEQ concentrations will often have a time coincidence with peak runoff in the whole catchment and thus will be diluted, as it was observed in FC4. This will lower the concentration to which organisms are exposed.

Our results are supported by the findings of Gall et al. [29] in their study on hormone export from a "tile-

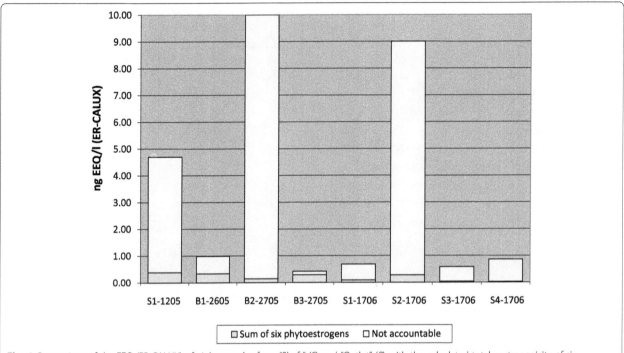

Fig. 6 Comparison of the EEQ (ER-CALUX) of eight samples from "Shaft" (*S*) and "Outlet" (*B*) with the calculated total oestrogenicity of six phytooestrogens measured in the sample extracts with LC/MS/MS

drained agroecosystem receiving animal wastes". They found that "higher hormone concentrations generally occurred during discrete periods of increased flow", "high flow rates often were associated with a disproportionately high hormone flux" and "hormone fluxes were highest during storm events that occurred shortly after animal waste applications".

Organisms (including their eggs, embryos and hatchlings) living in streams in agricultural areas with manure application and a high percentage of drained area may thus be exposed to numerous manure-borne EEQ peak concentrations per year. Fish eggs and fish embryos in the sediment would inevitably be exposed to such EEQ peak concentrations, because they are stationary. Brown trout (*Salmo trutta fario* L.) is a typical fish of small streams in the Swiss lowlands, whose embryogenesis may be affected by this. It spawns between October and January [31]. At the typical winter water temperatures of 4–5 °C, development of the eggs takes 90–100 days, and ends between January and April. Brown trout populations have been declining in Switzerland since 1980 [32].

Schubert et al. [33] examined the sensitivity of brown trout embryos (*Salmo trutta fario* L.) to "environmentally relevant concentrations" of E2 in the time between fertilisation and hatch (70 days). They exposed the fish embryos to transient E2 concentrations of 3.8 and 38.0 ng/l E2 for 2 h. Four scenarios were investigated: exposure (a)

directly after fertilisation, (b) at "eyeing stage", (c) weekly until hatch, and (d) bi-weekly until hatch. Their conclusion was that "even a single, transient E2 exposure during embryogenesis" has "significant effects on brown trout development".

In the Canton of Thurgau, manure is applied four to six times per year on managed pastures. The first application in March has a possible time coincidence with the hatching period of brown trout, while the last in October or November has a time coincidence with spawning. Since farmers have individual strategies for manure application, it is not practised in a synchronised manner, which potentially increases the number of EEQ peak concentrations in a given stream location. Development of brown trout embryos also falls into a season with low vegetation activity, low soil temperature and low general biological activity. Degradation of a postulated EEQ reservoir in the soil would therefore be slower than in spring or summer, and EEQ washout-events in connection with rainfall may even be more likely than in warmer seasons. If embryogenesis of brown trout should be negatively affected by manure-borne EEQ peaks, this may be an additional reason for declining brown trout populations in Switzerland.

There are also indications that pulses of estrogens in low concentrations can cause effects in juvenile or adult fish exposed to them. Hyndman et al. [34] examined the effects of differential timing of exposure with E2 on a range of fathead minnow (*Pimephales promelas*) biomarkers in a laboratory study. They found that the ability of treatment male fish to hold nest sites in direct competition with control males was sensitive to E2 exposure. Labadie and Budzinski [35] concluded that juvenile male turbots (*Psetta maxima*) are susceptible to hormonal imbalance as a consequence of short-term exposure to environmentally relevant 17α-ethinylestradiol (EE2) levels.

The postulated "long-term reservoir (…) in the soil" [19] will probably contain natural estrogens in conjugated and deconjugated forms, with specific REP for every substance. Urine of pregnant cattle for example contains estrone-3-sulphate (E1-3S), the dominating conjugated form of E1, which has a very low relative potency in the E-screen assay (0.000012, [10]). It can be expected that E1-3S will be deconjugated over time, but it is, to our knowledge, unknown how fast this process proceeds. If conjugated estrogens are "hydrolysed to their free forms in the environment, they could contribute additional estrogenic activity" [10]. Degradation of estrogens is reportedly slowed down by anaerobic conditions [17], sheep urine [15] and by DOC from manure [16]. It can be expected that these factors will influence the half-life-time of estrogen conjugates and thus the size and estrogenic activity of the "long-term reservoir" as well. This aspect deserves further investigation.

Effects observed in bioassays display the overall estrogenicity of a sample and cannot easily be associated to specific substances. In this study, we could not include chemical measurements of natural estrogens. Thus, a complete toxicity identification evaluation was not possible. Further research is therefore necessary to verify and understand the postulated "long-term reservoir" and link specific estrogens to the EEQ values measured with bioassays.

Methods

One drainage pipe from a cattle pasture was monitored for 73 days in spring 2010 (April 16–June 28). A set of physical and chemical parameters was monitored continuously (see "Contribution of phytoestrogens to total estrogenic activity" section). During three typical storm events and two reference periods without rainfall, water samples were collected in consecutive intervals at two sampling sites (see "Estrogenic activity of manure and drainage water samples" section). estrogenic activity in the samples was determined with the ER-CALUX® and the YES bioassays. Manure-related nutrients were determined using field equipment.

Sampling location

This study was performed at a 30-ha dairy farm in Guettingen, Switzerland, with about 60 dairy cows, one bull and ten calves. The stocking density of about 2 animal

units/ha was slightly above the Swiss average of 1.71 [36]. The study area is located on a gently north-sloping hillside facing the Lake of Constance. The soil developed on glacial till and can be characterised as well weathered, slightly acidic brown earth. It is influenced by groundwater in the lower parts of the horizon (C. Boesiger, ZHAW Bachelor's thesis 2010, unpublished), which was the reason for draining it. Guettingen lies 440 m above sea level and has an average precipitation of 916 mm per year (source: MeteoSwiss). The climate at Guettingen can be characterised as temperate oceanic to humid continental [37].

Monitoring and sampling was done in the pipe outlet draining the dairy cow pastures "West" and "East" (sampling site "Shaft", Fig. 1). In addition, samples were collected where the whole drainage system enters the stream Othmarsbach (sampling site "Outlet"). Drainage at "Shaft" originated exclusively from the two cow pastures. Water at "Outlet" was a mixture of drainage pipe water from "Shaft", road runoff and the much larger area of drained pastures uphill. It cannot be excluded that the water at "Outlet" contained traces of domestic wastewater. The exact contribution of drainage water from "Shaft" to the outflow at "Outlet" is unknown.

The GIS record of the drainage system shown in Fig. 1 is probably incomplete. According to W. Vogt, the drainage pipes collecting at "Shaft" (built around 1950) lie at an estimated soil depth of about 1–1.50 m and have a total length of about 40–50 m. Based on these incomplete data, the size of the catchment drained at "Shaft" was roughly estimated at about 500 m^2. Additional surveying was not possible within this project. Soil parameters were examined at the sampling point "Soil", between the two drainage pipes.

On average, manure is spread four to six times per year on the farm's pastures, using a drag hose device. Cow manure is sometimes mixed with pig and chicken manure from neighbouring farms. The grass of the pastures is harvested or used by direct grazing at different times of the year (personal communication of the farmer).

Sampling

As a general precaution, all material that came into contact with the samples was pre-rinsed three times with ultra-clean acetone (puriss.p.a., Sigma-Aldrich 00570). All dilutions were made with HPLC-grade water (J.T.Baker 4218). Contact of the sample with plastics was avoided as much as possible.

Manure samples were collected with a scoop on the day of manure application and transferred directly to 1-litre glass bottles. They were immediately deep-frozen until further processing in the lab.

During rainfall events, water samples were collected in time-dependant steps with auto-samplers (ISCO). Sampling generally started before the rainfall events, and continued for 24 h in consecutive sampling intervals. Sampler settings and sampling intervals are summarised in Table 7.

Five FC were conducted (see Fig. 2). FC1 (30.4./1.5.), FC3 (26.5./27.5.) and FC4 (17.6./18.6.) started at 16:00 and lasted for 24 h, equalling six sampling intervals at "Shaft", and three sampling intervals at "Outlet", respectively. FC2 (12.5., 11:30–15:30) and FC5 (24.6., 11:20–15:10) lasted for 4 h, or 1 sampling interval at each location. They were conducted during dry weather conditions.

After each sampling interval, the collected samples were removed from the auto-samplers. Three 600-ml subsamples were taken from each sample and stored in 1-litre glass bottles. The remaining volume was analysed for chemical parameters. The bottles were labelled with "S" (Shaft) or "B" (Outlet), the number of the interval, the number of the subsample and the date. All samples were deep-frozen at –18 °C within 2 h after collection.

Sample preparation and extraction

The frozen samples were transported to the lab and stored there at –18 °C until preparation.

For the extraction of manure samples, the slightly modified method of Zhao et al. [38] was used: manure samples were thawed in a water bath at 10 °C. They were agitated vigorously to resuspend particles. Subsequently, 20 ml of raw manure was taken and mixed with 80 ml of 1 M NaOH and allowed to settle for 30 min. Of the supernatant, 6 ml was transferred to a fresh vial with 6 ml of chloroform. The vial was vortexed two times for 20 s and phase separation was awaited. An aliquot of 5 ml of the aqueous phase was transferred to another vial, neutralised with 190 µl of acetic acid (90 %) and run through solid-phase extraction (SPE) as described below.

For the extraction of water samples, the samples were thawed in a water bath at 10 °C and immediately adjusted to pH 3 (±0.1) with 1 M HCl. Thereafter, 1.2 l of each sample was filtered through 1 µm glass fibre filters (Whatman GF/F). SPE Cartridges (200 mg LiChrolut EN, Merck) were preconditioned by subsequently adding 2 ml of hexane (>99 %, Sigma-Aldrich), 2 ml of acetone (puriss.p.a, Fluka), 3 × 2 ml of methanol (>99 %, Sigma-Aldrich) and 3 × 2 ml of HPLC-grade water (J.T. Baker), and by letting the solvents run through the cartridge by gravity. After the last preconditioning step, the lower valve of the SPE cartridge was closed; it was filled with

HPLC-grade water up to the upper rim and the sample bottle was connected with a Teflon-coated tube. The filtered samples were then pulled through the SPE cartridges under vacuum, within about 60–90 min. After completion, SPE cartridges were dried under N_2 flow (0.4 bar) and stored at –18 °C if necessary. Elution was conducted with 4 × 1 ml acetone (puriss.p.a, Fluka). The extracts were evaporated under N_2 flow, redissolved in 0.5 ml absolute ethanol (puriss.p.a, Sigma), portioned and stored in silanised amber vials (Supelco 27072-U). The portions were sent cooled and via express mail to BDS (0.2 ml, Amsterdam, The Netherlands) for the ER-CALUX® assay, and to the Ecotox Centre Eawag/EPFL (0.3 ml, Duebendorf, Switzerland) respectively, for the YES-assay. The weight of the samples was recorded at each step of the procedure, and concentration factors for each sample were calculated based on these records. The mean of the concentration factors was 1615 (standard deviation 330).

To assess the accuracy of the SPE, three procedural controls (K1–K3) were processed along with the manure samples (see Table 4). Nine procedural controls (HPLC-grade water without or with acetone, unspiked or spiked) were processed along with the water samples (Table 6). The extracts were analysed on estrogenic activity as described in "Analysis of estrogenic activity" section.

Analysis of estrogenic activity

The ER-CALUX is an "estrogen-receptor mediated, chemical-activated luciferase reporter gene-expression assay" based on human U2OS cells with an exogenous hERα receptor [39]. Estrogenic activity of a sample is quantified by using the amount of luciferase activity after 24 h of exposure.

In this project, all ER-CALUX measurements were commissioned to the ISO/IEC 17025-accredited company BioDetectionServices (BDS, Amsterdam, The Netherlands). BDS received 0.2 ml of water extracts (redissolved in ethanol) for analysis. These extracts were evaporated by them, redissolved in 25 µl DMSO and used for analysis in the ER-CALUX. All extracts and reference compounds were analysed in triplicates. In the ER-CALUX, "only dilutions that are negative in the cytotoxicity test" are "used for quantification of the response" [16]. Based on sample-specific SPE concentration factors provided by us, BDS reported (a) the calculated EEQ in the matrix (in ng 17β-estradiol equivalents per litre of water), (b) the

Table 7 Settings of the ISCO auto-samplers at "Shaft" and "Outlet"

Site	Sampler	Duration of sampling interval	Number of bottles in sampler	Volume per time step	Total sample volume after interval
"Shaft"	ISCO 6712 Portable Sampler	4 h	4	0.1 l/10 min	2.4 l
"Outlet"	ISCO GLS Compact Composite Sampler	8 h	1	0.125 l/25 min	3.0 l

measurement uncertainty and (c) the level of detection (LOD) for every batch of measurements. Data were regarded "quantifiable between the limit of quantification (LOQ) and the EC50", and "only results within this range are included in the final results" (BDS reports). At BDS, all measurements with a standard deviation higher than 15 % are repeated as part of the regular laboratory routine. Measurement uncertainty was reported as 26 % for all measurements. The LOD of the ER-CALUX differed between batches but was always 0.017 ng/l E2 or lower.

The YES, an estrogen-inducible expression system, is described in detail by Routledge and Sumpter [40]. In brief, the yeast (*Saccharomyces cerevisiae*) genome carries a stably integrated DNA sequence of the human estrogen receptor (hERα). Yeast cells also contain expression plasmids carrying estrogen responsive elements, regulating the expression of the reporter gene lacZ (encoding the enzyme β-galactosidase). Thus, when an active ligand binds to the receptor, β-galactosidase is synthesised and secreted into the medium, leading to a colour change of chromogenic substrate chlorophenol red β-d-galactopyranoside (CPRG) from yellow to red.

In this project, all YES measurements were commissioned to the Ecotox Centre Eawag/EPFL (Switzerland) and were measured as described by Rutishauser et al. [41]. Based on the sample-specific SPE concentration factors, the Ecotoc Centre reported (a) the calculated EEQ in the matrix (in ng 17β-estradiol equivalents per litre of water) and (b) the standard deviation for every batch of measurements. The LOD of the YES at the Ecotox Centre Eawag/EPFL has been reported as 0.02 to 0.1 ng/l E2 [42]. The yeast cells were provided by John Sumpter (Brunel University, Uxbridge, UK). The evaluation of the generated data by fitting a dose response curve was carried out with GraphPad Prism 5 Software (La Jolla, CA, USA). The results were expressed as EC50 (the concentration causing 50 % of the maximum effect) as well as EEQ (estrogen equivalent concentration). The fit provided the EC50 value and out of this, the EC10 and EEQ values were calculated.

Chemical analysis of phytoestrogens

Selected samples from FC2, FC3 and FC4 (S1-1205, B1-2605, B2-2705, B3-2705, S1-1706, S2-1706, S3-1806 and S4-1806) were analysed for phytoestrogens at Agroscope Reckenholz-Tänikon (ART) Research Station. For budget reasons, analysis of phytoestrogens was limited to six phytoestrogens common in Swiss rivers: daidzein, genistein, coumestrol, equol, formonetin and biochanin A (selection based on [43]). The frozen sample bottles were brought to ART, thawed, spiked with an isotope-labelled internal standard, solid-phase extracted and analysed with LC/MS/MS as described by Erbs et al. [44].

Calculated EEQs (calEEQ) of the analytically determined concentrations of the estrogenic compounds were determined by multiplying the concentration of each compound with its relative potency in the YES, and adding up the values for the compounds [5]. The numbers for the relative potencies of the phytoestrogens were taken from [23].

Chemical and physical parameters

All water samples were analysed in the field on selected chemical parameters using Hach-Lange test kits (Hach-Lange, Rheineck, Switzerland) and a portable Hach-Lange Xion spectrophotometer: NH_4-N (LCK 304), NO_3-N (LCK 339), PO_4-P (LCK 349), chloride (LCK 311), German degrees of hardness °dH (LCK 327). In addition, all samples were measured with a portable multi-probe Hach HQ40d (Hach-Lange, Rheineck, Switzerland) on electrical conductivity, pH and temperature. At "Shaft", electrical conductivity, pH, temperature and oxygen content were also recorded every 15 min with a Troll 9500 multi-parameter-probe (In-Situ Inc., Ft. Collins, CO, USA) from May 12 to June 28, 2010.

Monitoring of rainfall and outflow

Rainfall data were obtained from the SwissMetNet station at Guettingen. For measuring the drainage pipe outflow at "Shaft", the outflow was directed through a V-notch weir. Water levels behind the V-notch weir were measured every 30 min by a pressure transducer (Keller Drucktechnik, type: PR-36 X W, Winterthur, Switzerland) and transmitted once a day to a server using the cellular phone network. The outflow was calculated from a calibration equation, which was based on measurements at different outflows.

Soil parameters

The soil horizons were classified according to Swiss soil taxonomy [45]. Saturated hydraulic conductivity was determined with a constant head well permeameter in the field, using water stained with "Brilliant Blue" to visualise flow paths. Organic carbon content, grain size and volume of macropores (at 60 hPa) were determined on undisturbed samples in the lab. Soil moisture tension (SMT) was monitored with six Watermark probes (Irrometer, Riverside, CA, USA) dug into the ground, positioned in the middle of the three horizons (at 12.5, 25 and 50 cm depth) and measured every 60 min with a data logger. The manufacturer Irrometer considers a soil below 100 hPa as "saturated" and between 100 and 200 hPa as "adequately wet". Between 300 and 600 hPa, an agricultural soil should be irrigated, and above 1000 hPa, it is considered "dangerously dry for production" (Irrometer Company Inc., undated) [46]. More details on the determination of soil parameters are described in (C. Boesiger, ZHAW Bachelor's thesis 2010, unpublished).

Abbreviations

calEEQ: calculated estrogenicity; E1: estrone; E2: 17β estradiol; EE2: 17α-ethinylestradiol; EEQ: 17β estradiol equivalent; FC: field campaign; LOD: level of detection; LOQ: level of quantification; REP: relative estrogenic potency; SMT: soil moisture tension; YES: yeast estrogen screen.

Competing interests

The authors declare that they have no competing interests.

Authors' contributions

AS conceived the study, acquired funding, guided the study design and coordination, analysed and interpreted the data and co-authored the manuscript. PK managed the YES measurements, participated in data analysis and interpretation and co-authored the manuscript. MK participated in study design, data analysis and co-authored the manuscript. All authors read and approved the final manuscript.

Acknowledgements

Thank you to the Swiss Federal Office of the Environment FOEN for funding this study. W. and A. Vogt generously offered help and hospitality during the field campaign. O. Feurer, C. Boesiger and P. de Buren (all ZHAW) made the field campaign possible by being willing to be on the go, irrespective of the mood changes of the local weather. M. Kneubühl and L. Petit Matile helped to measure and interpret the data. T. Mengesha (EAWAG) was directly involved in the YES measurements. R. Schönenberger and M. Suter (EAWAG) and C. Hörger (ART) were responsible for LC/MS/MS measurements of selected samples. H. Besselink did the ER-CALUX measurements. M. Faden provided one of the two ISCO samplers. Güttingen municipality provided the GIS map of the drainage system. Meteorological data have been provided by MeteoSwiss, the Swiss Federal Office of Meteorology and Climatology. Cordial thanks go to all of you for your help in this field study.

Author details

[1]Institute of Natural Resource Sciences, Zurich University of Applied Sciences, P.O. Box CH-8200, Waedenswil, Switzerland. [2]Ecotox Centre Eawag/EPFL, P.O. Box 611, CH-8600 Duebendorf, Switzerland. [3]Kanton Thurgau, Amt für Umwelt, Abteilung Gewaesserqualitaet und -nutzung, Bahnhofstrasse 55, CH-8510 Frauenfeld, Switzerland.

References

1. Tyler CR, Routledge EJ. estrogenic effects in fish in English rivers with evidence of their causation. Pure Appl Chem. 1998;70:1795–804.
2. Jobling S, Nolan M, Tyler CR, Brighty G, Sumpter JP. Widespread sexual disruption in wild fish. Environ Sci Technol. 1998;32:2498–506.
3. Vethaak AD, Rijs GBJ, Schrap SM, Ruiter H, Gerritsen A, Lahr J. Estrogens and xeno-estrogens in the aquatic environment of the Netherlands occurrence, potency and biological effects, Dutch National Institute of Inland Water Management and Waste Water Treatment (RIZA) and the Dutch National Institute for Coastal and Marine Management (RIKZ). RIZA/RIKZ-report no. 2002.001, ISBN 9036954010. 2002.
4. Christiansen L, Winther-Nielsen M, Helweg C. Feminisation of fish—the effect of estrogenic compounds and their fate in sewage treatment plants and nature, Danish Environmental Protection Agency, Environmental Project No. 729. 2002.
5. Vermeirssen EL, Körner O, Schönenberger R, Suter MJ-F, Burckhardt-Holm P. Characterization of environmental estrogens in river water using a three-pronged approach: active and passive water sampling and the analysis of accumulated estrogens in the bile of caged fish. Environ Sci Technol. 2005;39:8191–8.
6. Burki R. 2006, Environmental estrogens as endocrine disruptors in Swiss rivers: Assessment of exposure, effects on brown trout and interaction with other stressors, Ph.D. Thesis, Philosophisch-naturwissenschaftliche Fakultät der Universität Bern, Switzerland.
7. Johnson AC, Williams RJ, Matthiessen P. The potential steroid hormone contribution of farm animals to freshwaters, the United Kingdom as a case study. Sci Total Environ. 2006;362(1–3):166–78.
8. Lange IG, Daxenberger A, Schiffer B, Witters H, Ibarreta D, Meyer HHD. Sex hormones originating from different livestock production systems: fate and

9. potential disrupting activity in the environment. Anal Chim Acta. 2002;473(1–2):27–37.
9. Hanselmann TA, Graetz DA, Wilkie AW, Szabo NJ, Diaz CS. Determination of steroidal estrogens in flushed dairy manure wastewater by Gas-chromatography-mass spectrometry. J Environmental Quality. 2006;35:695–700.
10. Gadd JB, Tremblay LA, Northcott GL. Steroid estrogens, conjugated estrogens and estrogenic activity in farm dairy shed effluents. Environ Pollut. 2010;158:730–6.
11. Alvarez DA, Shappell NW, Billey LO, Bermudez DS, Wilson VS, Kolpin DW, et al. Bioassay of estrogenicity and chemical analyses of estrogens in streams across the United States associated with livestock operations. Water Res. 2013;47(10):3347–336.
12. Peterson EW, Davis RK, Orndorff AK. 17b-estradiol as an indicator of animal waste contamination in mantled Karst aquifers, J. Environ. Qual. 2000; 29: 826-34.
13. Kjaer J, Olsen P, Bach K, Barlebo HC, Ingerslev F, Hansen M, et al. Leaching of estrogenic hormones from manure-treated structured soils. Environ Sci Technol. 2007;41:3911–7.
14. Lucas SD, Jones DL. Biodegradation of estrone and 17ß-estradiol in grassland soils amended with animal wastes. Soil Biol Biochem. 2006;38:2803–15.
15. Lucas SD, Jones DL. Urine enhances the leaching and persistence of estrogens in soils. Soil Biol Biochem. 2009;41:236–42.
16. Stumpe B, Marschner B. Dissolved organic carbon from sewage sludge and manure can affect estrogen sorption and mineralization in soils. Environ Pollut. 2010;158:148–54.
17. Ying GG, Kookana ARS, Dillon P. Sorption and degradation of selected five endocrine disrupting chemicals in aquifer material. Water Res. 2003;37(15):3785–91.
18. Schuh MC, Casey FXM, Hakk H, DeSutter TM, Richards KG, KhanE, Oduor P. An on-farm survey of spatial and temporal stratifications of 17ß-estradiol concentrations. Chemosphere, 2010, doi:10.1016/j.chemosphere.2010.10.093.
19. Shappell NW, Elder KH, West M, Estrogenicity and nutrient concentration of surface waters surrounding a large confinement dairy operation using best management practices for land application of animal wastes, Environ. Sci. Technol. 2010;44:2365–2371.
20. Matthiessen P, Arnold D, Johnson AC, Pepper TJ, Pottinger TG, Pulman KGT. Contamination of headwater streams in the United Kingdom by estrogenic hormones from livestock farms. Sci Total Environ. 2006;367(2–3):616–30.
21. Hooda PS, Edwards AC, Anderson HA, Miller A. A review of water quality concerns in livestock farming areas. Sci Total Environ. 2000;250(1–3):143–67.
22. Liechti P. Methoden zur Untersuchung und Beurteilung der Fliessgewässer. Chemisch-physikalische Erhebungen, Nährstoffe. Bern: Umwelt-Vollzug Nr. 1005. Bundesamt für Umwelt; 2010. p. 44.
23. Coldham NG, Dave M, Sivapathasundaram S, McDonnell DP, Connor C, Sauer MJ. Evaluation of a recombinant yeast cell estrogen screening assay. Environ Health Perspect. 1997;105(7):734–42.
24. Fent K. 2007, Ökotoxikologie – Umweltchemie, Toxikologie, Ökologie, Thieme Verlag, Stuttgart.
25. Zhao S, Zhang P, Melcer ME, Molina JF. Estrogens in streams associated with a concentrated animal feeding operation in upstate New York. USA Chemosphere. 2010;79:420–5.
26. Hanselman TA, Graetz DA, Wilkie AC, Manure-Borne Estrogens as Potential Environmental Contaminants: A Review, Environ. Sc. Technol. 2003; 37: 5471-8.
27. Zheng W, Yates S, Bradford SA. Analysis of steroid hormones in a typical dairy waste disposal system. Environ Sci Technol. 2008;42:530–5.
28. Schlenker G, Müller W, Birkelbach C, Glatzel PS. Experimental investigations into the effect of Escherichia coli and Clostridium perfringens on the steroid estrone. Berl Munch Tierarztl Wochenschr. 1999;112:14–7.
29. Gall HE, Sassman SA, Jenkinson B, Lee LS, Jafvert C. Hormone loads exported by a tile-drained agroecosystem receiving animal manure wastes, Hydrological Processes, 2013, doi:10.1002/hyp.9664.
30. Schuh MC, Casey FXM, Hakk H, DeSutter TM, Richards KG, Khan E, et al. Effects of field-manure applications on stratified 17β-estradiol concentrations. J Hazard Mater. 2011;192(2):748–52.
31. Riedl C, Peter A. Timing of brown trout spawning in Alpine rivers with special consideration of egg burial depth. Ecol Freshw Fish. 2013;22:384–97.
32. Burkhardt-Holm P, Giger W, Güttinger H, Ochsenbein U, Peter A, Scheurer K, et al. Where have all the fish gone? The reasons why the fish catches in Swiss rivers are declining. Environ Sci Technol. 2005;39:441A–7A.
33. Schubert S, Peter A, Schönenberger R, Suter MJ, Segner H, Burkhardt-Holm P. Transient exposure to environmental estrogen affects embryonic development of brown trout (Salmo trutta fario). Aquat Toxicol. 2014;157:141–9.

34. Hyndman KM, Biales A, Bartell SE, SchoenfusS HL. Assessing the effects of exposure timing on biomarker expression using 17b-estradiol, Aquatic Toxicology 2010; 96: 264–72.

35. Labadie P, Budzinski H. Alteration of steroid hormone profile in juvenile turbot (Psetta maxima) as a consequence of short-term exposure to 17alpha-ethinylestradiol. Chemosphere. 2006;64(8):274–1286.

36. Bundesamt für Statistik, 2015, Nutztierhalter und Nutztierbestände 1985–2013, http://www.bfs.admin.ch/bfs/portal/de/index/themen/07/03/blank/data/01/03.html (3.2.2015).

37. Peel MC, Finlayson BL, McMahon TA. Updated world map of the Köppen-Geiger climate classification. Hydrol Earth Syst Sci. 2007;11:1633–44.

38. Zhao Z, Fang Y, Love NG, Knowlton KF. Biochemical and biological assays of endocrine disrupting compounds in various manure matrices. Chemosphere. 2009;74:551–5.

39. Van Der Linden SC, Heringa MB, Man HY, Sonneveld E, Puijker LM, Brouwer A, et al. Detection of multiple hormonal activities in wastewater effluents and surface water, using a panel of steroid receptor CALUX bioassays. Environ Sci Technol. 2008;42:5814–20.

40. Routledge EJ, Sumpter JP. Estrogenic activity of surfactants and some of their degradation products assessed using a recom- binant yeast screen. Environ Toxicol Chem. 1996;15:241–8.

41. Rutishauser BV, Pesonen M, Escher BI, Ackermann GE, Aerni H-R, Suter MJ-F, et al. Comparative analysis of estrogenic activity in sewage treatment plant effluents involving three in vitro assays and chemical analysis of steroids. Environ Toxicol Chem. 2004;23:857–64.

42. Escher BI, Bramaz N, Quayle P, Rutishauser S, Vermeirssen ELM. Monitoring of the ecotoxicological hazard potential by polar organic micropollutants in sewage treatment plants and surface waters using a mode-of-action based test battery. J Environ Monit. 2008;10:622–31.

43. Hoerger C, Wettstein F, Hungerbühler K, Bucheli T. Occurrence and origin of estrogenic isoflavones in Swiss river waters. Environ Sci Tech. 2009;43(16):6151–7.

44. Erbs M, Hoerger C, Hartmann N, Bucheli TD. Quantification of six phytestrogens at the nanogram per liter level in aqueous environmental samples using 13C3-labeled internal standards. J Agric Food Chem. 2007;55:8339–45.

45. ART/ACW, (2008): Schweizerische Referenzmethoden der Forschungsanstalten Agroscope, Band 1, Methode KOM, E1.056.d, Version 2008.

46. Irrometer Company Inc. (undated), Irrometer/Watermark Sensor FAQ's, http://www.irrometer.com/faq.html#when (3.2.2015).

47. Bundesamt für Statistik, 2015, Bevölkerungsstand, http://www.bfs.admin.ch/bfs/portal/de/index/themen/01/02/blank/key/bevoelkerungsstand.html (3.2.2015).

48. Ying G-G, Kookana RS, Ru Y-J. Occurrence and fate of hormone steroids in the environment. Environ Int. 2002;28:545–51.

49. Brunner H, Conradin H, Gasser U, Kayser A, Lüscher P, Meuli R, et al. Klassifikation der Böden der Schweiz, Bodenkundliche Gesellschaft der Schweiz. 2010.

4

Effects of socio-economic status and seasonal variation on municipal solid waste composition: a baseline study for future planning and development

Ali Kamran[1*], Muhammad Nawaz Chaudhry[1] and Syeda Adila Batool[2]

Abstract

Background: The present study has highlighted the effects of seasonal variation and socio-economic status on generation and composition of municipal solid waste (MSW) in Shalimar Town (ST). The total amount of MSW generated in ST is estimated to be 927 tons per day per year. The average per capita rate of MSW collected in ST is 0.69 kg per day in all four seasons.

Results: No significant difference was found in overall waste generation; however, statistical analyses show the significant difference for food waste, paper and plastic ($p < 0.01$) among socio-economic groups and seasons. The results show that the lowest income group produces 0.39 kg per capita per day during winter months which is the minimum of MSW generated as compared to high (1.1 kg per capita per day) and middle (0.56 kg per capita per day) income groups in the same season.

Conclusions: It is also concluded that the low-income group produces the minimum of waste in each of the four seasons. In terms of breakdown of the MSW, organic waste is in the highest percentage (81%) followed by paper (5%), plastic (6%), glass (2%) and others (5%). Food waste is 84% of the entire generated MSW as well as having very low heating value of 5,642 J per g. Elemental and proximate analyses of mixed food waste had carbon 48.72%, nitrogen 2.41%, hydrogen 6.37%, sulphur 0.29 and oxygen 40.15% respectively.

Background

Waste management is a complex issue, which has to assess and aptly take into account the environmental impacts, technical aspects, implementation and operational costs of each specific treatment and disposal option [1–3]. The municipal solid waste (MSW) generation has radically increased with the increase in economic growth, population as well as with the upsurge in living standards in semi-developing countries. The generated municipal solid waste is not managed appropriately in most of the under-developed countries and causes stern issues with respect to collection, storage, and ultimate disposal of waste [4–6].

Public health and environmental risks are often caused by incongruous solid waste operations, storage, collection, and transportation and disposal practices [7].

For an apposite management of urban solid waste, knowledge of solid waste composition is exceptionally important [8, 9]. There are numerous ways to deal with the solid waste generated by human activities in cosmopolitan areas of the developing countries all over the world. MSW production is strongly influenced by geographical conditions [10], rapid urbanization and pattern of consumption [11]. Those responsible for the design and operation of MSW management systems must know the sources and composition of MSW generated and to set out goals for the collection in their borough. Information on the composition of MSW helps to define equipment needs, collection methodologies, management programs and plans. The waste management process can

*Correspondence: alikamran46@yahoo.com
[1] College of Earth and Environmental Sciences, University of the Punjab, Lahore, Pakistan
Full list of author information is available at the end of the article

help conserve resources and safeguard the environment [12]. The waste generation and management issues cause amassed effects on socio-economics, public health and aesthetics of many societies around the world [13–15].

Improper storage of waste as well as open dumping of waste in streets and on open plots is a grave concern [16, 17]. Unscientific and inapt management of MSW can cause different types of environmental pollution affecting public health, with MSW providing harborage for various forms of disease-carrying vectors such as flies, mosquitoes and rats [18]. Many have stated that increasing generation of MSW is causing environmental problems in urban areas of developing countries [10, 19]. The most common methods used for the disposal of MSW include open dumping, sanitary landfilling, composting and incineration [20]. City officials often devote derisory time or resources to the management of MSW in developing countries like Pakistan [5].

Description of study area

Lahore is the second largest city of Pakistan in terms of both population and land area. The city of Lahore [latitude 31°510′N, longitude 74°0.40′E, 711 feet (217 m) above sea level] experiences four distinct seasons, namely monsoon or rainy/wet season (July, August and September), winter (November, December and January), spring (February, March and April) and summer (May, June and October). Lahore is divided into nine administrative towns. ST is one of the administrative towns and it has a total population of 1,344,000. ST is further divided into 17 union councils. Table 1 shows the self-classified monthly income and average number of family members in ST.

Waste composition was measured during each of the following seasons:

- wet (July, August and September)
- winter (November, December and January)
- spring (February, March and April); and
- summer (May, June and October)

Description of MSW collection system in ST Lahore

More than half of the waste generated in ST is collected by the Lahore Municipal Waste Corporation (LMWC) of

Table 1 Self-classified monthly income and average number of family members in ST

Income groups	Average monthly income (Euro equivalent)	Average number of family members per house
High	>50,000/- Rs. (442.5 €)	5.2
Middle	>30,000/- Rs. (265.5 €)	5.8
Low	>15,000/- Rs. (132.7 €)	6.7

the City District Government Lahore (CDGL). LMWC is responsible for the collection of MSW from most parts of the city. LMWC is funded by the Government of the Punjab. Another important data is the composition of the MSW that is generated by the residents of three socio-economic levels in ST. In Lahore, a communal container collection system is mainly in practice. The total number of steel waste storage containers in ST is 62 out of which 40 are 5 m³ in size and 22 are of 10 m³ in size. These containers are placed at different locations of the town without considering their effectiveness and suitability. MSW from these containers is transported to the dump site which is located almost 10 km by road from the town center. The transportation of MSW is on a daily basis from containers placed in high-income areas and is on alternate days in low and middle income areas. A total number of 57 vehicles are used for the collection of MSW in ST. Most of these vehicles are not properly maintained and some of them often remain out of order. The diurnal collection operation of MSW is hampered due to worn out vehicles. Out of these 57, only 5 are fully contained compactor vehicles to control supplementary pollution during transportation and to improve efficiency of waste collection operations. MSW, generated in low and middle income areas, rarely, contains high-value recyclables as these are separated at source.

In Lahore, there are three dump sites. These dump sites are located in the out skirts of Lahore. The biggest among three is known as Mehmood Booti Dump Site (located in Mehmood Booti area near the River Ravi). It is situated within the flood plain between the channel belt and the first terrace of the River Ravi. The landfilling in and around Lahore is not an environmentally adequate option due to the geological factors and the price of land.

Objective

The objective of this study is to characterize and to quantify the MSW generated from three different income groups of ST, Lahore, during four different seasons of the year and among three different socio-economic levels (income groups). For planning a sustainable integrated municipal solid waste management (IMSWM), quality and quantity of MSW should be known. The study shall be helpful in minimizing the human health risks associated with waste management.

Methods

Before carrying out the actual sampling, numerous physical surveys of the area were carried out to fully understand the collection system of MSW in ST. During the surveys, it was also observed that scavenging is very active between 7:00 am to 9:00 am. Therefore, the sample collection was carried out early in the morning starting at 6:00 am.

Sample collection procedures

Sample collection was started from wet season in the month of July using ASTM Method D5231-92. On the basis of physical surveys, 11 union councils (UCs) were selected as population of these UCs comprised of all three socio-economic levels. Since the population of low-income groups is greater than both the high and middle income groups, therefore, for better statistical results of MSW composition, four steel containers were randomly selected from low-income areas and three steel containers each were selected from high and middle income areas. Hence, sampling was conducted from randomly selected ten steel containers from all three socio-economic levels. Samples were collected for a period of 1 week from open steel containers placed in each socio-economic area during all four seasons.

Collection system in ST

In poor countries like Pakistan, picking of waste is a source of income for poor people living in the third world countries [21, 22]. Therefore, the existing MSW collection system in ST can be abridged through a schematic flow sheet as presented in Figs. 1 and 2.

Number of samples and sample size

A total of 84 samples were collected from all three socio-economic levels for all seasons. The total sample size was 8,400 kg.

Segregation of collected samples

Municipal solid waste was originally classified into 48 possible fractions. As the authors are using EASEWASTE Life Cycle Assessment (LCA) computer based software model for future MSW management and planning. The 48 default material fractions are named according to the dataset on Danish household waste 2003 [23]. During the sampling phase many of the MSW fractions were found in smidgens and hence were not weighable. Finally, the 48 fractions were reduced to 13 reported fractions.

Quantifying the waste fractions

After the sorting process, waste fractions were weighed on a calibrated digital scale. The same process was followed for the next 6 days and same method was applied in all three socio-economic groups for four seasons during the whole year.

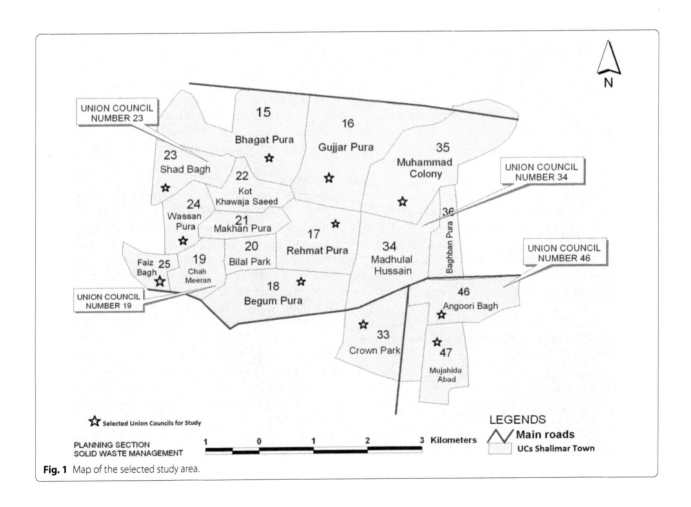

Fig. 1 Map of the selected study area.

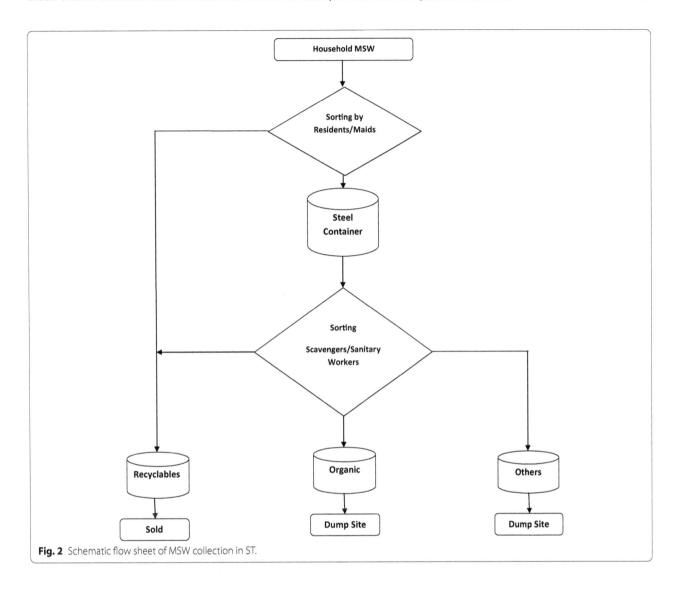

Fig. 2 Schematic flow sheet of MSW collection in ST.

Elemental and proximate analysis

Carbon (C), hydrogen (H), nitrogen (N) and sulphur (S) contents and oxygen (O) is determined by the difference using elemental analyser Elementar from Perkin-Elmer for elemental analysis. Gases resulting from combustion of sample in pure oxygen are measured by the instrument to find the content of the elements as per cent. Bomb calorimeter is used to determine the heating values [24]. Proximate analysis is performed according to the standard methods [25] which involve the determination of ash [26], fixed carbon, moisture [27] and volatile matter [28]. Upon drying the sample in an oven for 3–4 h at a temperature 103 ± 1°C, the moisture is determined by weight difference. Percentage of volatile matter is determines by the loss in weight of sample, when ignited in preheated muffle furnace at 950 ± 20°C. Finally, the fixed carbon is determined by adding the percentage values of ash,

moisture and volatile matter and the sum is subtracted from 100.

Statistical analysis

To check the efficacy of data, statistical software IBM® SPSS-16.0® used. Correlation between income groups and generated MSW was determined. Analysis of variance was carried out for MSW fractions and seasons of year.

Results and discussions

The average values of the MSW were calculated where data for the months of July, August and September represent the mean weighted values of wet or monsoon season when the rainy season increases the moisture with temperature of 25–35°C. Similarly, data for the months of November, December and January represent the mean weighted values of winter season when the temperature 3–20°C is lower. The months of February, March and April represent the mean weighted values of spring season when the temperature 12–25°C increases slightly. The months of May, June and October represent the mean weighted values of summer season when temperature 30–45°C increases and the weather is hotter. Table 2, shows the percentage of weekly generated waste for all seasons in three socio-economic levels of ST Lahore.

Chemical analysis

ASTM standard methods were performed to elemental and proximate analysis of food waste and average values both of analyses are presented in Table 3. Calorific values are not correspondent to that of fuel for combustion whereas, moisture values are very high. Having such high values of moisture and low heating value 5,642 J per g, introducing the modern waste management technologies like incineration and sanitary landfilling are not viable options.

Statistical analysis (correlation and analysis of variance)

Statistical analysis (ANOVA) shows significant difference between seasons for food waste, diapers, paper and plastic ($p < 0.01$) as shown in Table 4. A regular trend of decrease in average MSW generation rate from the higher income group to the lowest was observed. Data presented in Table 2, shows that showed that the MSW generation rate increases from spring season to wet season with highest values, then a steady decline have been observed from wet to winter season.

The increased financial status of households is likely to increase the overall MSW as the total amount of MSW and income groups was found to be significantly correlated, $r = 0.810$, $p = 0.001$. Sturdy correlation between economic categories and waste fractions has been found in food waste, paper, plastic, glass ($p < 0.001$) and yard waste and diapers ($p < 0.01$) respectively, as large amounts of these fractions are found in high-income group. Whereas, the fractions like stones, gravel, leather, rubber, textile, dust and straw, etc. present a negative correlation. The use of diapers increases in all income groups during the winter season because of the lower temperature, precluding frequent washing the infants. During monsoon season, the use of milk packed in tetrahedron packs (Tetra PAK) increases in all income levels because of humid weather. The data show no significant changes in batteries, glass, leather, and rubber during the whole year. Metals being high value recyclable are segregated at source. Textiles, including sewn clothes, are generally donated to the maids working at homes in high-income group, probably explaining why textile waste from this sector is generated in low numbers. In the middle and low-income groups, these numbers decrease further as old clothes and tailoring waste are used to make cushions. Therefore, it is revealed from the study that most of the collected generated waste in all income groups of ST is organic in nature. The results show that the collected hazardous waste is less in terms of quantity. As there is no transfer station facility in ST, the collected hazardous waste is dumped unscientifically at the dump site.

Table 2 Percentage of weekly waste generated for all seasons in three socio-economic levels of ST Lahore

Waste fractions	Wet (July/Aug/Sep)			Winter (Nov/Dec/Jan)			Spring (Feb/Mar/Apr)			Summer (May/June/Oct)		
	High	Middle	Low	High	Middle	Low	High	Middle	Low	High	Middle	Low
Food waste	52	78.1	80.1	42.4	69.7	70.6	55.8	74.6	74.8	53.2	75.2	82.3
Yard waste	11.2	5.2	0.001	5.6	2.7	0.004	15.2	3.7	0.005	13.1	3.8	0.01
Wood	0.6	0.2	tr	0.7	1.2	0.01	0.3	tr	tr	0.7	0.03	tr
Diapers	10.2	3.4	2.7	15.2	11.9	9.6	11.8	5.3	8.3	10.6	4.8	2.5
Paper & CB	8.5	3.7	1.1	10.6	5.8	2.4	7	4.7	0.97	9.7	3.9	1.2
Plastics	7.2	5.6	4.9	10.3	3.6	3.2	5.7	4.2	3.8	7.3	5.4	4.7
R & L	0.4	0.03	0.02	0.9	0.02	0.01	0.7	0.08	tr	0.6	0.2	tr
Metal	0.9	tr	tr	0.22	tr	tr	tr	tr	tr	tr	tr	tr
Glass	5.1	1.4	1.1	2.3	2.9	1.6	1.2	2.1	0.5	1.8	1.1	0.02
Textiles	0.8	0.5	0.2	2.7	0.5	0.4	0.2	2.7	0.6	0.13	2.7	0.5
Hazardous	0.09	0.09	0.6	0.7	0.03	0.3	0.04	0.3	0.5	0.06	0.13	0.7
Hygiene	0.03	0.01	tr	0.07	0.01	tr	0.01	tr	tr	tr	0.11	tr
Others	3.12	1.81	9.4	8.4	1.92	11.9	2.4	2.44	10.6	3.16	2.64	8.2
Total (%)	100	100	100	100	100	100	100	100	100	100	100	100

CB cardboard, *R & L* rubber and leather, *tr* trace.

Table 3 Elemental and proximate analyses

	Range	Average
Elemental analysis		
C (%)	46.20–49.16	48.72
N (%)	1.11–4.46	2.41
H (%)	5.68–6.62	6.37
S (%)	0.12–0.42	0.29
O (%)	36.43–41.87	40.15
C:N	10.35–43.20	28.64
Proximate analysis		
Ash (%)	0.76–2.38	1.34
Calorific value (J/g)	3,970–8,599	5,642
Fixed carbon (%)	3.42–8.09	5.41
Moisture content (%)	54.32–78.30	72.21
Volatile matter (%)	16.56–35.64	23.14

Table 4 ANOVA results for seasonal effects on waste generation

Item	df	F	Sig.
Food waste	8	0.248	0.860
Yard waste	8	0.199	0.894
Wood	8	0.987	0.446
Diapers	8	2.174	0.169
Paper & CB	8	0.171	0.913
Plastics	8	0.231	0.872
R & L	8	0.102	0.957
Metal	8	0.859	0.501
Glass	8	1.005	0.439
Textiles	8	0.241	0.865
Hazardous	8	0.040	0.988
Hygiene	8	0.450	0.724
Others	8	0.270	0.846

Socio-economic effect on waste generation

The trend of high waste generation in economically active areas is also experienced in high-income group of ST as it has been in Lahore [16, 29, 30], Abu Dhabi [31] and Morelia [32]. The food waste had the largest values in all the income groups. Figure 3 shows the mean values of different waste fractions generated in different socio-economic levels for the entire studied period.

The study further explains that the low-income group generates the least amount of MSW as compared to the other income groups. In ST, high-income group generates 43% of the entire waste generated while the contribution of middle and low-income groups is 31.4 and 25.8%, respectively.

In high-income group, the percentage of organic waste remains once again at the highest with 75% followed by paper and cardboard with 9% and plastics with 8%. Slight variations are found among the other different fractions of waste like glass, textiles and others for all socio-economic levels for all seasons. Similarly in middle and low-income groups, organic waste remains at the highest with 85 and 83% respectively.

The results also show that total amount of MSW collected is estimated to be 927 tons per day and the average collection rate in ST is 0.69 kg per capita per day. The collection rates of high, middle and low-income groups are 1.1 kg per capita per day, 0.56 kg per capita day and 0.39 kg per capita day respectively as shown in Fig. 4.

The study shows that MSW generated during winter season is significantly low as compared to the other seasons. It is because of the fact that low temperatures during the month of December and January reduce the consumption of fresh foods and drinks.

In Fig. 5 among the different fractions of collected organic waste, food waste is the most prominent with 84% followed by diapers and yard waste with 10 and 6% respectively.

Conclusions and recommendations

- Results show that carbon 48.72%, nitrogen 2.41%, hydrogen 6.37%, sulphur 0.29, oxygen 40.15%, and C and N ratio of 28.64 were the chemical composition of food waste fraction. Ash content in percentage is determined by the difference in weight by placing this moisture free sample in a muffle furnace at 580–600°C for 2 h.

- Physical survey conducted before the actual sampling, explicitly reconnoitered MSW collection system in ST Lahore. Samples of MSW were collected from steel containers placed in three different socio-economic levels in the town for four different seasons. Results show that average household collected waste is 0.69 kg per capita day and 81% of the collected waste is organic in nature.

- Food waste was 84% of total MSW with 72.21% moisture content and had a low heating value of 5,642 J/g. Seasonal variations in MSW are only significant for food waste, other organic and plastic.

- The results also show that there is positive correlation between the economic status and MSW generation ($p < 0.04$).

- Detailed composition of MSW was analysed initially using 48 fractions but sample collection was reduce to 13 fractions as most of the recyclables are separated by residents and maids/servants at source. The trend of high waste generation in economically active areas is also seen in the study area where high-income group generates 43% of entire waste generated in the town.

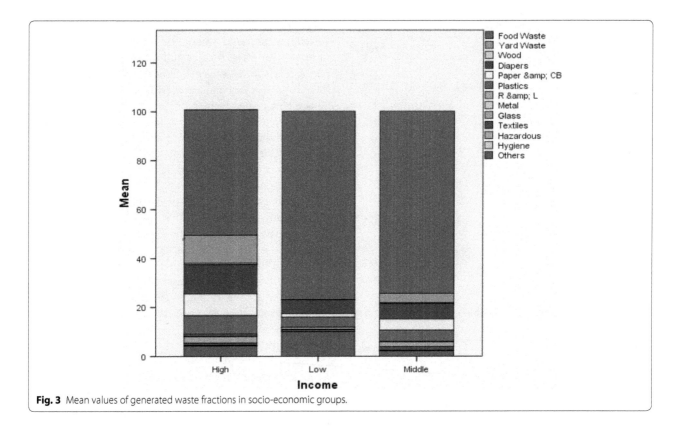

Fig. 3 Mean values of generated waste fractions in socio-economic groups.

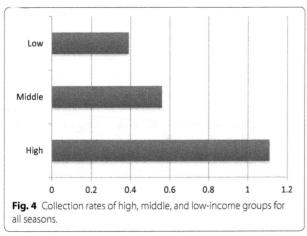

Fig. 4 Collection rates of high, middle, and low-income groups for all seasons.

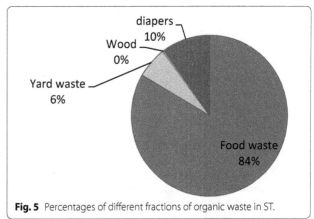

Fig. 5 Percentages of different fractions of organic waste in ST.

• Comparison of collected waste by composition and seasonal variation shows that in all three socio-economic levels slight variations are found among different fractions of organic waste like the use of diapers increases in winter season. For all studied seasons and economic levels, minimum waste is generated in the winter season as compared to the spring, summer and wet/rainy seasons.

• The upshot of this study suggested a methodology of waste composition survey, the summary of quantifi-cation and characterization of collected waste based on compostable and recyclable materials and the classification category list.

• Concerning MSW collection, transportation and dumping, the authors strongly suggest for a transfer station facility so that collected waste could be seg-regated on purely scientific grounds before the final dumping.

• The results and methodology are expected to be use-ful for authorities, planner and decision-makers of boroughs to develop an economically suitable and environmentally sustainable waste management system.

- Concerning further monitoring on MSW characterization and quantification for better results, the authors suggest that household survey and household sampling should also be conducted for different seasons. Moreover, household selection and sampling points must be prudently selected with different factors such as population density, household income, and level of urbanization.

Authors' contributions

All authors have read the manuscript thoroughly before submitting as a PhD student, 90% of the research is carried out the first/corresponding author AK and the co-authors have been guiding and supporting the corresponding author at every step of research as well as drafting and proof reading the manuscript. All authors read and approved the final manuscript.

Author details

[1] College of Earth and Environmental Sciences, University of the Punjab, Lahore, Pakistan. [2] Department of Space Science, University of the Punjab, Lahore, Pakistan.

Acknowledgements

We thank College of Earth and Environmental Sciences, University of the Punjab, Lahore, and the Higher Education Commission (HEC), Pakistan, for providing us the necessary facilities and financial support to conduct this study.

Compliance with ethical guidelines

Competing interests

The authors declare that they have no competing interests.

References

1. Kinnaman TC (2009) The economics of municipal solid waste management. Waste Manag 29(10):2615–2617
2. Consonni S, Giugliano M, Massarutto A, Ragazzi M, Saccani C (2011) Material and energy recovery in integrated waste management systems: project overview and main results. Waste Manag 31(9):2057–2065
3. Da Cruz NF, Simões P, Marques RC (2012) Economic cost recovery in the recycling of packaging waste: the case of Portugal. J Clean Prod 37:8–18
4. Al-Khatib IA, Monou M, Zahra ASFA, Shaheen HQ, Kassinos D (2010) Solid waste characterization, quantification and management practices in developing countries. A case study: Nablus district–Palestine. J Environ Manag 91(5):1131–1138
5. Batool SA, Chaudhry N, Majeed K (2008) Economic potential of recycling business in Lahore, Pakistan. Waste Manag 28(2):294–298
6. Burnley SJ (2007) A review of municipal solid waste composition in the United Kingdom. Waste Manag 27(10):1274–1285
7. World Resources Institute (1996) United Nations Environmental Program, United Nations Development Program. The World Bank, World Resources 1996–1997. The Urban Environment. Oxford University Press, Oxford. http://pdf.wri.org/worldresources1996-97_bw.pdf
8. McDougall FR, White PR, Franke M, Hindle P (2008) Integrated solid waste management: a life cycle inventory. Wiley-Blackwell, Oxford
9. Zeng Y, Trauth KM, Peyton RL, Banerji SK (2005) Characterization of solid waste disposed at Columbia Sanitary Landfill in Missouri. Waste Manag Res 23(1):62–71
10. Buenrostro O, Bocco G (2003) Solid waste management in municipalities in Mexico: goals and perspectives. Resour Conserv Recycl 39(3):251–263
11. Ashley R, Blackwood D, Souter N, Hendry S, Moir J, Dunkerley J et al (2005) Sustainable disposal of domestic sanitary waste. J Environ Eng 131(2):206–215
12. Sandulescu E (2004) The contribution of waste management to the reduction of greenhouse gas emissions with applications in the city of Bucharest. Waste Manag Res 22(6):413–426
13. Matthews E, Amann C, Bringezu S, Fischer-Kowalski M, Hüttler W, Kleijn R et al. (2000) The weight of nations. Material Outflows from Industrial Economies. World Resources Institute, Washington
14. Meyers GD, Mcleod G, Anbarci MA (2006) An international waste convention: measures for achieving sustainable development. Waste Manag Res 24(6):505–513
15. Louis GE (2004) A historical context of municipal solid waste management in the United States. Waste Manag Res 22(4):306–322
16. Batool SA, Chuadhry MN (2009) The impact of municipal solid waste treatment methods on greenhouse gas emissions in Lahore, Pakistan. Waste Manag 29(1):63–69
17. Gómez G, Meneses M, Ballinas L, Castells F (2008) Characterization of urban solid waste in Chihuahua, Mexico. Waste Manag 28(12):2465–2471
18. Kumar S, Bhattacharyya JK, Vaidya AN, Chakrabarti T, Devotta S, Akolkar AB (2009) Assessment of the status of municipal solid waste management in metro cities, state capitals, class I cities, and class II towns in India: an insight. Waste Manag 29(2):883–895
19. Pokhrel D, Viraraghavan T (2005) Municipal solid waste management in Nepal: practices and challenges. Waste Manag 25(5):555–562
20. Chen X, Geng Y, Fujita T (2010) An overview of municipal solid waste management in China. Waste Manag 30(4):716–724
21. Asim M, Batool SA, Chaudhry MN (2012) Scavengers and their role in the recycling of waste in Southwestern Lahore. Resour Conserv Recycl 58:152–162
22. Kamran A, Chaudhry MN, Batool SA (2015) Role of the informal sector in recycling waste in Eastern Lahore. Pol J Environ Stud 24(2):537–543
23. Riber C, Petersen C, Christensen TH (2009) Chemical composition of material fractions in Danish household waste. Waste Manag 29(4):1251–1257
24. ASTM STANDARD D5468-02 (2007) Standard test method for gross calorific and ash value of waste materials. ASTM International, West Conshohocken
25. ASTM Standard E870-82 (1998) Standard test method for analysis of wood fuels. ASTM International, West Conshohocken
26. ASTM Standard D1102-84 (2013) Standard test method for ash in wood. ASTM International, West Conshohocken
27. ASTM Standard E871-82 (2013) Standard test method for moisture analysis of particulate wood fuels. ASTM International, West Conshohocken
28. ASTM Standard E872-82 (2013) Standard test method for volatile matter in analysis of particulate wood fuels. ASTM International, West Conshohocken
29. Jadoon A, Batool SA, Chaudhry MN (2014) Assessment of factors affecting household solid waste generation and its composition in Gulberg Town, Lahore, Pakistan. J Mater Cycles Waste Manag 16:73–81
30. Hussain F, Chaudhry MN, Batool SA (2014) Assessment of key parameters in municipal solid waste management: a prerequisite for sustainability. Int J Sustain Dev World Ecol 21(6):519–525
31. Abu Qdais HA (2007) Techno-economic assessment of municipal solid waste management in Jordan. Waste Manag 27(11):1666–1672
32. Buenrostro O, Bocco G, Bernache G (2001) Urban solid waste generation and disposal in Mexico: a case study. Waste Manag Res 19(2):169–176

5

Acute toxicity of peroxy sulfonated oleic acids (PSOA) to freshwater aquatic species and sludge microflora as observed in laboratory environments

Stephan Solloch[1*], Nathan Pechacek[2], Bridget Peterson[2], Magdalena Osorio[2] and Jeffrey Caudill[3]

Abstract

Background: Peroxy sulfonated oleic acids (PSOA) is a novel surfactant peracid. The commercialapplications of PSOA result in the chemical primarily being disposed of via industrial waste water effluent.Given this manner of disposal, it is important to understand the aquatic hazards of the chemical to betterassess the risk posed to aqueous environments. Acute aquatic toxicity laboratory experiments wereperformed to evaluate aquatic hazards and were conducted according to standard OECD test guidelineswith rainbow trout (*Oncorhynchus mykiss*), water fleas (*Daphnia magna*) and algae (*Pseudokirchneriellasubcapitata*). In addition, microbial toxicity was evaluated in activated sludge obtained from a domesticsewage treatment facility.

Results: Lethal concentration in 50 % of test species (LC_{50}) and effect concentration in 50 % of test species (EC_{50}) values for PSOA ranged from 0.75 to 5.44 mg/L, representing a relatively small range spanning less than an order of magnitude. No observed effect concentration (NOEC) and lowest observed effect concentration (LOEC) ranges were also relatively small, with ranges of 0.25–1.66 and 0.5–3.6 mg/L, respectively. The EC_{50}, LOEC and NOEC values for microbial toxicity were 216, 60 and 20 mg/L, respectively. Predicted no effect concentrations (PNEC) for aqueous media were based on the 96-h LC_{50} (0.75 mg/L) for *O. mykiss*, the organism displaying the greatest sensitivity to PSOA. These values were derived for freshwater, marine water and intermittent releases to water and ranged from 7.5×10^{-5} to 7.5×10^{-3} mg/L. A sewage treatment plant PNEC of 2 mg/L was derived based on an activated sludge 3-h NOEC of 20 mg/L.

Conclusion: These values, along with the anticipated environmental fate and transport for PSOA, were considered in assessing the overall aquatic risk posed by this chemical. Despite the relatively high acute aquatic hazards for PSOA, environmental modeling suggests the overall risk of PSOA to aqueous environments is low based on its anticipated uses. This conclusion is consistent with the significant processing of industrial wastewater by onsite or municipal wastewater treatment facilities prior to release to the environment.

Keywords: Laboratory aquatic study, Acute aquatic toxicity, Peroxy sulfonated oleic acids (PSOA), PNEC, *Oncorhynchus mykiss*, *Daphnia magna*, *Pseudokirchneriella subcapitata*

Background

Peroxy sulfonated oleic acids (PSOA) is the common name of the substance "reaction product of sulfonated oleic acid potassium salt, hydrogen peroxide and sulfuric acid" (IUPAC name) (see Fig. 1). The official CAS entry is "9-Octadecanoic acid (9Z)-, sufonated, oxidized, potassium salts" (CAS number 1315321-94-8). As described in Pechacek et al. [13], PSOA is an organic peroxide that is characterized by the presence of one or more oxygen–oxygen bonds and is derived by reacting sulfonated oleic acid and hydrogen peroxide under acidic conditions

*Correspondence: stephansolloch@web.de
[1] Ecolab, Monheim, Germany
Full list of author information is available at the end of the article

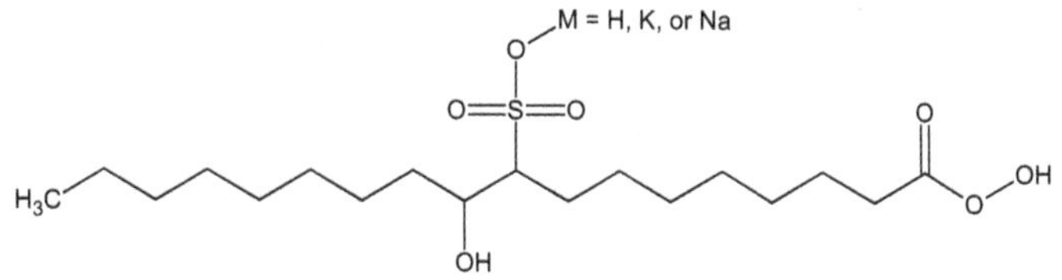

Fig. 1 Chemical structure of PSOA

[13, 14]. The oleic acid used in the production of PSOA is derived from renewable feedstocks such as plant oils and animal tallow [13]. Due to the variability inherent in the natural feedstock used to obtain oleic acid, PSOA subsequently falls into the regulatory definition of an unknown or variable composition, complex reaction products or biological materials (UVCB) substance [5, 13]. The resulting equilibrium mixture contains peroxy sulfonated acids, hydrogen peroxide, water and residual acid.

PSOA is a novel surfactant peracid developed to alleviate challenges with current peracid technology. It is commercially used as an antimicrobial, bleaching or coupling agent. Food and beverage industries primarily use PSOA as a coupling agent to clean and sanitize processing equipment [13]. PSOA is commonly available commercially as a concentrate due to its ability to maintain equilibrium longer as a concentrate than in diluted solutions. When PSOA comes into contact with organic matter such as milk or blood, it rapidly decomposes to its precursor reactants. Additionally, it rapidly decomposes when dried on a surface due to the water loss that shifts the equilibrium of the PSOA mixture [7]. PSOA is readily soluble in water with an aqueous solubility of ≥ 43.8 g/L at approximately 20 °C [15]. However, it is hydrolytically unstable and significant hydrolysis occurs at environmentally relevant pH values (4, 7 and 9) [2].

Disposal of PSOA primarily occurs in industrial wastewater effluent. Given this method of disposal, it is important to characterize the aquatic hazards of PSOA. This publication summarizes the results of four studies: a 96-h continuous flow study with rainbow trout (Oncorhynchus mykiss), a 48-h immobilization test with Daphnia magna, a 72-h algal growth inhibition test with P. subcapitata and a 3-h respiration inhibition test (ASRIT) with activated sludge. Given the known environmental fate and transport of PSOA in aqueous environments, the overall goal is to discuss the aquatic hazards identified in their appropriate risk context.

Methods

Test item
100 % UVCB substance PSOA was synthesized by Ecolab chemists and served as the test article. PSOA was stored at -20 °C and thawed prior to preparation of stock solutions, and was described as a viscous, pasty white or yellowish liquid. The PSOA used in all described tests originated from the same manufactured batch.

Experimental organisms
To determine the aquatic toxicity of PSOA, acute toxicity experiments were performed on three different freshwater species: rainbow trout (O. mykiss), water flea (D. magna) and algae (P. subcapitata). Acute lethality was evaluated for the fish (96 h), acute immobilization for the D. magna (48 h) and effects to growth rate and yield for algae (72 h). In addition, the toxicity of PSOA to indigenous microflora found in activated sludge of a domestic sewage treatment facility was assessed.

Fish
Fish obtained from Selcoth Fish Farm in Moffat, Scotland were used for the initial and repeat range finding tests, while fish from Brow Well Fisheries Ltd., Skipton, England were used in the definitive test. Fish measured 4–6 cm in length and were free from any apparent malformation or poor health. All fish were acclimated to laboratory conditions for a minimum of 12 days prior to commencement of the study. During holding, fish were fed daily. Twenty-four hours prior to initiation of the study, food was withheld from the fish and no food was provided during the study.

Daphnia magna
Daphnia were obtained from the Laboratory of Hydrobiology (Central Agricultural Office, Directorate of Plant and Soil Protection), Hungary. All daphnids were female and under 24 h old. Daphnia were bred and acclimatized in the laboratory under similar temperatures, light conditions and water quality as were used in the definitive test. Daphnia were fed a centrifuged green alga suspension during holding and no food was provided during the study.

Algae
Algae were supplied by the SAG: Collection of Algal Cultures, Institute of Plant Physiology, University of Göttingen, Germany. The stock cultures were small algal

cultures that were planted on agar regularly and transferred to fresh medium at least once every 2 months. The pre-culture was intended to give an amount of algae suitable for the inoculation of test cultures. The pre-culture was prepared with Algal Mineral Salts Culture Medium, incubated under the conditions of the definitive test and used when growing exponentially, which normally occurred after an incubation period of 3 days.

Activated sludge

Activated sludge was obtained from Totnes Sewage Works, Totnes, Devon, UK 1 day prior to the start of the definitive test. This sludge was used predominantly in the treatment of domestic sewage. The activated sludge was settled and the supernatant decanted. The settled sludge was incubated with 50 mL of OECD synthetic sewage feed per liter of sludge per day and aerated at room temperature until testing commenced. The OECD synthetic sewage feed was prepared as described in OECD Test Guideline 209 [12].

Aquaria and water quality
Fish

Fish were kept in 12-L (initial and repeat range finding tests) and 17-L (definitive test) tanks of reconstituted freshwater (RFW) according to the formula recommended by OECD Guideline 203 [9]. High-grade salts were dissolved in 500 L of deionized, reverse osmosis-grade water and this solution was metered into a flow of deionized water to produce reconstituted water. Reconstituted water was stored at 12 °C and constantly aerated. For the initial range finding test, the pH of the holding and test medium was adjusted to 5.0–6.5. For the repeat range finding test and the definitive test, a reduced pH was not required and the holding and test medium were maintained at a pH range of 6.0–8.5. Test vessels were kept within a temperature-controlled laboratory with the aim of achieving a temperature in the range of 13–17 °C. A light cycle of 16 h of light and 8 h of darkness was in operation throughout the test. Artificial daylight fluorescent tubes provided illumination.

Daphnia magna

Daphnia were kept in beakers containing approximately 40 L test medium. Reconstituted water (ISO medium according to OECD 202) was prepared by adding 25 mL from each of four stock solutions to 1 L of water [10]. It had an approximate total hardness of 249 mg/L $CaCO_3$, a dissolved oxygen concentration of 8.31–8.81 mg/L, a pH of 7.50–8.08 and a temperature range of 19.6–20.5 °C. The temperature of the climate chamber was in the range of 19.2–20.6 °C, with 16-h light and 8-h dark cycles. Artificial daylight fluorescent tubes provided illumination.

Algae

Algae were kept in 250-mL Erlenmeyer flasks with 100 mL test medium. Water temperature in the flasks was in the range of 22.4–23.0 °C while the climate chamber had a temperature range of 22.1–23.9 °C. The pH was measured at the beginning and end of the study for each test concentration and the control, and it ranged from 5.99 to 8.57. Algal flasks were continuously illuminated. The light intensity at the position occupied by the algal cultures in the flasks was approximately 7947 lux, which was maintained by fluorescent lamps with a spectral range of 400–700 nm. The light intensity between test flasks did not vary in excess of ±15 %.

Activated sludge

The activated sludge contained a total filterable solid concentration of 8307 mg/L as determined on the day of testing. The pH of the sludge was measured as 5.5 and adjusted to 7.3 by the addition of 2 M sodium hydroxide. One drop of antifoaming agent, antifoam B emulsion, was dispensed in each flask after initial foaming was observed during aeration. The antifoam B emulsion contained 10 % active silicone. Flasks containing activated sludge were aerated at a temperature of 20 ± 2 °C.

Experimental designs
Fish

Preliminary range finding testing was conducted to determine the appropriate concentrations of the test item for the definitive testing. Weighed amounts of PSOA (5.01, 10.12, 20.15 and 50.3 mg) were individually added to 1 L of RFW and ultrasonicated for 30 s to ensure solubility and homogeneity to create four nominal concentrations: 5, 10, 20 and 50 mg/L. The pH values for each solution were determined. Based on these values, the initial range finding test was conducted at 20 mg/L, as this test concentration had a pH (5.78) similar to the pH to which the fish were acclimating. The initial range finding test was conducted over a 96-h period under semi-static conditions, at nominal concentrations of 2.5, 5, 10 and 20 mg/L with an untreated control. Three fish per tank were used, with one tank per concentration and one control. Due to overt effects observed within 18 min at all test concentrations, the study was terminated and all fish were humanely euthanized to avoid additional stress. A repeat range finding test was conducted at 0.002, 0.02, 0.2 and 2 mg/L with an untreated control. Once again, three fish per tank with one tank per concentration and control were used. Test solutions were prepared daily by dilution of a 20 mg PSOA/L stock solution with RFW. Fish were transferred to freshly prepared test solutions at 24-h intervals.

The definitive 96-h test with fish was conducted at nominal PSOA concentrations of 0.0625, 0.125, 0.25,

0.5 and 1 mg/L with a solvent and untreated control and continuous flow conditions. Seven fish were randomly added to each tank at the start of the exposure phase. For determination of the PSOA concentration, samples were taken from the testing concentrations and the controls at -24, 0, 24 and 96 h. For each test concentration, RFW was passed from a header tank into a pre-exposure glass mixing vessel containing a magnetic follower. The nominal flow rate of RFW to each mixing vessel was 35 mL/min which equated to five volume replacements per tank every 24 h. The water was mixed with a continuous flow of stock solution of PSOA in acetone or acetone alone for the solvent control. The stock solutions were contained in Becton–Dickinson syringes equipped with a 16-gauge Teflon syringe infusion tube and delivered using Medfusion Syringe Infusion Pumps, Model 2001. Stock solutions at the appropriate concentration were delivered to the mixing vessel at a rate of 0.21 mL/h where they were mixed with the dilution water. The test solutions flowed through tubing directly to the appropriate tank at a continuously controlled flow rate. Excess solution was siphoned off with an overflow tube. The delivery of solutions to the solvent and non-solvent control tanks was identical to the treated tanks but excluded PSOA. The solvent control received acetone only from the infusion pump, which was mixed with RFW and the non-solvent control received only RFW. Dose solution delivery rates were confirmed daily for each tank by collecting the outflow of dose solution over a timed interval and measuring the volume. Temperature, pH, conductivity, dissolved oxygen concentration and hardness of test water were also monitored throughout the study. During the in-life phase of the test, fish were observed at 1, 4, 24, 48, 72 and 96 h. Incidence of fish death was recorded and abnormalities noted. At the conclusion of the study, the length and weight of surviving fish were recorded.

Daphnia magna

For the range finding test for *Daphnia*, nominal PSOA concentrations of 0.01, 0.1, 1, 10 and 100 mg/L were prepared by appropriate dilution of a stock PSOA solution. The stock solution consisted of 100 mg PSOA/L and was prepared by dispersing the test item in ISO medium and then shaking for 30 min. Non-dissolved test material was separated by centrifugation. For the range finding test, ten daphnids for each test concentration and control were exposed for 48 h. Two replications of this test were conducted.

Based on the observations for the range finding test, five test concentrations in a geometric series with a separation factor of two and one control was used in the 48-h definitive test. The nominal concentrations were 0, 2.5, 5.0, 10.0, 20.0 and 40.0 mg/L. Preparation of the stock PSOA solution and test solutions occurred in the same manner as the range finding test. For determination of the PSOA concentration, samples were taken from the testing concentrations and the control at the start and end of each water renewal period. A semi-static water renewal method was chosen with a renewal frequency period of 24 h. Twenty *Daphnia* were exposed to each test concentration, with the twenty animals divided into four groups of five. Each group resided in approximately 40 mL of test medium. *Daphnia* were observed at 24- and 48-h intervals and were considered immobile when they were unable to swim after 15 s.

Algae

A range finding test was conducted in which algae were exposed for 72 h to nominal concentrations of 0.01, 0.1, 1, 10 and 100 mg/L. A PSOA stock solution and test solutions were prepared for the range finding and definitive tests as described in "Daphnia magna" section. Based on cell number counts from the range finding study, six test concentrations in a geometric series with a separation factor of two and one control was used in the 72-h definitive test. The nominal concentrations were 0, 0.4, 1.0, 2.6, 6.4, 16.0 and 40.0 mg/L. Introduction of algae into the 250-mL Erlenmeyer flasks occurred at 0 h by inoculation of 0.1 mL algal biomass (10^7 algal cells/mL) into 100 mL test solutions. The algal cells were taken from an exponentially growing pre-culture established 4 days prior to the start of the definitive study. The initial cell density was about 10^4 cells/mL in each test flask. There were three replicates per test concentration and six replicates in the untreated control. During incubation the flasks were stored on an orbital shaker and continuously shaken. Algal cell numbers and morphology were assessed at 24, 48 and 72 h. Cell number was determined by manual cell counting using a microscope with a counting chamber. Morphology was assessed microscopically.

Activated sludge

The definitive study ran in two sets of 3-h exposure periods in the course of 1 day. The first set consisted of three PSOA concentrations, five reference flasks and four control flasks. The second set consisted of two PSOA concentrations, five reference flasks and three control flasks. Nominal concentrations of 6, 20, 60, 200 and 600 mg PSOA/L were prepared in replicates of five, in addition to a control. Flasks of the reference substance, 3,5-dichlorophenol (3,5-DCP) were prepared at nominal concentrations of 3.2, 10, 32 and 100 mg/L. 3,5-DCP was used as a reference substance given its known inhibitory effect on respiration, as well to ensure that the batch of sludge used in the test showed a representative level of sensitivity. An abiotic flask at 100 mg/L and a control were also prepared.

Each flask contained 3.2 mL of synthetic sewage and 18 mL of activated sludge to give a final solids concentration of 1500 mg/L (with the exception of the abiotic flask), an appropriate quantity of either PSOA or 3,5-DCP stock solution and reverse osmosis water to give a final flask contents volume of 100 mL. After foaming was observed in the first aerating flasks, one drop of antifoaming agent was added to each test flask. The pH of each flask was measured at the beginning and end of the test. The pH of the PSOA stock solution was adjusted to 6.6 before use and the reference substance stock solution had a pH of 7.3.

Flasks were established in batches of six and aerated at 20 ± 2 °C for 3 h. Each batch included a control flask and five test or reference substance flasks. The temperatures of the flask contents were measured at the end of the 3 h aeration using a mercury-in-glass thermometer. The respiration rate of each flask was measured after 3 h and compared with the mean respiration rates of the control flasks. The rate of oxygen uptake was measured in glass sample tubes into which microcathode oxygen electrodes were inserted. The electrodes were connected to an interface unit, which converted the current produced by the electrodes into dissolved oxygen readings. These readings were transferred to a computer that calculated the respiration rate in each flask over the linear part of the curve and compared it to the mean of the control cultures. The rates of oxygen uptake were expressed as mg/L/h.

Analytical measurements
Fish
An analytical procedure for the determination of PSOA in Tap Water formulations has been developed and validated by Charles River [3]. Formulation prepared at 2.5 µg/mL was found to be not stable when stored at ambient laboratory temperature in the dark. Formulation prepared at 150 µg/mL was found to be stable for 12 h when stored at ambient laboratory temperature in the dark. Due to the instability of the PSOA formulations in Tap Water the stability of PSOA in acetone was investigated. The formulation and analytical procedures were found to be satisfactory for a formulation prepared at 1.00 mg/mL in acetone which could be used for a flow through test in the ecotoxicology study.

In the range finding tests, analytical samples were collected in duplicate from the test concentrations at −24, 0, 24 and 96 h to assess stability. For the definitive test, duplicate samples were collected at the start of the study and after 72 h. Samples were collected from expired test media after 24 and 96 h. Sample analysis generally occurred on the same day as sampling. In cases where this was not feasible, the samples were stored under suitable conditions until analysis could take place.

Daphnia magna
Analytical samples were collected from the test concentrations and the control at the start and at the end of each water renewal period. Samples were analyzed directly using a photometric method in a manner similar to the fish study.

Algae
Analytical samples were collected from the test concentrations and the control at the start of the test and 24-h intervals thereafter. For these daily measurements, one extra replicate was used in each test concentration that was treated the same as the replicates used for determination of the algal cells. In a manner similar to the fish and *Daphnia* studies, samples were analyzed directly after sampling using a validated photometric method.

Activated sludge
No analytical measurements were taken during the study. Only nominal concentrations were used.

Computational and statistical analysis
Fish
Median lethal concentrations (LC_{50}) values were estimated by taking the arithmetic mean of the 0 and 100 % mortality concentrations. No observed effect concentration (NOEC) values were based on both mortality and observed effects.

Daphnia magna
NOEC, lowest observed effect concentration (LOEC) and effective concentrations for the 50th and 100th percentile values (EC_{50} and EC_{100}) were determined for the definitive test. NOEC, LOEC and EC_{100} were identified directly from the data, while EC_{50} values were calculated by probit analysis with 95 % confidence limits.

Algae
The inhibition of algal growth was determined from the average specific growth rate and yield using the following equations:

Average specific growth rate (μ) : μ_{i-j}
$$= \frac{\ln X_j - \ln X_i}{t_j - t_i} + (\text{day}^{-1})$$

where μ_{i-j} is average specific growth rate from time i to j; X_j is biomass at time j; X_i is biomass at time i.

Percent inhibition of growth rate ($\%I_r$) : % I_r
$$= \frac{\mu_C - \mu_T}{\mu_c} \times 100$$

where μ_C is mean value for average specific growth rate in the control group; μ_T is average specific growth rate for the treatment replicate.

Percent inhibition in yield ($\%I_y$) : $\% I_y$

$$= \frac{Y_C - Y_T}{Y_C} \times 100$$

where Y_C is mean value for yield in the control group; Y_T is value for yield for the treatment group.

Mean values and standard deviations of cell concentrations were calculated for each treatment at 0, 24, 48 and 72 h. The percent inhibition of growth rate (μ) and yield (y) was also calculated using Excel for Windows software. The EC_{50} values for growth rate and yield and their associated confidence limits were calculated using probit analysis based on measured geometric mean concentrations. For the determination of NOEC and LOEC, the calculated mean growth rate and yield at the test concentrations were tested on significant differences to the control values by Bonferroni t test. TOXSTAT software was used to evaluate the normal distribution of the rate and its homoscedasticity. Using EC_{50}, NOEC and LOEC values were identified for algal growth rate and yield.

Activated sludge

The respiration of the flasks dosed with the PSOA or 3,5-DCP were expressed as percentages of the respiration rate of the control flasks and were derived as follows:

Percent inhibition : $\% I$

$$= \left[1 - \left[\frac{\text{Respiration rate of test flask}}{\text{Mean respiration rate of control flasks}}\right]\right] \times 100$$

The effective concentrations (EC) for the 20th, 50th and 80th percentiles, as well as the NOEC for respiration inhibition, along with their 95th confidence intervals, were calculated by linear interpolation using the US EPA program ICPIN (Version 2, June 1993).

Results
Range finding tests
Fish

In the repeat range finding test, all fish in the highest dose group (2 mg/L) experienced difficulties in respiration within 1 h. Therefore, all fish were removed and humanely euthanized to avoid additional stress. Fish at all other test concentrations (0.002, 0.02 and 0.2 mg/L) appeared active and healthy throughout the exposure period with no abnormal effects or behavior. Measurement of all test solution quality parameters verified that pH, temperature, conductivity and dissolved oxygen concentration remained within acceptable limits throughout the duration of the test: pH range of 6.52–6.94,

temperature range of 14–16.8 °C, conductivity range of 180.3–208 μS/cm and dissolved oxygen concentration range of 95–100 % air saturation value.

Daphnia magna

In the preliminary range finding test the *Daphnia* were considered immobile if they were unable to swim after 15 s. There were no immobile *Daphnia* observed at 48 h in the 0.01, 0.1 and 1 mg/L test groups. Six out of ten *Daphnia* were considered immobile in the 10 mg/L test group and all ten *Daphnia* were considered immobile in the 100 mg/L test group.

Algae

In the preliminary range finding test for algae, the nominal concentrations used were untreated, 0.01, 0.1, 1, 10 and 100 mg/L. The average of cell numbers at 72 h ($\times 10^4$ cells/mL) was 67, 65, 61.5, 59, 43 and 1.2, respectively.

Definitive tests
Fish

The definitive test results for fish are based on nominal concentrations of PSOA given that the test sample analysis confirmed PSOA levels were maintained at 99–100 % of nominal concentrations over the study duration. One dead fish was removed after 21 h of exposure to the highest dose (1 mg/L). The remaining fish in the same tank appeared lethargic. Two more dead fish were removed after 22.75 and 24 h, respectively. By 50.75 h, the remaining four fish were dead. After 72 h at the second highest dose (0.5 mg/L), all fish appeared to be respiring at an increased rate. All fish in both controls and at 0.0625, 0.125 and 0.25 mg/L appeared active and healthy throughout the test period. Fish surviving to 96 h were within the range 4.02–5.23 cm in length and 0.6984–1.7030 g in dry weight. No concentration-dependent growth effects were observed. Measurement of all test solution quality parameters verified that pH, temperature, conductivity and dissolved oxygen concentration remained within acceptable limits throughout the duration of the test: pH range of 6.41–7.15, temperature range of 16.0–16.9 °C, conductivity range of 194.8–208 μS/cm and dissolved oxygen concentration range of 74–97 % air saturation value.

Daphnia magna

The nominal concentrations used to test *Daphnia* were 0, 2.5, 5.0, 10.0, 20.0 and 40.0 mg/L with corresponding measured geometric mean concentrations of 0, 1.28, 2.69, 6.89, 15.61 and 32.91 mg/L, respectively. One hundred percent (100 %) of the daphnids were immobilized at 24 and 48 h at the two highest test concentrations (i.e., 15.61

and 32.91 mg/L). At 6.89 mg/L, 90 % of the daphnids were immobilized at 24 h and 100 % were immobilized at 48 h. At 2.69 mg/L, no immobilization was observed at 24 h; however, 33 % of the daphnids displayed immobilization at 48 h. No immobilization was observed at 1.28 mg/L or the control for the 24- and 48-h observations.

Algae
The nominal concentrations used to test algae were 0.4, 1.0, 2.6, 6.4, 16.0 and 40 mg/L. Geometric mean exposure concentrations were calculated for all nominal test concentrations except for the lowest two (i.e., 0.4 and 1.0 mg/L). The lowest two test concentrations were not within the measurable range of the photometric analytical method for the entire duration of the study and were not included in estimating EC_{50} values. For the third lowest test concentration, 2.6 mg/L, one measured concentration was below the limit of quantification (LOQ) of 0.25 mg/L and was assigned a value of one-half the LOQ ($0.5 \times LOQ$) in calculating the geometric mean concentration. For the nominal concentrations of 2.6, 6.4, 16.0 and 40 mg/L, geometric mean exposure concentrations of 0.56, 1.66, 3.60 and 8.25 mg/L were calculated. In terms of the control, the algal cell density increased from a nominal level of 1×10^4 cells/mL at 0 h to a mean value of 9.433×10^5 cells/mL at 72 h, representing sufficient algal growth to pass the validity criteria of the assay. Microscopic evaluation of the treated algal cells showed thinner and smaller cells at the two highest test concentrations (3.60 and 8.25 mg/L) relative to the control. No algal effects were observed at the two lowest test concentrations.

Activated sludge
The validity criteria of a respiration rate was met by all control test flasks (at least 20 mg oxygen per one gram of activated sludge in an hour), with measured values ranging from 33.0 to 39.8 mg O_2/g/h. The reference substance 3,5-DCP caused substantial inhibition of the respiration rate with a mean 3-h EC50 value estimated to be 2.6 mg/L. The percent inhibition for 3,5-DCP at 3.2, 10, 32 and 100 mg/L were 53.9, 76.1, 90.2 and 93.3, respectively. This was within the expected range of 2–25 mg/L, which indicated the sludge was responding accordingly and confirmed the viability of the sludge microflora [12]. The respiration rates in all the control flasks were within 15 % of each other. Therefore, the mean respiration rate of the control flasks associated with the reference substance and the relevant test concentration flasks was used to calculate the percent inhibition. The respiration rate of the abiotic control was negligible throughout the study.

Identification of point-of-departure values
Fish
Point-of-departure (POD) levels (i.e., LC_{50}, LOEC, NOEC) for the definitive fish study are noted in Fig. 2. The study concentrations noted for the fish are the nominal concentrations. As noted in "Fish" section, mortality was observed at the highest test concentration (i.e., 1 mg/L). At the next lower concentration, 0.5 mg/L, an increased respiration rate was observed but no mortality. For the next lower concentration, 0.25 mg/L, no effects were observed. Based on these observations, the fish 96-h LOEC and NOEC are 0.5 and 0.25 mg/L, respectively. Due to the lack of fractional mortality, the 96-h LC_{50} was estimated using the arithmetic mean of the 0 and 100 % mortality concentrations resulting in a value of 0.75 mg/L with 95 % confidence intervals (CI) of 0.5–1 mg/L.

Daphnia magna
Effect levels for the definitive Daphnia study are noted in Fig. 2. The study concentrations noted for the Daphnia effective levels are the analytically confirmed test concentrations rather than nominal concentrations. The Daphnia 48-h NOEC was 1.28 mg/L as all organisms appeared to be swimming normally at this concentration and no immobilization effects were observed. The lowest 48-h test concentration resulting in immobilization effects was 2.69 mg/L (i.e., LOEC of 2.69 mg/L). The concentration resulting in 100 % immobilization was 6.89 mg/L (i.e., 48-h EC_{100} of 6.89 mg/L). The 48-h EC_{50} was calculated to be 3.05 mg/L (95 % CI 2.57–3.92 mg/L) (Fig. 2).

Algae
Effects levels for the definitive algal study are noted in Fig. 2. No effects were observed over 72 h at 0.56 and 1.66 mg/L (i.e., 72-h NOEC of 1.66 mg/L). Significant growth inhibition was observed over 72 h at ≥3.60 mg/L (i.e., 72-h LOEC of 3.60 mg/L). The 72-h EC_r50 (i.e., EC_{50} for growth) was estimated to be 5.46 mg/L (95 % CI 4.98–5.99 mg/L) and 5.42 mg/L (95 % CI 4.77–6.16 mg/L) using the third and fourth highest test concentrations, respectively. The geometric mean of the EC_{r50} values is 5.44 mg/L. The 72-h EC_{y50} (i.e., EC_{50} for yield) was estimated to be 2.79 mg/L (95 % CI 2.62–2.97 mg/L) and 2.35 (95 % CI 2.08–2.66 mg/L) using the third and fourth highest concentrations, respectively. The geometric mean of the EC_{y50} of values is 2.56 mg/L.

Activated sludge
The NOEC and LOEC were determined to be 20 and 60 mg/L, respectively. The 3-h EC_{20}, EC_{50} and EC_{80} values were calculated to be 46 mg/L (95 % CI 37–73 mg/L), 216 mg/L (95 % CI 187–252 mg/L) and 504 mg/L (95 %

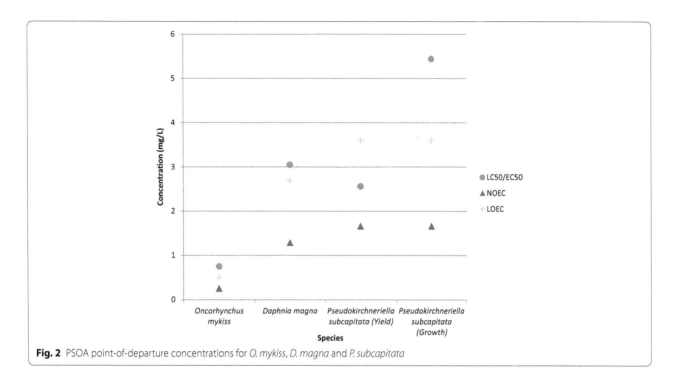

Fig. 2 PSOA point-of-departure concentrations for *O. mykiss*, *D. magna* and *P. subcapitata*

CI 480–525 mg/L), respectively. All POD values are based on nominal concentrations.

Discussion

PSOA was developed as a surfactant peracid to alleviate challenges with current peracid technology. Given its commercial applications, the primary disposal mode for PSOA is via industry wastewater effluent, which raises the importance of understanding whether the substances pose a risk to the aquatic environment. Evaluation of acute aquatic toxicity endpoints is a critical first step in evaluating the aquatic hazards of PSOA for risk assessment given its primary mode of disposal.

The aquatic toxicity studies described in this article were performed according to standard OECD test guidelines, where applicable. Given the reactivity of PSOA, significant effort was expended to confirm nominal test concentrations with analytical measurements, with the exception of the activated sludge assay. In reviewing the acute toxicity results, LC_{50} and EC_{50} values for fish, *D. magna* and algae ranged from 0.75 to 5.44 mg/L, representing a relatively small range of less than an order of magnitude. However, the EC_{50} value calculated for sludge was appreciably higher at 216 mg/L. It is unclear why the aquatic species were more sensitive to PSOA than the sludge microflora; however, the rapid reactivity and subsequent attenuation of PSOA once it comes into contact with organic matter may help explain this observation.

In terms of acute NOECs (0.25–1.66 mg/L) and LOECs (0.5–3.6 mg/L) for the aquatic organisms, a similar observation of the LC_{50}/EC_{50} ratio occurred; the NOEC and LOEC ranges were less than an order of magnitude apart. In terms of activated sludge, the NOEC (20 mg/L) and LOEC (60 mg/L) values were higher than the corresponding values for the test aquatic species. The small ranges for the POD values for the aquatic species indicates that PSOA has a relatively steep dose–response curve and rapid transitions from non-toxic concentrations to toxic levels over a small concentration range.

The mode-of-action for PSOA is not known, but of its four primary structural elements (i.e., hydroperoxide structure, acid component, sulfonate moiety and monohydroxy structure), the hydroperoxide structure is generally considered to be the defining component of the toxicity of PSOA [13]. In assessing the acute aquatic and microbial toxicity hazards of PSOA, insight can be garnered from evaluating acute toxicity studies conducted with similar hydroperoxides. Peracetic acid (PAA, CASRN 79-21-0) and *tert*-butyl hydroperoxide (TBHP, CASRN 75-91-2) are two representative hydroperoxides that can be used for comparison purposes (results shown in Table 1). From the table, it appears that PSOA has similar potency to PAA in terms of acute toxicity to *O. mykiss* but is slightly less potent to *D. magna* and *P. subcapitata* than PAA and appreciably less toxic to microorganisms. In comparison to TBHP, PSOA displayed

Table 1 Comparison of acute aquatic toxicity and microbial toxicity point-of-departure values for PSOA, PAA and TBHP

Species	Peracid chemistry					
	PSOA		PAA[a]		TBHP[b]	
	NOEC (mg/L)	LC$_{50}$/EC$_{50}$ (mg/L)	NOEC (mg/L)	LC$_{50}$/EC$_{50}$ (mg/L)	NOEC (mg/L)	LC$_{50}$/EC$_{50}$ (mg/L)
Oncorhynchus mykiss (96-h)	0.25 ($N=1$)	0.75 ($N=1$)	0.16–1.5 ($N=4$)	0.91–2 ($N=4$)	29.8 ($N=1$)	29.61–56.88 ($N=2$)
Daphnia magna (48-h)	1.28 ($N=1$)	3.05 ($N=1$)	0.035–<1 ($N=7$)	0.035–1.1 ($N=7$)	7 ($N=1$)	14.1 ($N=1$)
Pseudokirchneriella subcapitata (72-h growth rate)	1.66 ($N=1$)	5.44 ($N=1$)	0.084 ($N=1$)	0.035–0.86 ($N=2$)	0.22 ($N=1$)	1.5 ($N=1$)
Pseudokirchneriella subcapitata (72-h yield)	1.66 ($N=1$)	2.56 ($N=1$)	<1 ($N=1$)	<1 ($N=1$)	0.22 ($N=1$)	0.8 ($N=1$)
Activated sludge (3-h)	20 ($N=1$)	216 ($N=1$)	16.7 ($N=1$)	5.1–38.6 ($N=2$)	Not reported	17 ($N=1$)[c]

[a] Referenced in OECD [11]

[b] Referenced in ECHA [6]

[c] 30-min value rather than 3 h

greater toxicity to *O. mykiss* and *D. magna* but lower toxicity to *P. subcapitata* and sludge microorganisms. These comparisons are limited given the paucity of studies conducted for PSOA and TBHP, particularly relative to PAA. It is noted that much of the available aquatic information for peracids are based on industry-sponsored studies summarized in publicly available databases. There is relatively little published information for peracid in the peer-reviewed scientific literature. Therefore, a discussion on species sensitivity would be based on speculation regarding toxicity such as metabolic capacity of certain aquatic species that allow the bioactivation of PSOA resulting in, e.g., oxidative stress. Further the discussion above about the mode-of-action shows that structural similar chemicals show different species sensitivity regarding toxicity.

The data for *O. mykiss* were used to derive an aquatic health benchmark for PSOA for use in a risk assessment given that this species displayed the greatest sensitivity to PSOA. The 96-h LC$_{50}$ of 0.75 mg/L for *O. mykiss* was used to derive an aquatic predicted no effect concentration (PNEC) for PSOA. The activated sludge 3-h NOEC of 20 mg/L was used to derive a health benchmark for sewage treatment plants. Table 2 provides a description of the aquatic and sewage treatment PNEC values for PSOA.

As noted earlier, the disposal of PSOA via industry wastewater effluent requires identification of aquatic hazards of this chemical. The acute aquatic and microbial toxicity testing of PSOA enables identification of these hazards, as well as derivation of various environmental health benchmarks (i.e., PNECs). However, another critical facet to assessing the overall risk of PSOA to aquatic environments is to consider its environmental fate and transport.

The stability of PSOA in aqueous environments under environmentally relevant pH values has been assessed

following OECD Test Guideline 111 [2]. At pH levels of 4, 7 and 9, PSOA had hydrolyzed >10 % after 5 days with an estimated half-life of <1 year [2]. Based on these results, PSOA is considered hydrolytically unstable at environmentally relevant pH values. Experience with PSOA indicates that its stability in water is influenced by its aqueous concentration, with low ppm levels of PSOA showing less stability than higher concentrations (data not shown). This is important considering that in commercial applications of PSOA, the chemical is diluted prior to application and wastewater effluent containing PSOA is likely to be mixed with other facility wastewater lacking PSOA, further diluting the overall PSOA concentration prior to discharge and potentially accelerating chemical hydrolysis. Therefore, PSOA is not expected to persist in the environment.

Another important consideration is the microbial degradation potential of PSOA in aqueous environments. In an assessment following OECD Test Guideline 301B, PSOA biodegraded 56 % after 28 days with 40 % biodegradation observed by the Day 10 observation point (LAUS [8]. Though PSOA did not meet the OECD criteria for consideration as a "readily biodegradable" substance under the conditions tested, it did demonstrate appreciable aqueous biodegradability.

To gauge the hydrophilic nature of PSOA, EPI Suite software [KOWWIN Program (v1.68)] was used to estimate the Log Kow values for two of the primary components of PSOA, 10-hydroxy-9-sulfooctadecaneperoxoic acid and 10-hydroxy-9-sulfooctadecanoic acid [16]. Based on this work, a representative Log Kow of 3.12 was identified. The adsorption coefficient (Log Koc) can be used to estimate PSOA's ability to bind to suspended sediment in water. Testing conducted in accordance with OECD Test Guidelines 121 identified Koc values ranging from 0 to 190,546 (Log Koc of un-retained to 5.28)

Table 2 Aquatic and sewage treatment plant PNEC for PSOA

Environmental compartment	PNEC (mg/L)	Remarks and justification
Freshwater—PNEC aqua (freshwater)	0.00075[a]	Assessment factor: 1000 PNEC aqua (freshwater) equals lowest short-term LC50 value of 0.75 mg/L (fish) divided by assessment factor of 1000
Marine water—PNEC aqua (marine water)	0.000075[a]	Assessment factor: 10,000 PNEC aqua (marine water) equals lowest short-term LC50 value of 0.75 mg/L (fish) divided by assessment factor of 10,000
Intermittent releases to water—PNEC aqua (intermittent releases)	0.0075[a]	Assessment factor: 100 PNEC aqua (intermittent releases) equals lowest short-term LC50 value of 0.75 mg/L (fish) divided by assessment factor of 100
Sewage treatment plant—PNEC (STP)	2	Assessment factor: 10 PNEC STP equals 3 h NOEC of 20 mg/L divided by assessment factor of 10

[a] PSOA is not anticipated to be hydrolytically stable at these environmental concentrations

[1]. To identify a representative value in this range, a Koc weighted average based on the percent peak area of the test solutions was calculated to be 1.06, resulting in a Log Koc estimate of 0.024. This representative value indicates a relatively low sediment binding potential overall for the PSOA mixture; however, select minor fractions of the PSOA mixture may readily bind sediment.

Based on the laboratory tests conducted for PSOA, it is suggested that the environmental aqueous fate of this substance would be to exist largely unbound in the water column with hydrolytic and biological factors degrading the chemical and limiting its availability to aquatic organisms. The rate of degradation may vary appreciably depending on the initial concentration of PSOA in the water column, and for some fractions of this mixture, persistence in the aqueous environment is possible. Additionally, a minor fraction of the PSOA mixture does displays appreciable sediment affinity which is anticipated to attenuate its toxicity and could increase its persistence in the aqueous environment.

As part of the requirements under the Registration, Evaluation, Authorization and Restriction of Chemicals (REACH), an environmental risk assessment was conducted using EUSES software to determine if the proposed uses of PSOA posed an unacceptable risk to aqueous environments. All risk characterization ratios (RCRs), which are ratios of the PNEC to the estimated environmental concentration (i.e., estimated environmental concentration/PNEC), were <0.01, indicating no unacceptable risk to the environment [4]. Therefore, despite notable acute aquatic hazards for PSOA when considering the acute toxicity values, no unacceptable risk to the aqueous environment is anticipated based on its use and other factors. This aligns with the qualitative assessment that PSOA in industry wastewater effluent would be further treated and diluted via a municipal waste water treatment facility prior to release to aquatic

environments. Such treatment and dilution is anticipated to render PSOA below concentrations that pose a risk to native fauna or flora.

Conclusion

PSOA is a novel surfactant peracid whose commercial use will result in disposal via industrial wastewater effluent. Given this manner of disposal, it is important to understand the aquatic hazards of the chemical to better assess the risk posed to aqueous environments. Laboratory testing of PSOA for representative freshwater fish, invertebrate and algal species displays a degree of toxicity that appears to align with other peracid chemistries, such as peracetic acid. However, PSOA displayed appreciably lower microbial toxicity in activated sludge relative to other peracids. Despite the relatively high acute aquatic hazards for PSOA, environmental modeling suggests the overall risk of PSOA to aqueous environments is low based on its anticipated uses. Such a conclusion aligns with the significant processing of industrial wastewater by municipal wastewater treatment facilities prior to release to the environment.

Authors' contributions

This paper is part of a global registration project of PSOA manufactured by Ecolab. All authors have made substantial contributions to conception and design, acquisition of data, analysis, discussion and interpretation of data. They have been involved in drafting the manuscript or revising it critically for important intellectual content; and they have given final approval of the version to be published. BP contributed authoring the "Methods" section of the paper, as well as reviewing the Introduction, Results and Discussion. JC contributed authoring the "Results" section and reviewed the "Background" and "Methods" sections. MO wrote the introduction, assisted with the "Results" section, and peer reviewed the entire article. NP contributed authoring and reviewing to all sections of the paper. SS contributed authoring and reviewing to all sections of the paper. He was the study monitor and managed the authoring team.

Authors' information

All authors are working at Regulatory Affairs and are member of Ecolab's Global Corporate Toxicology Team. BP is Associate Regulatory Specialist working on hazard communication. JC (M.S.) is Regulatory Specialist 1 working on hazard communication. MO (B.A. in Biology, Society and Environment, Associate) is Regulatory Specialist working on hazard communication. NP (M.S., DABT) is Principal Regulatory Specialist working as subject matter expert in toxicology. SS (Food Chemist, Dr. rer. nat.) is working as subject matter expert in food and environmental safety. He is member of the GDCh and is doing the post-graduate study to become a certified ecotoxicologists established by the SETAC-GLB/GDCh.

Author details
[1] Ecolab, Monheim, Germany. [2] Ecolab, St. Paul, MN, USA. [3] Ecolab, Naperville, IL, USA.

Acknowledgements
The authors acknowledge Charles River Laboratories for conducting the acute fish study, Toxicoop-ZRT for conducting the acute *D. magna* and algae studies, and LAUS GmbH for conducting the activated sludge study. The authors also acknowledge Kathryn Sande for her input during the development of the manuscript. This study is based on regulatory scientific work which is an inherent part of the post-graduate study for certified ecotoxicologists, established by the SETAC-GLB/GDCh in 2005. The corresponding author wants to thank SETAC-GLB/GDCh for providing the opportunity to receive advanced training in ecotoxicology.

Compliance with ethical guidelines

Competing interests
The authors are employed by Ecolab. PSOA is manufactured by Ecolab. Therefore, none of the authors have competing interests.

References
1. Charles River Laboratories (2011) Product chemistry of PSOA: determination of adsorption coefficient. Study No. 219311. Edinburgh, United Kingdom
2. Charles River Laboratories (2012). Product chemistry of PSOA: determination of hydrolysis as a function of pH. Study No. 219295. Edinburgh, United Kingdom
3. Charles River Laboratories (2012b) Validation of methodologies for the formulation and analysis of PSOA in tap water. Study No. 428044. Edinburgh, United Kingdom
4. Ecolab (2013) Chemical safety report, reaction product of sulfonated oleic acid potassium salts, hydrogen peroxide and sulfuric acid. CAS No. 1315321-94-8
5. European Chemicals Agency (ECHA) (2011) Guidance for identification and naming of substances under REACH and CLP, ECHA-11-G-10-EN
6. European Chemicals Agency (ECHA) (2013) Registered substances, *tert*-butyl hydroperoxide. CAS No. 75-91-2
7. Hilgren J (2014) Microsoft Power Point Presentation—PSOA: new data supporting inert petition 26 February 2014
8. LAUS GmbH (2011) Determination of the aerobe ready biodegradability of PSOA (peroxy sulfonated oleic acids) in the CO_2 evolution test following OECD 301B resp. EUC.4.C. Study No. 11051701G605. Kirrweiler, Germany
9. Organization of Economic Co-operation and Development (OECD) (1992) Fish, acute toxicity test. Test Number: 203. OECD guidelines for the testing of chemicals, Section 2, OECD Publishing. doi:10.1787/9789264069961-en
10. Organization of Economic Co-operation and Development (OECD) (2004) *Daphnia* sp. acute immobilisation test. Test Number: 202. OECD guidelines for the testing of chemicals, Section 2, OECD Publishing. doi:10.1787/9789264069947-en
11. Organization of Economic Co-operation and Development (OECD) (2008) Screening information dataset (SIDS) dossier, peracetic acid. CAS No. 79-21-0
12. Organization of Economic Co-operation and Development (OECD) (2010) Activated sludge, respiration inhibition test (carbon and ammonium oxidation). Test Number: 209. OECD guidelines for the testing of chemicals, Section 2, OECD Publishing. doi:10.1787/9789264070080-en
13. Pechacek N, Laidlaw K, Clubb S, Aulmann W, Osorio M, Caudill J (2013) Toxicological evaluation of peroxy sulfonated oleic acid (PSOA) in subacute and developmental toxicity studies. Food Chem Toxicol 62:436–447
14. Sanchez M, Meyers TN (1996) Kirk-Othmer Encyclopedia of Chemical Technology, 4th edn. vol 18. Wiley, Inc., pp 230–310
15. TOXI-COOP—ZRT (2011) Determination of water solubility of PSOA. Study No. 675-107-2971, Balatonfüred, Hungary
16. TOXI_COOP—ZRT (2011) Determination of the partition coefficient of PSOA (peroxy sulfonated oleic acids). Study No. 675.194.2970. Balatonfüred, Hungary

6

Predictions of Cu toxicity in three aquatic species using bioavailability tools in four Swedish soft freshwaters

S. Hoppe[*], M. Sundbom, H. Borg and M. Breitholtz

Abstract

Background: The EU member countries are currently implementing the Water Framework Directive to promote better water quality and overview of their waters. The directive recommends the usage of bioavailability tools, such as biotic ligand models (BLM), for setting environmental quality standards (EQS) for metals. These models are mainly calibrated towards a water chemistry found in the south central parts of Europe. However, freshwater chemistry in Scandinavia often has higher levels of DOC (dissolved organic carbon), Fe and Al combined with low pH compared to the central parts of Europe. In this study, copper (Cu) toxicities derived by two different BLM software were compared to bioassay-derived toxicity for *Pseudokirchneriella subcapitata*, *Daphnia magna* and *D. pulex* in four Swedish soft water lakes.

Results: A significant under- and over prediction between measured and BLM calculated toxicity was found; for *P. subcapitata* in three of the four lakes and for the daphnids in two of the four lakes. The bioassay toxicity showed the strongest relationship with Fe concentrations and DOC. Furthermore, DOC was the best predictor of BLM results, manifested as positive relationships with calculated LC_{50} and NOEC for *P. subcapitata* and *D. magna*, respectively.

Conclusion: Results from this study indicate that the two investigated BLM softwares have difficulties calculating Cu toxicity, foremost concerning the algae. The analyses made suggest that there are different chemical properties affecting the calculated toxicity as compared to the measured toxicity. We recommend that tests including Al, Fe and DOC properties as BLM input parameters should be conducted. This to observe if a better consensus between calculated and measured toxicity can be established.

Keywords: Copper, Soft freshwater, Sweden, Bioavailability tools, DOC, Al, Fe

Background

When implementing water quality criteria (WQC) for metals in aquatic environments it is crucial to consider water chemistry parameters such as hardness, concentration of humic substances and pH, since these strongly affect the speciation of many metals [1–4]. For instance, cations (e.g. Ca^{2+} and Mg^{2+}) compete with other free metal ions (M^{n+}) for biotic ligands, such as gill membranes, thereby reducing their uptake [2, 5, 6]. This protective effect is less pronounced in soft than in hard waters due to lower Ca^{2+} levels [7]. Total and dissolved organic carbon (TOC/DOC), especially the humic fractions, generally reduce metal toxicity by binding M^{n+} into less bioavailable, high molecular weight complexes [5, 6, 8]. The interaction of metals and organic carbon is also dependent on ambient pH, as hydrogen ions can compete with the M^{n+} for binding sites, presenting further challenges for accurately predicting the toxicity of metals in soft acidic freshwaters. In the process of setting WQC it is central to try to mimic organism exposure in natural environments [9], which can be difficult, given the vast natural variability within and among natural aquatic ecosystems. To facilitate this process, Biotic ligand models (BLMs) using freshwater chemical characteristics and chemical equilibrium calculations combined

*Correspondence: Sabina.hoppe@aces.su.se
Department of Environmental Science and Analytical Chemistry (ACES),
Stockholm University, 106 91 Stockholm, Sweden

with organism biology can be used to calculate site-specific predicted no-effect concentrations (PNEC), environmental quality standards (EQS) and effect concentrations (LC/EC_{50}) [6, 9–12]. For a reliable use in environmental risk assessment, BLMs use a variety of physico-chemical parameters, of which TOC, alkalinity and pH are the three most important for assessing metal toxicity [1, 13, 14]. In current BLMs, the calibration range is targeted towards the majority of south central European freshwaters [15]. However, in Scandinavia, freshwaters often have low levels of Ca^{2+} due to the dominating soft water qualities, combined with low pH values and high levels of DOC, Fe and Al [16]. According to Swedish national lake surveys, the median Ca^{2+} concentration is around 2 mg/L and the 90 percentile, 10 mg/L ($n = 56,000$), which consequently can be characterised as soft- to ultra-soft waters, often leaving them outside of the BLM calibration range [15, 17]. Studies have shown that chronic BLM-predicted Cu and Zn toxicity can be underestimated in freshwaters with low pH and elevated levels of TOC, Al and Fe [18, 19], a water quality commonly found in Sweden.

We hypothesize that the usage of BLMs outside of their targeted calibration range will also result in less reliable results compared to bioassay studies considering acute toxicity. To test this hypothesis, Cu toxicity was measured for three aquatic species: *Daphnia magna, D. pulex* and *Pseudokirchneriella subcapitata*, in water from four soft water lakes with different water chemistry. These lakes were chosen as representatives from the Swedish national monitoring program as they together represent the four most common Swedish water types (Älgsjön: dark humic forest lake, St. Envättern: uncontaminated lake with relatively high TOC, Fiolen: nutrient rich forest lake and Abiskojaure: nutrient poor clear water lake with low primary production). Results from the bioassays were compared with BLM-derived toxicities for chosen organisms and waters to see if there were any divergences between measured and calculated results.

Results

Chemical properties

Table 1 presents the most important chemical characteristics of the four studied lakes, including Ca, Mg, Na, K,

Table 1 Properties of the lake waters used in this study: lake pH and DOC concentrations were determined in the test water used for the bioassays

Lake	Abiskojaure ($n = 32$)	Fiolen ($n = 58$)	St. Envättern ($n = 56$)	Älgsjön ($n = 58$)
RT90 *X–Y*	758,208–161,749	633,025–142,267	655,587–158,869	655,275–153,234
Latitude	68.3067°N	57.0917°N	59.0948°N	59.0948°N
Longitude	18.6550°E	14.5317°E	16.3693°E	16.3693°E
Catchment	Above treeline, tundra vegetation	Coniferous forest, some agriculture	Coniferous forest	Coniferous forest, some wetlands
Lake area (km^2)	2.8	1.6	3.7	0.35
Max depth (m)	35	10	11	7.7
pH[a]	7.6	6.5	6.5	6.5
DOC (mg/L)[a]	0.88	7.0	10	17
Ca (mg/L)	4.5 ± 1.46	2.9 ± 0.22	3.4 ± 0.19	5.6 ± 0.93
Mg (mg/L)	0.70 ± 0.21	1.0 ± 0.1	0.85 ± 0.06	1.9 ± 0.30
Na (mg/L)	1.0 ± 0.53	3.9 ± 0.27	2.2 ± 0.09	3.2 ± 0.54
K (mg/L)	0.59 ± 0.13	1.5 ± 0.15	0.29 ± 0.02	0.85 ± 0.15
SO_4 (mg/L)	4.3 ± 1.35	6.2 ± 0.68	5.9 ± 0.56	5.6 ± 1.31
Cl (mg/L)	1.1 ± 0.87	6.0 ± 0.51	2.8 ± 0.17	2.8 ± 0.42
Alk (mg $CaCO_3$/L)	10 ± 4.43	3.0 ± 0.76	3.0 ± 0.62	13 ± 3.31
Hardness (mg $CaCO_3$/L)	13	9.8	10	18
Cu (µg/L)*	0.91	0.74	0.38	0.78
Fe (µg/L)[a]	2.01	33	31	393
Cd (µg/L)[a]	0.012	0.034	0.006	0.006
Zn (µg/L)[a]	0.5	5	1.6	0.8
Pb (µg/L)[a]	0.005	0.1	0.06	0.08
Al (µg/L)[a]	1.5	49.7	33.7	56.9

Major ion concentrations represent multi-annual means of all data 2000–2009 from the Swedish national monitoring program. Trace metal concentrations (0.22 µm filtered) were determined in the *Daphnia* test waters at test 0 and 48 h, the numbers presented is at test 48 h

* Original concentration without any addition, detection (± 0.08)

[a] Analysed from bioassays at the end of 48 h

SO_4, Cl, alkalinity (multi annual levels), pH, DOC, hardness as well as total dissolved (<0.2 μm) concentrations of Cu, Fe, Cd, Zn, Pb and Al sampled at the test start. The variability in pH, alkalinity and hardness is quite small among the lakes but they differ widely in DOC (0.9–17 mg/L), Fe (2–393 μg/L) and Al (1.5–57 μg/L) concentrations. Direct optical and integrated isotopic measurements suggest that the lakes also differ in the quality and origin of DOC. The four lakes fall within the typical ranges of Swedish lakes for these auxiliary optical and isotopic variables (Fig. 1). Älgsjön, a humic forest lake, displays not only the highest concentrations of DOC, Fe and Al, but also deviates most from the other three lakes in the ratio between absorbance (420 nm, 5 cm) and DOC, a measure of the carbon-specific colour of dissolved organic matter, and carbon-specific fluorescence (CSF), a measure of the relative fluorophore abundance, suggesting a different chemical composition of DOC in this lake as compared with the others. The fluorescence index (FI) and fish $\delta^{13}C$, indicative of the organic precursors of DOC, also varied among the lakes. The patterns indicate a higher portion of terrestrial carbon sources in Älgsjön compared to St. Envättern and especially to Fiolen and Abiskojaure (Fig. 1).

BLM calculations vs. bioassays

Both calculated and bioassay *Daphnia* LC_{50} values were lowest and highest in Abiskojaure and Älgsjön, respectively. The same was noticed for the algae concerning calculated NOECs and bioassay EC_{50} values (Table 2). These two lakes also differed most with respect to DOC, Fe and Al among the four investigated lakes in this study. BLM calculations resulted in higher LC_{50} compared to the measured (bioassays) ones in three of the four lakes for daphnids (bioassays with *D. pulex* was only available for two lakes due to cultivating problems), and higher BLM NOEC compared to measured EC_{50} for *P. subcapitata* in all four lakes (Table 2). The difference between calculated and measured toxicity for *D. magna* was statistically significant for Abiskojaure ($p = 0.025$) and Älgsjön ($p < 0.001$), however, differing only by a factor 2 and 1.3, respectively. For the alga, the difference was ranging from a factor 2.5 in Älgsjön to 9.3 in St. Envättern, proving to be significant in three of the four lakes ($p < 0.001$). The only exception was Älgsjön ($p = 0.365$) where BLM suggested higher toxicity to *D. magna* than the bioassays. Calculated BLM LC_{50} values for Abiskojaure and St. Envättern were high enough too corresponded to bioassay LC_{100} (LC_{99} *D. magna* Abisko: 13.2 μg Cu/L; St. Envättern: 59.5 μg Cu/L) rather than a 50 % mortality. In contrast, for Älgsjön, the calculated LC_{50} value was below the measured value with a factor of 0.7. BLM calculated PNEC values (Table 2) based on the HC_5 concentration

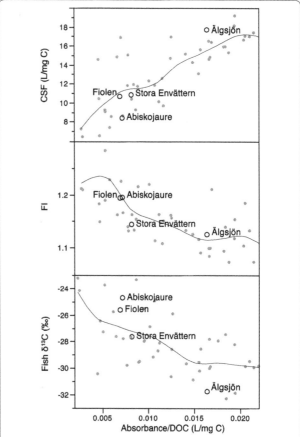

Fig. 1 DOC properties: stable carbon isotope signatures (δ13C, multiannual mean) in perch muscle tissue, water fluorescence index (FI) and carbon specific fluorescence (CSF-) in Swedish lakes. The four lakes used in this study highlighted as *circles*

varied between 4.9 μg Cu/L (Abiskojaure) and 31.4 μg Cu/L (Älgsjön). For two of the four lakes, Abiskojaure and St. Envättern, the PNEC value was on par with or exceeded the measured algal EC_{50} value for growth inhibition.

Influence by water chemistry

Although BLM yielded higher values than bioassays, in 8 of 10 cases, the general direction of the regression slopes was coherent between BLM and bioassay estimates for the key chemistry variables. In an attempt to find which factors that could explain the observed systematic difference, the relationship between water chemistry and the ratios between BLM and bioassay results was plotted. The difference between BLM and bioassays was related to the ratio of dissolved metals and DOC, e.g. the ratio between Fe and Al and DOC (Figs. 2, 3). The correlation between estimated toxicity and all available variables was examined and a subset of these is visualized in Fig. 2. DOC was the best

Table 2 Test and modelled results: bioassay (mean and SD values) and BLM results (µg Cu/L) for the crustacean and algae species compared as well as the eventual significance (p value)

	Abiskojaure	Fiolen	St. Envättern	Älgsjön
D. magna				
LC_{50}	8.53 ± 1.16	34.9 ± 1.78	34.3 ± 2.67	128 ± 23
BLM LC_{50}	17.4	40.0	61.7	92.5
p value LC50/BLM	0.025*	1.0	0.48	<0.001*
D. pulex				
LC_{50}	7.6 ± 1	30.4 ± 2	–	–
BLM LC_{50}	10.2	19.6	30.3	45.3
p value LC50/BLM[a]				
P. subcapitata				
EC_{50}	1.4 ± 0.2	27.4 ± 2.5	20.8 ± 0.4	111 ± 8.7
BLM NOEC	9.8	127	185	287
p value EC50/BLM	<0.001*	<0.001*	<0.001*	0.365
PNEC	4.9	14.5	20.6	31.4

As the BLM used for the algae only can produce NOEC values this was compared to the bioassay EC_{50} value

* Denotes a statistical significant difference (p < 0.05); – was not tested

[a] Was not tested due to lack of statistic material

predictor of BLM calculated toxicity, whereas Fe, followed by DOC was the best predictor of bioassay results. Other key water chemistry parameters, such as pH and hardness, showed no clear relationship with toxicity for the tested waters. Compound and optical variables was also investigated and it was found that the molar sum of Al and Fe was a better predictor of bioassay toxicity for both test species than Fe was alone and CSF appeared to correlate strongly with bioassay toxicity (Fig. 2).

Discussion
Usage of BLMs in soft freshwaters
In this study, we applied two currently available BLM on four soft water lakes, over a wide range of DOC (0.8–17 mg C/L), Al and Fe concentrations in order to define Cu toxicity in typical Swedish freshwaters. The performance of these BLM software versions for waters outside of the intended calibration range was examined by comparing model output to measured acute Cu toxicity for *P. subcapitata*, *D. magna* and *D. pulex* (species only tested in two lakes). The *Daphnia* results showed that BLM v.2.2.3 both significantly over- and under-predicted Cu toxicity. In the lake where BLM overestimated toxicity, i.e. lake Älgsjön, a higher proportion of allochtonous natural organic

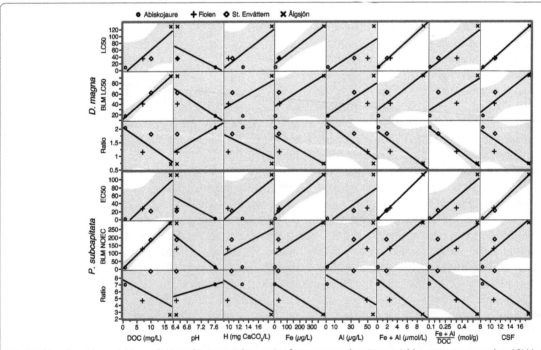

Fig. 2 Water chemistry variables vs. BLM results: scatter-plot matrix of some water chemistry variables versus measured and BLM-derived toxicity indices for two species in four lakes. The *third* and *sixth rows* show the ratios between the *two rows* above. The *lines* are fitted linear regression lines and the *shadowed areas* represent the 95 % confidence interval of the *fitted line*

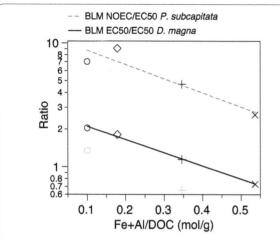

Fig. 3 ANCOVA results for BLM vs. test results: Lines fitted by ANCOVA. The *lines* depict the relationship for the ratio between measured and BLM calculated toxicity (measured LC_{50}, EC_{50}) as well as the ratio between the molar sum of metals (Fe, Al) and DOC in four lakes. *Red line* ratio of LC50's for *D. magna*; *blue line* ratio between NOEC and EC50 for *P. subcapitata*. The *greyed-out symbols* indicate LC-ratios for *D. pulex* that was exposed only to water from two lakes. *D. pulex* data was not part of the ANCOVA. The effect (slope) is statistically significant but the slopes does not significantly differ between the two species (Table 3)

matter (NOM), as compared to the other three lakes, was found, which indicates that lakes with these NOM properties could be less sensitive to Cu toxicity than lakes with a more autochthonous carbon source. However, for this model, both the over- and underestimation for the daphnids was within a factor of two, which could be considered a fairly good agreement. As for the algae, BLM v.0.0.0.17, underestimated the toxicity with a factor of 2.5 to 9.3, being significant in three out of four lakes (Table 2).

The lakes used in this study represent four typical Swedish freshwater lakes: *Älgsjön*: Dark humic forest lake with a high TOC, some months over 30 mg C/L, *St. Envättern*: uncontaminated lake with relatively high TOC, *Fiolen*: Nutrient rich forest lake; *Abiskojaure*: nutrient poor clear lake with low TOC and primary production. The lakes are used as reference lakes in the Swedish national monitoring program and are not considered to be affected by any local anthropogenic activities in their catchments, apart from some forestry. The only influence comes from long-range airborne pollutants. The levels of toxic trace metals (Pb, Cd, Cu, Zn) in the waters are low (Table 1), and should not affect the test organisms or compete with the added Cu for humic or biotic ligands. Concentrations of inorganic Al forms which are known to cause toxic effects on fish and invertebrates [20, 21] were found to be very low (<3 µg/L). This indicates that it is unlikely that toxic Al fractions contributed to any direct negative effects in this study.

Differences between the models

The NOEC-values (BLM v.0.0.0.17) exceeded the bioassay EC_{50}-values for the algae, with a factor of 2.5–9.3. They also exceeded the calculated PNEC-values by a factor of 2.2 in Abiskojaure up to a factor of 9.1 in Älgsjön and St. Envättern. Since the Cu NOEC value reflects the concentration where there should be no negative effect on the algal growth rate, it is notable that the NOEC value was higher than the measured EC_{50}, i.e. the estimated Cu concentration for 50 % growth reduction. However, since the use of NOEC values in ecotoxicology is strongly questioned by international standardisations bodies [22], the BLM software should use an EC_x endpoint instead. BLM v.0.0.0.17 overall gave a larger deviation between calculated and measured values than BLM v.2.2.3. BLM v.2.2.3, which uses a conversion factor to calculate chronic toxicity, produced results for the two *Daphnia* species within a factor of two. Thus, in these soft waters BLM v.2.2.3 calculated more similar results compared to the bioassays. This since BLM v.0.0.0.17 calculated algae NOEC values (where no growth inhibition is expected), which were higher than the measured EC_{50} values (concentration where there was a 50 % growth inhibition). Furthermore, even though BLM v.2.2.3 is an acute model, the approach of using the conversion factor seems to be working, at least for the investigated waters.

Influences by DOC, Fe and Al on toxicity

The soft water lakes used for this study covered a wide range in DOC, whereas other key variables such as pH and hardness were rather comparable among the lakes. As the hardness and alkalinity were similar there was no clear difference in the BLM performance to predict Cu-toxicity that could be related to the variation in these variables (Fig. 2). It is apparent that calculated effect concentrations derived by using these BLMs agree better with the measured effect concentrations at elevated Al+Fe/DOC-ratios. In other words, these BLMs or more specifically the speciation model WHAM V (Windermere Humic Acid model) which is incorporated into BLM 2.2.3, will produce more accurate calculations of metal speciation when there is a higher amount of metal ions that can bind to the NOM ligands present in the water. This has previously been shown [23] for both Cu and Cd, although, the problem was more pronounced for Cu. This relationship was also statistically significant when Fe was included and the p values decreased as more metals were added (Table 3). Up to a certain degree, the more metals per unit carbon that is present in the water, the better the agreement is between the bioassay and BLM results. An analogous relationship has previously been reported by Tipping [23, 24] who found that earlier versions of the WHAM V, which is used in BLM 2.2.3,

Table 3 ANCOVA results: ANCOVA table testing the effect of the ratios between metals and DOC (independent covariate) on the log-transformed ratios between toxicity indices estimated by BLM and bioassays (dependent variable) for two species _D. magna_ and _P. subcapitata_ (nominal factors)

	Al/TOC	Fe/TOC	$\sum$FeAl/TOC (mol/g)	$\sum$Metals/TOC (mol/g)
R^2	0.739	0.945	0.980	0.985
Root mean square error	0.590	0.270	0.163	0.142
Analysis of variance				
$F_{3,7}$ ratio	3.77	23.04	65.59	87.37
Prob > F	0.1164	0.0055	0.0007	0.0004
Effect tests:				
Species	0.03	0.002	0.0003	0.0002
log(Me/TOC)	0.5493	0.0144	0.0018	0.0010
Species x log(Me/TOC)	0.93	0.79	0.80	0.72
Estimated slope of covariate ± SE				
log(Me/TOC)	−0.069 ± 0.11	−0.046 ± 0.01	−2.531 ± 0.34	−2.655 ± 0.31

See Fig. 3 for further details

calculated reliable simulations for copper titrations of humic and fulvic acids when the [Cu]/DOC ratios were high, but performed less well when the [Cu]/DOC ratio was low.

DOC properties

The important aspect of the dissolved organic matter (DOM) quality is not considered in the two BLM softwares used in this study. This could be an important factor as DOM properties have been shown to influence both speciation [23–29] and toxicity [30] of Cu. Älgsjön, where BLM calculations overestimated the Cu toxicity compared to bioassays, differs from the other lakes concerning the ratio between water colour (absorbance of filtered water at 420 nm) and DOC. This water colour ratio was found to be similar among three of the lakes, 0.007, 0.008 and 0.009 in Abiskojaure, Fiolen, and St Envättern, respectively, whereas Älgsjön had a higher ratio of 0.018 (Fig. 1). The latter value, to some extent, likely reflects the higher relative Fe concentration in this lake. However, spectrofluorometric and isotopic data also suggest that there are qualitative differences in the DOC among the lakes which can influence the DOC complexing capacity for trace metals. In particular, the CSF showed a correlation with Cu toxicity. As a measure of the fluorophore density, CSF increase with the degree of aromatization which at current wavelength settings is indicative of humic substances. Fiolen and St. Envättern had very similar CSF as well as toxicity indices derived by both BLM and bioassays, while Fiolen in terms of C origin instead was very close to Abiskojaure, the lake most sensitive to Cu. Fiolen and Abiskojaure were estimated to have a more autochthonous DOC than St. Envättern and especially Älgsjön were characterized by allochtonous DOC

(Fig. 1). Within the narrow ranges of the other parameters no clear linear patterns could be seen.

Fe and Al

In the bioassays, varying concentrations of Fe and Al may influence the toxicity of Cu, while BLM-calculated toxicity will not be affected, as these metals are not among the model input parameters. For this reason, it is not surprising that the BLM estimates were best correlated with DOC, as DOC supposedly is one of the most important input variables in current BLM versions. However, bioassay results suggested that Fe, especially in combination with Al, is as important as DOC for the Cu toxicity to the tested organisms (Fig. 2). An additional explanation for the overestimation of toxicity in Älgsjön could be connected to the high Fe and DOC content in the lake, creating humus-iron aggregates as well as inorganic colloidal and particulate Fe-forms, such as ferrihydrite, which are fairly adsorptive for other trace metals, including Cu [31]. Even though the water was filtrated, aggregates could have been formed during the time between the filtration and the test start. These strong adsorbents may influence the Cu-speciation and could contribute to a decreased concentration of free Cu-ions in the dissolved bioavailable phase, which then would decrease the toxicity.

When comparing calculated with measured results, the magnitude and direction of the plotted slopes (Figs. 2, 3) were consistent for both the daphnids and the alga, suggesting that the bioavailability of Cu to these organisms is regulated by the same mechanisms. The degree of underestimation, defined as the ratio between BLM and measured toxicity was, for both the alga and the daphnids, best correlated with the molar ratio between the sum of metals (Fe and Al) and DOC (Fig. 2). As Fe had the strongest correlation when single variables were considered, Fe was in this case apparently the more important of the two (Fig. 2). The slope patterns are similar for the tested species, as shown by the ANCOVA where the interaction term was non-significant (Fig. 3; Table 3). It has previously been shown [24] that both Al and Fe can compete with Cu for binding sites at the DOC, especially in cases with low Cu levels, affecting the speciation in the water.

A few other studies have indicated that BLM could underestimate Cu toxicity for *P. subcapitata* (factor of 2–4), chronic toxicity for *D. magna* (factor 8) and toxicity to larval fathead minnows in soft waters [8, 14, 18, 19, 32]. Some of these studies suggest that the underestimation by BLM on Cu toxicity could be explained by the model not sufficiently accounting for Fe or Al. Indicating that these metals are competing with Cu for the binding sites on humic substances, preventing Cu from binding to the humic ligands and instead increase the bioavailability.

Conclusion

This study shows that using current BLM software (i.e. v.0.0.0.17 and v.2.2.3) outside of their intended calibration range could prove problematic since the models differed in their ability to accurately predict toxicity. The significant difference between measured and calculated toxicity was found to be between a factor of 0.7 up to 9.3. The toxicity measured in the bioassays showed the strongest relationship with concentrations of Fe, Al and DOC. In the lake where BLM overestimated Cu toxicity, different properties as regard to DOC origin was found, indicating that qualitative DOC properties can affect the performance of the BLM calculations. As our results are indicative towards Al, Fe and NOM properties influencing BLMs ability to calculate Cu toxicity we strongly recommend that studies considering the implementation of these parameters are conducted. DOM composition is extremely complex and not easily implemented in operative models like BLM; still, if the important aspects of DOC can be condensed into simple and cost effective optical properties like carbon-specific fluorescence or absorbance, as our results indicate, it could be worthwhile to further investigate.

Methods
Sampling and treatment
The selected lakes; Abiskojaure, Fiolen, St. Envättern and Älgsjön are included in the Swedish national monitoring program, sampled 8–10 times/year for the last 20 years, and are consequently well characterized both chemically and biologically. The lakes vary regarding chemistry as well as catchment properties (Table 1). All of the lakes have low hardness (9–19 mg $CaCO_3$/L) and

circumneutral pH (6.5–7.6), and differ widely in TOC (0.8–17 mg/L). The water used in this study was sampled 2011. The pH values of these lakes were kept in the neutral range. Water from Abiskojaure was collected by staff at Abisko Scientific Research station (Royal Swedish Academy of Science), using polycarbonate water sampler from the ice, in March–April, and sent to ITM (Dept. of Applied Environmental Science). Water from Fiolen was similarly collected by staff at the County Administrative board in Växjö and sent to ITM. These two waters were filtered upon arrival, in a clean room, within 3 days after sampling. Water from Älgsjön and St. Envättern were sampled by the authors and filtered in situ or the next day in a clean room. To avoid contamination of test waters, acid cleaned (0.1 M HCl) 5 L polypropylene containers were used. All filtrations were performed on-line by pumping water through acid cleaned (0.1 M HCl) plastic tubes (Masterflex tubing, silicone) and 0.22 μm acid cleaned (0.1 M HCl) polypropylene capsule filters (Calyx capsule). The filtered water was placed in a dark cold room at 6 °C for 1 month until time of analysis and bioassay testing.

Water chemical analyses
Physico-chemical measurements were performed by the Swedish University of Agricultural Science (SLU) (major ions), ACES (trace metals, O_2, pH and fluorescence) and Stockholm Water Company (Stockholm Vatten) (TOC). Oxygen saturation levels and pH were determined using a SympHonic SP90N5 multimeter and a Radiometer pHM82 Standard. Trace metal levels were analysed both at the start of the bioassays and after 48 h using an ICP-MS Thermo X-series II (inductively coupled plasma mass spectrometry). The ICP-MS accuracy was checked using certified reference water (SLRS-4 riverine water; Cu: certified 1.81 ± 0.08 μg/L) during each test batch. The labile inorganic Al fraction in the lake waters was determined by cation exchange spectrophotometry [33]. TOC was determined using a Shimadzu TOC-5050, and DOC was assumed to be 90 % of TOC concentrations. DOC is often considered to differ approximately 5 % from TOC [34, 35], however, since there could be aggregates present, due to the water standing for one month prior to test start, 10 % was chosen. Fluorescence emission spectra of the lake waters were determined by a spectrofluorometer (JASCO FP-777) equipped with a 150 W xenon lamp and monochromator. Excitation wavelength was set to 370 nm and intensities were corrected for inner filtering effects and blanks. The fluorescence spectra were used to determine the carbon-specific fluorescence (CSF, ratio between max fluorescence and TOC concentration) and the fluorescence index (FI, the ratio between intensities at 450 and 500 nm) [36]. Both CSF and FI are

optical measures of DOC properties that can give e.g. the aromaticity and origin of dissolved organic matter [34]. Stable carbon isotope signatures (δ^{13}C) in fish (Sundbom, unpublished data) were measured by the Stable Isotope Facility, University of California at Davis, using isotope ratio mass spectrometry (PDZ Europa 20-20, Sercon Ltd., Cheshire, UK). This measurement has previously been shown to provide an indication of the amount of allochtonous carbon found in the water, meaning that at a higher δ^{13}C (more negative) the DOC often is of allochtonous origin. All chemical analyses were conducted following quality assurance routines specified in the accreditation of the laboratories.

Bioassays
Daphnia spp

Daphnia tests were conducted using both *D. magna* and *D. pulex*. The *D. magna* clone was provided by University of Göteborg in Sweden, where it has been cultured since 1979. The clone originated from a small lake in Bohuslän in the south west of Sweden. The *D. pulex* clone (PA$_{33}$: Portland arch) was provided by University of Indiana, USA. Daphnids were cultured in groups of 15-20 animals in 3 L clear glass beakers containing 2.5 L of M7 medium. The M7 medium was renewed on a weekly basis and the animals were fed a mixture of *Monoraphidium contortum* (800 µL) and *P. subcapitata* (6.8 mL), approximately 0.2 mg C/day/*Daphnia*. The light cycle was 16:8 h light/darkness. The condition of the *Daphnia* culture was tested with $K_2Cr_2O_7$ in synthetic soft water media (M7) and the LC$_{50}$ found to be within the recommended range [37]. One month before test start, the M7 culture medium was exchanged to lake water in order for the daphnids to acclimatize. After one month the new cultures were in good shape and no visible differences compared to the mother culture were observed, except for Abiskojaure where the animals were slightly longer and thinner. Experimental setup was designed according to OECD Test Guideline No. 202 [38]. Briefly, neonates (<24 h) were exposed to five different concentrations of copper dissolved in 50 mL lake water; Abiskojaure: 0;4.6; 7; 11; 17; 25 µg Cu/L, Fiolen: 0;13; 20; 30; 60; 90 µg Cu/L, St. Envättern: 0; 12; 24; 35; 50; 70 µg Cu/L and Älgsjön: 0; 80; 120; 180; 270; 405 µg Cu/L, in acid cleaned pre-conditioned beakers during a 48 h period. Since the waters differed in DOC concentrations and had different metal background concentrations, the added metal concentrations differed between the studied lakes. For each concentration, 20 daphnids divided into 4 groups of 5 individuals were used. The number of immobile daphnids was recorded at 24- and 48 h. Dissolved oxygen, pH and trace metals (Fe, Al, Zn, Pb, Cd and Cu) were measured at start and end of experiments.

Pseudokirchneriella subcapitata

Micro algae tests were performed using a clone of *P. subcapitata*, which has been cultured at Stockholm University since 1975. It is grown in 250 mL culture flasks during constant light in 20 % Z8 culture medium [39]. Growth inhibition by Cu was tested according to OECD [40] during 72 h at three different concentrations for all the lake waters; Abiskojaure: 0, 0.5, 1.5 and 4.5 µg Cu/L, Fiolen and St. Envättern: 0, 5, 15 and 45 µg Cu/L; Älgsjön: 0, 50, 150 and 250 µg Cu/L. However, fluorescence-inferred biomass was used as endpoint instead of dry weight biomass according to modifications in Nyholm [41]. Prior to exposure, glass test tubes (5 mL) were acid cleaned (0.1 M HCl), preconditioned with copper (0–130 µg Cu/L) and enriched with nutrients (50 µg P/L and 5 mg N/L) [41]. Before bioassay start the Z8 medium was replaced with fresh test solution in the test tubes and the algae added. The test tubes (4 concentrations*6 replicates) were held at 20 °C with constant light (60 µE × m^2 × s^{-1}) for 72 h [40]. After 72 h algal density (chlorophyll a 440–460 nM: blue and 685 nM: red) was measured by fluorometry (10-AU Fluorometer Turner Designs).

Data analysis and BLM

The statistical software PROBIT v. 2.3 was used to calculate bioassay LC- and EC-values as well as their 95 % confidence intervals. Biotic Ligand Model v.2.2.3 [42] was used to calculate LC$_{50}$ results for *D. magna* and *D. pulex*, whereas BLM v 0.0.0.17, based on the Cu-VRA document [43], was used to calculate no observed effect concentration (NOEC) values for the alga and PNEC values for the four lakes. BLM-calculated PNEC values were based on HC$_5$ concentration, i.e. where 95 % of test organisms included in the model database will not be affected from the Cu concentration. These HC5 curves are based on species sensitivity distributions (SSDs) for those species that are included in the model's database [44]. The BLM v.2.2.3 was used since it has a wide calibration range and uses a conversion factor to transform acute to chronic data. The BLM v 0.0.0.17 was chosen since it can predict PNEC values as well as toxicity to *P. subcapitata*. These BLM softwares use the chemical equilibrium model WHAM V [45] to calculate Cu speciation data for LC/EC and PNEC/NOEC values [4, 46]. Neither of these software include DOC origin or the input parameters Fe and Al when calculation Cu toxicity. Differences between modelled and measured toxicity were tested using a one-way ANOVA combined with a post hoc test (Tukey or Dunnett C) (SPSS v. 18). A series of ANCOVAs were applied to ratios between measured and BLM toxicity for *D. magna* and *P. subcapitata*, both species in the same model. The modeled and measured toxicity indices have different definitions for the two species, and hence ratios differ considerably. To obtain approximately equal variances for the two species, the ratios were log-transformed before ANCOVA analyses.

Abbreviations

ANCOVA: analysis of covariance; ANOVA: analysis of variances; BLM: biotic ligand model; DOM: dissolved organic matter; PNEC: predicted no effect concentration; NOEC: no effect concentration; NOM: natural organic matter; LC: lethal concentration; EC: effect concentration; TOC: total organic carbon.

Authors' contributions

SH has design the study, preformed it, assembled and interpreted the data material as well as written the manuscript. MS has analysed the data with ANCOVA as well as contributed in writing the manuscript. HB has been involved during the interpretation of the results as well as in drafting the manuscript and MB has been involved in designing the study, interpret the results as well as writing the manuscript. All authors read and approved the final manuscript.

Acknowledgements

This work was supported by FORMAS (2006-638). The authors would like to thank Stockholm Vatten for analysing the TOC samples, Karin Ek for *Daphnia* tests, Pär Hjelmquist for fluorometric analyses as well as Karin Holm and Jörgen Ek for ICP-MS analyses.

Competing interests

The authors declare that they have no competing interests.

References

1. Deleebeeck NME, De Schamphelaere KAC, Janssen CR (2007) A bioavailability model predicting the toxicity of nickel to rainbow trout (*Oncorhynchus mykiss*) and fathead minnow (*Pimephales promelas*) in synthetic and natural waters. Ecotoxicol Environ Saf 67:1–13
2. Deleebeeck NME, Muyssen BTA, De Laender F, Janssen CR, De Schamphelaere KAC (2007) Comparison of nickel toxicity to cladocerans in soft versus hard surface waters. Aquat Toxicol 84:223–235
3. Paquin PR, Zoltay V, Winfield RP, Wu KB, Mathew R, Santore RC, Di Toro DM (2002) Extension of the biotic ligand model of acute toxicity to a physiologically-based model of the survival time of rainbow trout (*Oncorhynchus mykiss*) exposed to silver. Comp Biochem Physiol C: Toxicol Pharmacol 133:305–343
4. Santore RC, Mathew R, Paquin PR, Di Toro D (2002) Application of the biotic ligand model to predicting zinc toxicity to rainbow trout, fathead minnow, and *Daphnia magna*. Comp Biochem Physiol C: Toxicol Pharmacol 133:271–285
5. Meyer JS, Santore RC, Bobbitt JP, Debrey LD, Boese CJ, Paquin PR, Allen HE, Bergman HL, Di toro DM (1999) Binding of nickel and copper to fish gills predicts toxicity when water hardness varies, but free-ion activity does not. Environ Sci Technol 33:913–916
6. Santore RC, Di Toro DM, Paquin PR, Allen HE, Meyer JS (2001) Biotic ligand model of the acute toxicity of metals. 2. Application to acute copper toxicity in freshwater fish and *Daphnia*. Environ Toxicol Chem 20:2397–2402
7. Kozlova T, Wood CM, McGeer JC (2009) The effect of water chemistry on the acute toxicity of nickel to the cladoceran *Daphnia pulex* and the development of a biotic ligand model. Aquat Toxicol 91:221–228
8. Boeckman CJ, Bidwell JR (2006) The effects of temperature, suspended solids, and organic carbon on copper toxicity to two aquatic invertebrates. Water Air Soil Pollut 171:185–202
9. De Laender F, De Schamphelaere KAC, Verdonck FAM, Heijerick DG, Van Sprang PA, Vanrolleghem PA, Janssen CR (2005) Simulation of spatial and temporal variability of chronic copper toxicity to *Daphnia magna* and *Pseudokirchneriella subcapitata* in Swedish and British surface waters. Human Ecol Risk Assess 11:1177–1191
10. Bossuyt BTA, De Schamphelaere KAC, Janssen CR (2004) Using the biotic ligand model for predicting the acute sensitivity of Cladoceran dominated communites to copper in natural surface waters. Environ Sci Technol 38:5030–5037
11. Di Toro DM, Allen HE, Bergman HL, Meyer JS, Paquin PR, Santore RC (2001) Biotic ligand model of the acute toxicity of metals: 1. Technical basis. Environ Toxicol Chem 20:2383–2396
12. Meylan S, Behra R, Sigg L (2004) Influence of metal speciation in natural freshwater on bioaccumulation of copper and zinc in periphyton, A microcosm study. Environ Sci Technol 38:3104–3111
13. De Schamphelaere KAC, Heijerick DG, Janssen CR (2003) Refinement and field validation of a biotic ligand model predicting acute copper toxicity to *Daphnia magna*. Comp Biochem Physiol C: Toxicol Pharmacol 134:243–258
14. Sciera KL, Isely JJ, Tomasso JR, Klaine SJ (2004) Influence of multiple water-quality characteristics on copper toxicity to fathead minnows (*Pimephales promelas*). Environ Toxicol Chem 23:2900–2905
15. Hoppe S, Gustafsson J-P, Borg H, Breitholtz M (2015) Evaluation of current copper bioavailability tools for soft freshwaters in Sweden. Ecotoxicol Environ Saf 114:143–149
16. FOREGS (2011). http://weppi.gtk.fi/publ/foregsatlas/text/Ca.pdf
17. Wilander A, JohnsonRK, Goedkoop W, Lundin L (1998) Riksinventering 1995. En synoptisk Studie av vattenkemi och bottenfauna i svenska sjöar och vattendrag. Naturvårdsverket, rapport 4813
18. De Schamphelaere KAC, Vasconcelos FM, Heijerick DG, Tack FMG, Delbeke K, Allen HE, Janssen CR (2003) Development and field validation of a predictive copper toxicity model for the green alga *Pseudokirchneriella subcapitata*. Environ Toxicol Chem 22:2454–2465
19. De Schamphelaere KAC, Janssen CR (2004) Development and field validation of a biotic ligand model predicting chronic copper toxicity to *Daphnia magna*. Environ Toxicol Chem 23:1365–1375
20. Campbell PGC, Stokes PM (1985) Acidification and toxicity of metals to aquatic biota. Can J Fish Aquat Sci 42:2034–2049
21. Andrén CM, Rydin E (2012) Toxicity of inorganic aluminium at spring snowmelt-In-stream bioassays with brown trout (*Salmo trutta* L.). Sci Total Environ 437:422–432
22. Jager T (2012) Bad habits die hard: the NOEC's persistence reflects poorly on ecotoxicology. Environ Toxicol Chem 31:228–229
23. Tipping E (1998) Humic ion-binding Model VI: an improved description of the interactions of protons and metal ions with humic substances. Aquat Geochem 4:3–48
24. Tipping E, Rey-Castro C, Bryan SE, Hamilton-Taylor J (2002) Al(III) and Fe(III) binding by humic substances in freshwaters, and implications for trace metal speciation. Geochim Cosmochim Acta 66:3211–3224
25. Al-Reasi HA, Wood CM, Smith DS (2011) Physicochemical and spectroscopic properties of natural organic matter (NOM) from various sources and implications for ameliorative effects on metal toxicity to aquatic biota. Aquat Toxicol 103:179–190
26. Baken S, Degryse F, Verheyen L, Merckx R, Smolders E (2011) Metal complexation properties of freshwater dissolved organic matter are explained by its aromaticity and by anthropogenic ligands. Environ Sci Technol 45:2584–2590
27. Chappaz A, Curtis J (2013) Integrating empirically dissolved organic matter quality for WHAM VI using the DOM optical properties: a case study of Cu–Al–DOM Interactions. Environ Sci Technol 47:2001–2007
28. Mueller KK, Lofts S, Fortin C, Campbell PGC (2012) Trace metal speciation predictions in natural aquatic systems: incorporation of dissolved organic matter (DOM) spectroscopic quality. Environ Chem 9:356–368
29. Tipping E, Lofts S, Sonke JE (2011) Humic Ion-Binding Model VII: a revised parameterisation of cation-binding by humic substances. Environ Chem 8:225–235
30. Richards JG, Curtis PJ, Burnison BK, Playle RC (2001) Effects of natural organic matter source on reducing metal toxicity to rainbow trout (*Oncorhynchus mykiss*) and on metal binding to their gills. Environ Toxicol Chem 20:1159–1166

31. Gamble DS, Schnitzer M (1973) The chemistry of fulvic acid and its reactions with metal ions. In: Singer PC (ed) Trace metals and metal-organic interactions in natural waters. Ann Arbor Science Publishers Inc, Ann Arbor, pp 265–302

32. De Schamphelaere KAC, Janssen CR (2004) Effects of dissolved organic carbon concentration and source, pH, and water hardness on chronic toxicity of copper to *Daphnia magna*. Environ Toxicol Chem 23:1115–1122

33. Andrén C, Rydin E (2009) Which aluminium fractionation method will give true inorganic monomeric Al results in freshwaters (not including colloidal Al)? J Environ Monit 11:1639–1646

34. Ivarsson H, Jansson M (1993) Regional variation of dissolved organic matter in running waters in central northern Sweden. Hydrobiologia 286:37–51

35. Mattsson T, Finér L, Kortelainen P, Sallantus T (2003) Brookwater quality and background leaching from unmanaged forested catchments in Finland. Water Air Soil Pollut 147:275–297

36. McKnight DM, Boyer EW, Westerhoff PK, Doran PT, Kulbe T, Andersen DT (2001) Spectrofluorometric characterization of dissolved organic matter for indication of precursor organic material and aromaticity. Limnol Oceanogr 46:38–48

37. Persone G, Jenssen C (1994) Third practical training course in aquatic toxicity testing. The laboratory for biological research in aquatic pollution, University of Ghent, Belgium

38. OECD (2004) OECD guidelines for the testing of chemicals/section 2: effects on biotic systems, Test No. 202: *Daphnia* sp. Acute Immobilisation Test

39. SIS (2005) Water quality—freshwater algal growth inhibition test with unicellular green algae (ISO 8692:2004)

40. OECD (2006) OECD guidelines for the testing of chemicals/section 2: effects on biotic systems, Test No. 201: Alga, Growth Inhibition Test

41. Nyholm N (1985) Response variable in algal growth-inhibition tests—biomass or growth-rate. Water Res 19:273–279

42. Hydroqual (2007) Biotic ligand model version 2.2.3. http://www.hydroqual.com/wr_blm.html

43. European Copper Institute (2007) Voluntary risk assassment of Copper, Copper II Sulphate Pentahydrate, Copper(I)oxide, Copper(II)oxide, Dicopper chloride trihydroxide, European Union Risk Assessment Report. http://echa.europa.eu/copper-voluntary-risk-assessment-reports

44. Wheeler JR, Grist EPM, Leung KMY, Morritt D, Crane M (2002) Species sensitivity distributions, data and model choice. Mar Pollut Bull 45:192–202

45. Tipping E (1994) WHAM—a chemical-equilibrium model and computer code for waters, sediments, and soils incorporating a discrete site electrostatic model of ion-binding by humic substances. Comput Geosci 20:973–1023

46. Paquin PR, Gorsuch JW, Apte S, Batley GE, Bowles KC, Campbell PGC, Delos CG, Di Toro DM, Dwyer RL, Galvez F, Gensemer RW, Goss GG, Hogstrand C, Janssen CR, McGeer JC, Naddy RB, Playle RC, Santore RC, Schneider U, Stubblefield WA, Wood CM, Wu KB (2002) The biotic ligand model: a historical overview. Comp Biochem Physiol C: Toxicol Pharmacol 133:3–35

7

Rainfed winter wheat cultivation in the North German Plain will be water limited under climate change until 2070

Nikolai Svoboda[*], Maximilian Strer and Johannes Hufnagel

Abstract

Background: We analysed regionalised ECHAM6 climate data for the North German Plains (NGP) in two time slots from 1981 to 2010 and 2041 to 2070.

Results: The annual mean temperature will increase significantly (by about 2 °C) that will result in shorter growing periods since the sum of degree days until harvest will be reached earlier. Even if the amount of total precipitation does not change there appears to be a shift towards increased winter precipitation and thus noticeable reduced summer precipitation.

Conclusions: Through the example of winter wheat we show a future limitation of water availability if yields are to be maintained or even increase.

Keywords: Summer rainfall, Growing period, Resource efficient production systems

Background

Water is fundamental to plant growth, so the impact of climatic water availability on crop production is significant. Extreme yield drops in Europe in 2003 (loss of 13 billion Euros) were associated with an environmental temperature increase of nearly 6 °C above the long-term mean and below average precipitation of approximately 300 mm [17]. Many authors [9, 11, 12] show there is a general increase in winter precipitation, visible in predicted climate data. Meinke et al. [12] show an increase in winter precipitation with regional climate models, for North Germany, of +22 %, but a decrease in summer of −17 %. Thus, we could expect reduced summer rainfall and consecutively increased risk of yield losses due to increased water deficit of field crops. Aim of this study is to evaluate if there may arise serious problems and answer the following questions:

1. Is there a relevant change by comparing the status quo with current climate projections?

2. Is there a shift towards winter rainfall in the NGP, and in particular in the study regions, as predicted in the literature?

3. Is there a trend to decreased and less steady rainfall during the summer growing period of winter wheat visible when evaluating current climate projections?

Methods
Study area

The North German Plain (NGP) covers the administrative units of Schleswig–Holstein, Mecklenburg Vorpommern, Lower Saxony, Brandenburg and parts of Saxony-Anhalt. As described in Dickinson [4] most of the area is less than 100 m in altitude, and only its zones of low hills reach more than 200 m. Surface deposits are the results of glaciation. The general climate follows a gradient of increasing continentality from west (oceanic) to east (sub-continental). The mean annual temperature is comparable across the NGP but the western part has a temperature range, from annual minimum to annual maximum, of 16.4 °C and the eastern part a range of 18.5 °C. The western part of the NGP has a precipitation of 600–800 mm per year, while the eastern part

*Correspondence: Nikolai.Svoboda@zalf.de
Institute of Land Use Systems, Leibniz Centre for Agricultural Landscape Research, Eberswalder Straße 84, 15374 Müncheberg, Germany

has a smaller total of 500–600 mm [4]. Main field crops in terms of acreage in the NGP are winter wheat, winter rape, silage maize and winter rye. In the present study, Diepholz (DH) as the most western and Oder-Spree (OS) as the most eastern regions were investigated (Fig. 1). DH has a long-term (1981–2010) mean temperature of 9.6 °C and 719 mm of precipitation (Fig. 2a). OS has 9.6 °C and 568 mm in long term (Fig. 2b). In 2003, precipitation in DH was measured at 523 mm and 434 mm in OS, respectively. In DH, 2003 was the year with the lowest precipitation during the observation period (1981–2010).

Crop

Winter wheat is the most important crop in the NGP and, matching with Boogaard et al. [2], the dominant crop of Europe in terms of acreage. In DH, 16 % of all cropping area is winter wheat (WW). In OS, the share is 7 %. The sowing date (JD_s) is September 15 as common in the NGP. Due to temperature as the main driver for physiological processes [1], the harvest date of winter wheat is essentially determined by cumulated temperature (heat sum), expressed in degree days (DD) [8]. Growth of winter wheat depends strictly on the air temperature [18].

Modelling the harvest date and growing period

The duration of the growing period (V_{per}) is determined by:

$$V_{per} = JD_h + \left(365\frac{1}{4} - JD_s\right) \tag{1}$$

Fig. 1 The North German Plain (*grey*) and the study regions Diepholz (*left*) and Oder-Spree (*right*)

with JD_h [day of year (DOY)] and JD_s [DOY] being the harvest and the sowing date, and $365\frac{1}{4}$ denoting 365 days per year and 366 in the leap year, respectively. The harvest date JD_h is defined by

$$JD_h = i_{GDD=T_h}, \tag{2}$$

with $i_{GDD=T_h}$ being the iterator i of growing degree days (GDD) reaching the threshold (T_h). Growing degree days is determined as:

$$GDD = \sum_{i=1}^{GDD=T_h} \begin{cases} (T_{mean_i} - T_{base}), & T_{mean_i} \geq T_{base} \\ (T_{mean_i} - T_{base}) = 0, & else \end{cases}, \tag{3}$$

where T_{mean}, T_{base}, and T_h are the daily mean daily temperature, base temperature ($T_{base} = 2.5$ °C, [15]: root growth (3 °C) and shoot growth (2 °C)), and threshold temperature as a fit parameter. The same value ($T_h = 2100$ °C) was used for both study sites. We determined—based on harvest and sowing date—the vegetation days (V_{day}) as the number of days with temperatures above base temperature during growing period (V_{per}). Therefore, we derived the equation

$$DD_M = \sum_{j=JD_s}^{JD_h} \begin{cases} V_{Day} + 0, & T_{MAV_j} > T_{base} \\ V_{Day} + 0, & else \end{cases} \tag{4}$$

where T_{MAV} denotes the simple moving average of the mean daily temperature given by

$$T_{MAV_n} = \frac{T_{MAV_{(n-2)}} + T_{MAV_{(n-1)}} + T_{MAV_n} + T_{MAV_{(n+1)}} + T_{MAV_{(n+2)}}}{n} \tag{5}$$

Iterators are j and n.

Time slots

Time period analysed within this study is from 1981 until 2070. Within this period we selected two representative time slots of 30 years each. First slot is from 1981 to 2010 representing the status quo and delineates the reference period. The second slot is from 2041 to 2070 representing the future. Differences between the time slots indicate a possible climate change.

Climate—recent climate

Scenario weather data for representative weather stations are available with daily values for the model regions in the NGP. These data are the result of fitting "Statistical regionalization model: STAR" [13] to recent measured data of the appropriate weather stations. STAR scenario data (SCEN: 1981–2010) then match the observed values for each study area in terms like mean monthly precipitation, temperature and solar radiation. To exclude model

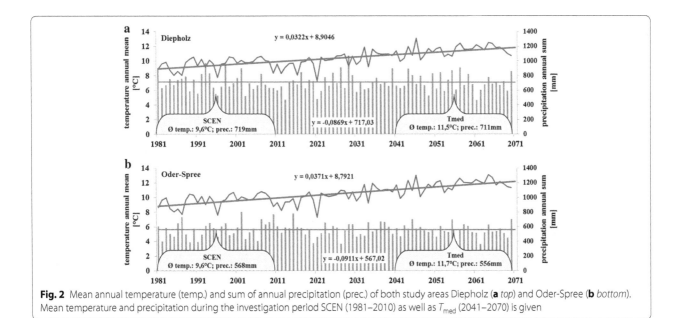

Fig. 2 Mean annual temperature (temp.) and sum of annual precipitation (prec.) of both study areas Diepholz (**a** *top*) and Oder-Spree (**b** *bottom*). Mean temperature and precipitation during the investigation period SCEN (1981–2010) as well as T_{med} (2041–2070) is given

bias when comparing status quo with future climate data, all following evaluations of the status quo were based on the scenario (SCEN) climate data.

Climate—climate change (CC) scenarios for future climate prediction

The results for the current condition were compared to projected weather data driven by the output of general circulation models (GCM) run under representative concentration pathway 8.5 (RCP 8.5). Collective climate models were used for analysis and prediction of climate change. Collective climate models include 21 GCM; all were driven by the scenario RCP 8.5. For the present study, we have selected 3 out of 21 GCM on the basis of their temperature gradient: (a) Minimum mean temperature increase ($T_{min} \rightarrow$ INM-CM4, Russia, +1 °C until 2070). (b) Medium mean temperature increase ($T_{med} \rightarrow$ ECHAM6, MPI Hamburg, Germany, +2 °C). (c) Maximum mean temperature increase ($T_{max} \rightarrow$ ACCESS1.0, CSIRO-BOM, Australia, +3 °C). The regionalisation of the GCM output was realised by the STAR model.

First of all, we need to define which aspects of climate change are relevant concerning crop production in general. Thus, in this study, the relevant climate change intends relevant for cropping winter wheat and includes in particular evaluations during the growing period and this period in parts.

Winter rainfall in our context is defined by DIN 4049 where the hydrological year (H_a) runs from 1 November of year one to 31 October of the following year. The winter season includes the months of November to April; the summer season includes the months of May to October.

The second benefit is the start and end of hydrological winter (H_W) that reflects start and end of leaching period in the NGP. Calculating this way enables us to analyse the winter rainfall during the typical leaching period and the summer rainfall from the end of the leaching period during summer until the harvest date, respectively.

Since rainfall during the growing period (P_{veg}) is not a meaningful parameter for analysing possible water deficit of winter wheat, we introduced the precipitation during main growing period (P_{m-veg}) as a parameter of interest (beginning of possible water deficit due to emptying the soil water storage with the beginning of hydrological summer); P_{m-veg} is defined by the amount of precipitation measured from May 1 (assumed end of leaching period due to the beginning of significant transpiration) until harvest date.

Statistical analysis

All data were evaluated using the R software package R Core Team [16].

Results

Model fit

Pre-tests showed that, the regional data (scenario) agree with respect to their general temperature trend, their variability and their precipitation with the climate data of the weather stations (observed) in the regions (data not shown). Harvest dates were reasonably well predicted by our simple model. The mean observed harvest dates of the study region Diepholz over 21 years were day 216 while the model underestimates by 3 days. The same good model fit could be shown for the Oder-Spree region where the observed mean harvest date was 214 and the modelled was 214. Annually simulated as well as observed harvest dates are presented in Fig. 3a, b.

Shift towards winter rainfall

The mean precipitation during the hydrological year (H_a: 1 May until 31 April) for the reference period (SCEN: 1981–2010) is 705 mm for DH and is 566 mm for OS. In the future (2041–2070) precipitation during H_a ranges between 683 and 711 mm for DH and between 512 and 570 mm for OS depending on the scenario (T_{min}, T_{med}, T_{max}). Thus, there is no significant change in annual precipitation while comparing the reference with the future period. Compared to the very little alteration of total amount of precipitation, standard deviation (as an indicator of constancy) of mean precipitation decreases in DH from 122 (SCEN) to 97 (T_{med}) and in OS from 101 (SCEN) to 67 (T_{med}) when comparing recent with future time period (Table 1). Mean precipitation during hydrological winter (H_W) during the SCEN period is 331 mm in DH and 246 mm in OS. Within the future time slot T_{med} DH has a mean H_W precipitation of 387 mm and OS 296 mm, respectively. The share of precipitation during H_W (H_a/H_W) in the SCEN period is for DH 0.47 and for OS 0.44 and for the future time slot in T_{med} for DH 0.54 and for OS 0.53. The scenario T_{max} delivered comparable results, while in the T_{min} scenario the share ranges between 0.50 (DH) and 0.48 (OS).

Harvest date

The mean harvest date within the SCEN period lies between the 3 and 5 August while for the T_{med} period earlier dates between 3 July and 30 June were calculated (Fig. 4a, b). Evaluating T_{min}, the harvest date is earlier than in SCEN but later than T_{med} (13 and 14 July). Much earlier is the H_{day} when dealing with the T_{max}: 19 and 21 June.

Growing period

Length of the growing period is strongly correlated with the harvest date. The growing period of winter wheat (V_{per}) during the reference period (SCEN) is 324 days in DH and 323 days in OS while the vegetation days during the V_{per} is 267 in DH and 246 in OS (Table 2). Therefore, in DH 57 cold days (days with less than 2.5 °C within the V_{per} as an indicator for the frequency of the interruption of biomass accumulation) and in OS 77 cold days were detected during the 1981–2010 period. In the future (2041–2070) period (T_{med}) the V_{per} is shorter by 33 days and 34 days in DH and OS, respectively, when compared to SCEN. The cold days in the T_{min} were reduced to 30

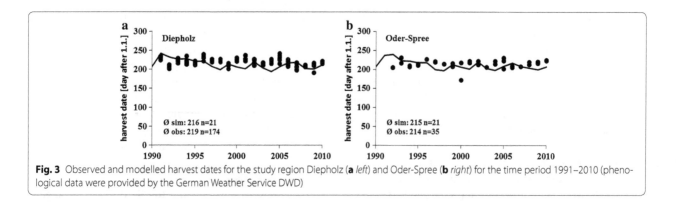

Fig. 3 Observed and modelled harvest dates for the study region Diepholz (**a** *left*) and Oder-Spree (**b** *right*) for the time period 1991–2010 (phenological data were provided by the German Weather Service DWD)

Table 1 Precipitation in the study regions Diepholz (DH) and Oder-Spree (OS) differentiated according to annual precipitation (1.1.–31.12.), precipitation during hydrological year (1.10.–31.9.), hydrological winter (1.10.–31.4.) and the share of precipitation during hydrological winter (H_W)

	Annual precipitation		Hydrological year		Hydrological winter		Share of H_W	
	SCEN	T_{med}	SCEN	T_{med}	SCEN	T_{med}	SCEN	T_{med}
DH								
P (mm)	709	711	705	711	331	387	0.47	0.54
SD (mm)	132	112	122	97	63	50		
OS								
P (mm)	572	556	566	556	246	296	0.44	0.53
SD (mm)	104	80	101	67	55	43		

SCEN represents the recent time period (1981–2010), T_{med} the future time period (2041–2070), *SD* is the standard deviation

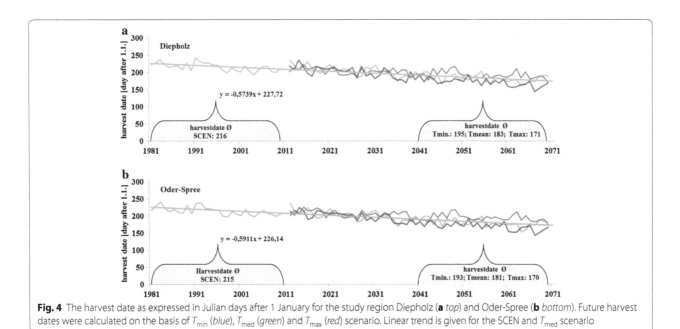

Fig. 4 The harvest date as expressed in Julian days after 1 January for the study region Diepholz (**a** *top*) and Oder-Spree (**b** *bottom*). Future harvest dates were calculated on the basis of T_{min} (*blue*), T_{med} (*green*) and T_{max} (*red*) scenario. Linear trend is given for the SCEN and T_{med} scenario

Table 2 Growing period (days) of winter wheat (V_{per}) as defined by the delimiters sowing and harvest date for the study regions Diepholz (DH) and Oder-Spree (OS)

	SCEN	T_{med}
DH		
V_{per}	324	291
V_{day}	267	261
OS		
V_{per}	323	289
V_{day}	246	246

SCEN represents the recent time period (1981–2010), T_{med} the future time period (2041–2070)

Table 3 Precipitation during main growing period (P_{m-veg})

	SCEN	T_{med}
DH		
P_{m-veg} (mm)	197	115
OS		
P_{m-veg} (mm)	171	98

SCEN represents the recent time period (1981–2010), T_{med} the future time period (2041–2070)

(DH) and 42 (OS). Within the T_{max} a minimum of cold days of 19 (DH) and 30 (OS) was counted.

Rainfall during main growing period and potential drought
During the main growing period (P_{m-veg}: 1 May until harvest) 197 mm were measured in DH and 171 mm in OS, Table 3). For the T_{med} scenario less rainfall during P_{m-veg} 115 to 98 mm was calculated (Fig. 5a, b). Similar results for P_{m-veg} can be shown for T_{min} (149 mm in DH and 136 mm in OS) and for T_{max} (78 mm in DH and 68 mm in OS).

Discussion
Model fit
When comparing the data from the weather stations during the reference period with the modelled STAR outcome no significant differences are noticed. This is in

good agreement of Gerstengarbe et al. [7] who compared STAR with the current climatology of selected regions all over Germany. Gallardo et al. [5] show similar results while analysing an ensemble of 15 regional climate models nested into six GCM. They found differences depending on the region and the investigated model. Our simple model for calculating the harvest date reasonably well predicts the mean harvest date over a long period of 30 years. For some years the prediction is less precise. For this reason, we have based all results to the long term.

Shift towards winter rainfall
The shift towards winter rainfall with +7 % in DH and +9 % in OS is less pronounced than reported in many studies [9, 11, 12]. That may be because of the different period (hydrological vs. calendric) selected on the one hand and the different period of time (1981–2010) in total. Badeck et al. [1] suggested that a fraction of uncertainty may arise due to the time frame analysed. Comparing the mean annual precipitation of calendric against hydrologic year in the present time period, DH

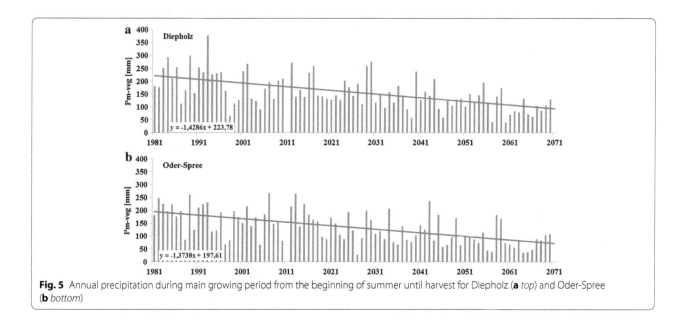

Fig. 5 Annual precipitation during main growing period from the beginning of summer until harvest for Diepholz (**a** *top*) and Oder-Spree (**b** *bottom*)

shows with 709 mm compared to 711 mm only little difference. However, OS reflects similar results on a lower level (572 to 566 mm). Kozuchowski and Degirmendizc [10] analysed long time weather data in different regions in Poland and found that regional differences are widespread. Following this, it may be possible, that the regions investigated in the present study may have a different shift than the mean of the NGP. Further studies should clarify the situation.

Harvest date

Patil et al. [14] found evidence that increased temperature led to earlier harvest date; the same effect we discovered for both regions. Depending on the scenario (T_{min}, T_{med}, T_{max}) the harvest date will be three (T_{min}), five (T_{med}) or six (T_{max}) weeks earlier than today. For Southern Sweden, Eckersten [6] has also found earlier harvest dates for winter wheat along with rising temperatures, while the yields stayed the same or decreased.

Growing period and rainfall during growing period

While comparing the growing period of winter wheat (V_{per}) in SCEN (1981–2010) with the V_{per} in T_{max} (2041–2070), there is a reduction of 45 (14 %) days in both regions. These findings correspond with Brown and Rosenberg [3] who calculated the length of the growing season of winter wheat in North America with different GCM. They pointed out that with increasing temperature the potential of water stress may arise. Reciprocal to the growing days we calculated the so-called cold days, with less than 2.5 °C, during the growing period. The amount of cold days decreased by >60 % to 19 days in the T_{max}

scenario. Walther et al. [19] discovered a comparable trend for frost days when analysing recent data of southern Switzerland. This could be relevant for vernalisation. Porter and Gawith [15] reported the optimal temperature for vernalisation process of winter wheat is between 3.8 and 6.0 °C, while in this study 2.5 °C [18] was taken to define cold days. Further regional adopted climate evaluations have to take care of optimal parameters. Under current conditions, 32 % (DH) to 36 % (OS) of the precipitation within the growing period comes during the main growing period from beginning of hydrological summer to harvest date. We observed a distinct shift of the precipitation towards the period in which the wheat plant does not require a lot of water (sowing until 1 March).

Conclusion and outlook

It became clear that there is a relevant difference comparing the status quo with current climate projections for the NPG. We found clear indications that the available precipitation during main growing period of winter wheat will decrease. Effects on yield have to be investigated using an appropriated plant soil model. While total annual rainfall does not change significantly a strong shift towards winter precipitation becomes evident. Possible consequences (e.g. nutrient leaching, erosion, need of introduction of catch crops) have to be evaluated in further studies.

Authors' contributions

NS developed the design of the model and the study in total, evaluated the results and drafted the manuscript. JH participated in the study, coordinated and helped to draft the manuscript. MS participated in the design of the study and performed R coding. All authors read and approved the final manuscript.

Acknowledgements

This project was supported by the German Ministry of Research (BMBF). Project: NaLaMa-nT, FKZ 033L029. The PIK (Potsdam Institute for Climate Impact Research) is gratefully acknowledged for providing the climate data.

Competing interests

The authors declare that they have no competing interests.

References

1. Badeck FW, Bondeau A, Böttcher K, Doktor D, Lucht W, Schaber J, Sitch S (2004) Responses of spring phenology to climate change. New Phytologist 162:295–309
2. Boogaard H, Wolf J, Supit I, Nimeyer S, van Ittersum M (2013) A regional implementation of WOFOST for calculating yield gaps of autumn-sown wheat across the European Union. Field Crop Res 143:130–142
3. Brown R, Rosenberg N (1999) Climate change impacts on the potential productivity of corn and winter wheat in their primary United States growing regions. Climatic Change 41:73–107
4. Dickinson RE (1961) A general and regional geography. E. P. Dutton and Company, Germany
5. Gallardo C, Gil V, Hagel E, Tejeda C, de Castro M (2013). Assessment of climate change in Europe from an ensemble of regional climate models by the use of Köppen - Trewartha classification. Int J Climatol
6. Eckersten H (2011). Climate change scenarios, crop production, length of growing period. Risk assessment/risk management, forecasting pests and diseases of field crops in a changing climate—Control strategies for pests, diseases and weeds
7. Gerstengarbe FW, Werner P, Österle H, Burghoff O (2013) Winter storm- and summer thunderstorm-related loss events with regard to climate change in Germany. Theor Appl Climatol 114:715–724
8. Goudriaan I, Laar HV (1994) Modelling potential crop growth processes textbook with exercises. Kluwer Acad. Publishers, Dordrecht
9. Grimm NB, Chapin FS, Bierwagen B, Gonzalez P, Groffman PM, Luo Y, Melton F, Nadelhoffer K, Pairis A, Raymond PA, Schimel J, Williamson CE (2013) The impacts of climate change on ecosystem structure and function. Frontiers Ecol Environ 11:474–482
10. Kozuchowski K, Degirmendzic J (2005) Contemporary changes of climate in Poland: trends and variation in thermal and solar conditions related to plant vegetation. Pol J Ecol 53:283–297
11. Lotze-Campen HCLDA, Noleppa S, Rock J, Schuler J, Uckert G (2009). Klimawandel und Kulturlandschaft Berlin. Technical report, Senatsverwaltung für Stadtentwicklung, Abteilung I, Gemeinsame Landesplanung Berlin-Brandenburg, Berliner Forsten, Berliner Stadtgüter GmbH
12. Meinke I, Gerstner E, von Storch H, Marx A, Schipper H, Kottmeier C, Treffeisen R, Lemke P (2010) Regionaler Klimaatlas Deutschland der Helmholtz-Gemeinschaft informiert im Internet über möglichen künftigen Klimawandel. Mitteilungen DMG, 02/2010, pp 5–7. http://www.norddeutscher-klimaatlas.de/klimaatlas/2071-2100/jahr/durchschnittliche-temperatur/norddeutschland.html (in German). Accessed 05 Aug 2015
13. Orlowsky B, Gerstengarbe F-W, Werner PC (2008) A resampling scheme for regional climate simulations and its performance compared to a dynamical RCM. Theor Appl Climatol 92:209–223
14. Patil RH, Laegdsmand M, Olesen JE, Porter JR (2010) Growth and yield response of winter wheat to soil warming and rainfall patterns. J Agric Sci 148:553–566
15. Porter JR, Gawith M (1999) Temperatures and the growth and development of wheat: a review. Eur J Agron 10:23–36
16. R Core Team (2013) R: A language and environment for statistical computing. Vienna, Austria, R Foundation for Statistical Computing. Available at http://www.R-project.org. Accessed 05 Aug 2015
17. Tubiello FN, Soussana J-F, Howden SM (2007) Crop and pasture response to climate change. Proc Natl Acad Sci 104:19686–19690
18. Waloszczyk K (1995) Einfluss von Lufttemperatur und Bestandesdichte auf das Wachstum von Winterweizen von Aufgang bis Vegetationsbeginn im Frühjahr. Arch Agron Soil Sci 39:379–387
19. Walther GR, Post E, Convey P, Menzel A, Parmesan C, Beebee TJC, Fromentin JM, Hoegh-Guldberg O, Bairlein F (2002) Ecological responses recent climate change. Nature 416:389–395

8

Efficiency of advanced wastewater treatment technologies for the reduction of hormonal activity in effluents and connected surface water bodies by means of vitellogenin analyses in rainbow trout (*Oncorhynchus mykiss*) and brown trout (*Salmo trutta f. fario*)

Anja Henneberg* and Rita Triebskorn

Abstract

Endocrine effects in the aquatic environment are in the focus of scientists and media along with debates on the necessity of further steps in wastewater treatment. In the present study VTG responses were compared to evaluate upgrades at wastewater treatment plants (WWTPs). We investigated several advanced sewage treatment technologies at two WWTPs connected to the Schussen, a tributary of Lake Constance, for the reduction of hormonal activity: (1) a powdered activated charcoal filter at the WWTP Langwiese; and (2) a combination of ozonation, sand filter, and granulated activated carbon filter at the WWTP Eriskirch. Rainbow trout and brown trout were either directly exposed to the effluents in aquaria or cages, or in a bypass system flown through by surface water of the Schussen. As a reference, trout were kept in bypass aquaria at the Argen River, which is less influenced by micropollutants. As a biomarker for estrogenicity, we analyzed the yolk precursor protein vitellogenin in immature rainbow trout and brown trout and in trout larvae (100 days post-fertilization) prior to and after the upgrade with the new technologies. Trout of different ages and species were used to detect differences in their sensitivity. At both bypass stations, larvae of brown trout showed significantly higher vitellogenin levels prior to the upgrade compared to negative control levels. Female brown trout exposed at the bypass station downstream of the WWTP showed decreased vitellogenin levels after the upgrade. In 1-year-old immature trout directly exposed to the respective effluents, no significant effects of the upgrades on vitellogenin levels were found. In general, larger effects were observed in brown trout than in rainbow trout, indicating that they are more sensitive test organisms.

Keywords: Endocrine disruption, Micropollutants, Wastewater treatment plant, Fish, Vitellogenin

Background

Endocrine disruptors (EDs) are hormonally active chemicals which are able to influence the endocrine system of organisms by mimicking or repressing the body's own hormones. EDs are a very diverse group of chemicals including, for example, ingredients of personal care products, pharmaceuticals containing steroid hormones, pesticides, plasticizers, dioxins, furans, phenols, alkylphenols, polychlorinated biphenyls, and brominated flame retardants [1, 2]. Still more endocrine-active chemicals were identified over the last years. The priority list of the European Commission contains 564 chemicals that had been suggested by various organizations and published papers as being suspected EDs [1].

*Correspondence: anja.henneberg@gmail.com
Animal Physiological Ecology, University of Tübingen, Auf der Morgenstelle 5, 72076 Tübingen, Germany

Because the aquatic environment is an important sink for natural and anthropogenic chemicals [3], the release of pollutants including EDs into surface waters via wastewater treatment plants (WWTPs) has come into the focus of scientists, authorities, and the public. Today, most wastewater is treated before it is released into bodies of water, but many studies show that not all hazardous chemicals, especially EDs, can be completely removed by routine wastewater treatment (see, e.g., [4]). Therefore, the discharge of wastewater treatment plants into recipient rivers is a main source for EDs to enter the aquatic environment. The level of pollution in rivers is particularly high if the catchment area is highly populated, has industry, or agriculture. Because wastewater can contribute up to 50 % and more of the flow of a river in months with low water [3], the released chemicals can play an important role for the occurring biota. For example, steroid estrogens, like the pharmaceutical ethinyl estradiol (EE2), are known to be extraordinarily active in fish at low to sub-ng/L concentrations [5, 6], and are found in many WWTP effluents at effect concentrations [7, 8].

This raises the question whether we should eliminate more pollutants, especially EDs, to improve wastewater quality. Whereas, for example, the Swiss Federal Government started projects introducing a tertiary treatment step at many of its WWTPs, the discussion whether

additional wastewater treatment technologies are ecologically worthwhile is still ongoing [9].

The present study is part of the "SchussenAktiv*plus*" project in the Lake Constance area investigating differently sized WWTPs which were equipped with additional wastewater treatment techniques [10]. Two of them (WWTP Langwiese and WWTP Eriskirch) are in the focus of the present study. To characterize the efficiency of technologies newly introduced at these WWTPs, we investigated vitellogenin (VTG) in juvenile male and female trout as well as in trout larvae as a biomarker of estrogenicity [5, 11–16]. VTG is an egg yolk precursor protein which is normally only produced by female fish. It is estrogen-dependent and EDs can act on hepatic receptors to induce the synthesis of VTG in males and juveniles [11, 17]. We compared VTG levels of trout that were exposed (1) directly to the conventional and modified effluent in aquaria connected to the effluents; (2) upstream and downstream the effluent prior and after the WWTP upgrade; and (3) in bypass systems downstream the WWTP and at a reference river prior and after the WWTP upgrade. Figure 1 gives an overview of these three approaches. It also shows the two WWTPs with their new technologies and summarizes the exposure experiments in the years 2013 and 2014 (for detailed information see methods section).

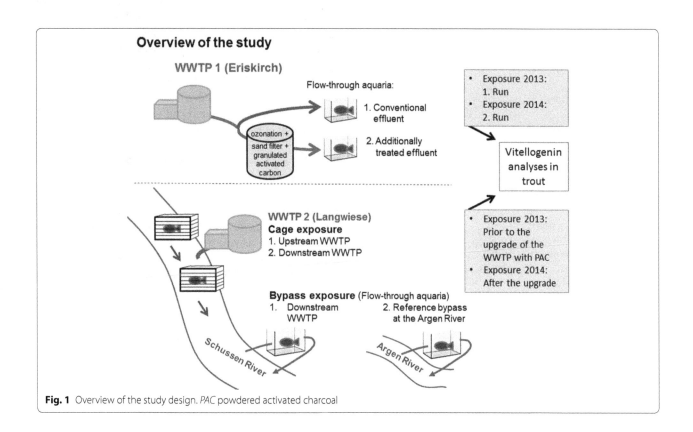

Fig. 1 Overview of the study design. *PAC* powdered activated charcoal

Results and discussion
Exposure experiments at the WWTPs

In rainbow trout exposed at the conventional and modified effluent at the WWTP Eriskirch, VTG levels in females varied between treatments (conventional and additionally treated effluent) and years (exposure in 2013 and 2014), whereas VTG levels in males were constantly low or even non-detectable (significant differences could not be determined) in both years (Fig. 2).

The increased VTG level in females in 2013 after the exposure to additionally treated wastewater might be due to the altered composition of the effluent in 2013 compared to 2014 with more ozone used in 2013 compared to 2014. This possibly could have resulted in the formation of by-products with estrogenic activity [18]—thus leading to higher VTG levels in females. The lacking reactions in male fish, however, indicate that the wastewater at the WWTP Eriskirch was not highly estrogenic in general. In line with this, chemical analyses of the effluent showed only low concentrations of estrogen-active substances (Bisphenol A: 39–110 ng/L in the conventional effluent and 11-160 ng/L after the additional treatment) or concentrations below the detection limit (EE2 > 1 ng/L) [19].

In contrast to our results, a study with crucian carp showed that VTG levels in immature female and male carps were reduced when the wastewater was treated with ozone [20]; however, the carps already had higher VTG induction in the normal effluent compared with controls and our trout did not show higher VTG levels in the normal effluent compared with the negative controls.

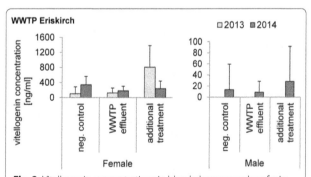

Fig. 2 Vitellogenin concentrations in blood plasma samples of rainbow trout exposed at the WWTP Eriskirch in aquaria connected to the conventional effluent or to the additionally treated effluent in 2013 and 2014; means and standard deviation (SD) are shown. Analyzed by Biosense rainbow trout vitellogenin ELISA kit. N-numbers 2013 females: negative control $n = 9$, WWTP effluent $n = 6$, additional treatment $n = 8$; males: negative control $n = 1$, WWTP effluent $n = 7$, additional treatment $n = 3$. No significant differences with Steel–Dwass test; $p > 0.05$. N-numbers 2014 females: negative control $n = 6$, WWTP effluent $n = 5$, additional treatment $n = 4$; males: negative control $n = 13$, WWTP effluent $n = 6$, additional treatment $n = 10$. No significant differences with Steel–Dwass test; $p > 0.05$. No significant differences between years; $p > 0.05$

Furthermore, we observed slightly higher VTG levels in females of our negative control in 2014 compared to 2013. Differences in VTG baseline levels in negative controls between the years 2013 and 2014 were probably due to the slower fish growth in the laboratory in 2013. In 2013, the mean weight of rainbow trout was 16.3 g ± 2.7 SD and, in 2014, the mean weight was 89 g ± 21.8 SD (Fig. 3). The brown trout showed similar results (Fig. 4). The gonadal development depends on the size of a fish. The bigger the fish the more developed are its gonads, and developed gonads are associated with higher VTG concentrations because the gonads induce the VTG synthesis in liver cells via hormones [21].

The results of the caging experiments performed upstream and downstream of the WWTP Langwiese showed no evidence of estrogenic disruption in males, neither before nor after the upgrade (Fig. 5). Chemical analyses found no EE2 in the effluent (detection limit 1 ng/L), but in vitro tests revealed estrogenic potentials prior to the upgrade [22]. In females, slightly, but not significantly higher VTG levels were measured upstream the WWTP in both years. Lower values downstream might possibly be caused by the combined activity of estrogenicity, anti-estrogenicity, and androgenicity which were all detected in parallel in in vitro bio tests [22].

In summary, the results of our exposure experiments at the two WWTP effluents made evident that, in contrast to other studies which showed an induction of VTG by wastewater in juvenile, sexually immature, and male trout [11, 17, 23], even the conventional effluents of these WWTPs did not lead to increased VTG levels. This speaks for the high efficiency of the already established technologies at these two WWTPs, which, like most of the other larger WWTPs connected to tributaries of Lake Constance, are already equipped with a flocculation sand filter as a final cleaning step.

Exposure experiments at the bypass stations at the Schussen and the Argen River
Rainbow trout

In the two bypass systems, at the Schussen downstream the WWTP Langwiese and at the reference river Argen, no VTG induction became evident in male fish, neither before nor after the upgrade of the WWTP Langwiese with the powdered activated charcoal filter (Fig. 6a). The VTG levels in females were highly variable; however, the highest percentages production in relation to the levels in the respective negative control fish were found in trout exposed at the bypass at the Schussen prior to the WWTP upgrade (Fig. 6b). Significant differences, however, did not occur.

Juvenile rainbow trout which hatched at the bypass stations and were continuously exposed there afterwards showed neither before nor after the upgrade any induction

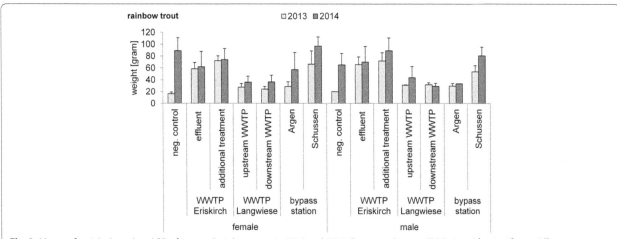

Fig. 3 Means of weight (gram) and SD of exposed rainbow trout in 2013 and 2014. For *n*-numbers see Table 1 and for significant differences see Table 2

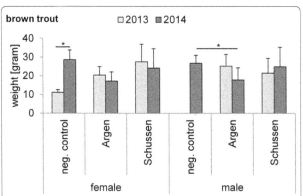

Fig. 4 Means of weight (gram) and SD of exposed brown trout in 2013 and 2014. For *n*-numbers see Table 3. Significant differences with the Tukey–Kramer HSD test: females: neg. control 2013—neg. control 2014 $p = 0.0226$ and males 2014: neg. control—Argen $p = 0.0498$ (*Asterisks* significant differences; *p < 0.05)

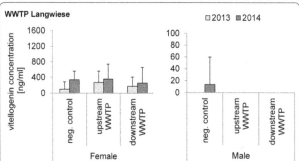

Fig. 5 Vitellogenin concentrations in blood plasma samples of rainbow trout exposed in 2013 and 2014 in cages upstream and downstream of the WWTP Langwiese; means and SD are shown. Analyzed with Biosense rainbow trout vitellogenin ELISA kit. *N*-numbers 2013 females: negative control $n = 9$, upstream WWTP $n = 15$, downstream WWTP $n = 7$; males: negative control $n = 1$, upstream WWTP $n = 2$, downstream WWTP $n = 4$. No significant differences with Steel–Dwass test; p > 0.05. *N*-numbers 2014 females: negative control $n = 6$, upstream WWTP $n = 9$, downstream WWTP $n = 13$; males: negative control $n = 13$, upstream WWTP $n = 11$, downstream WWTP $n = 8$. No significant differences with Steel–Dwass test; p > 0.05. No significant differences between years; p > 0.05

of VTG. In contrast to that, Stalter et al. showed a significant increase in the VTG concentrations using yolk sac rainbow trout which were directly exposed to WWTP effluents for 60 days [17]. We used river water instead of effluent and the results of our other experiments revealed only a weak estrogenic pollution, explaining why we did not find increased VTG levels in juveniles.

These results for juvenile rainbow trout coincide with data for male fish, both indicating that neither at the Schussen downstream the WWTP nor at the Argen River are rainbow trout affected by estrogen disruptors.

Brown trout

In 2013, prior to the upgrade, we found no significant differences in VTG levels in female and male brown trout

exposed at the bypass stations (Fig. 7a). In 2014, after the upgrade, female brown trout showed significantly lower VTG levels at the Schussen (downstream WWTP Langwiese), whereas males showed no significant differences (Fig. 7a). Note that VTG levels of brown trout from different years cannot be compared because semi-quantitative VTG kits (semi-quantitative Salmonid (*Salmoniformes*) biomarker ELISA from Biosense) were used, implying that values are only comparable within one kit. This is the reason why the absolute values are also presented as relative values to the respective negative control

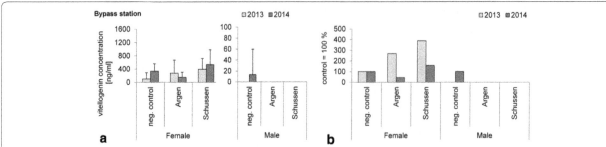

Fig. 6 **a** Vitellogenin concentrations in blood plasma samples of rainbow trout exposed in 2013 and 2014 at the bypass stations; means and SD are shown. Analyzed by Biosense rainbow trout vitellogenin ELISA kit. *N*-numbers 2013 females: negative control *n* = 9, Argen *n* = 8, Schussen *n* = 4; males: negative control *n* = 1, Argen *n* = 5, Schussen *n* = 9. No significant differences with Steel–Dwass test; *p* > 0.05. *N*-numbers 2014 females: negative control *n* = 6, Argen *n* = 8, Schussen *n* = 6; males: negative control *n* = 13, Argen *n* = 1, Schussen *n* = 16. No significant differences with Steel–Dwass test; *p* > 0.05. No significant differences between years; *p* > 0.05. **b** Values of **a** relative to negative control. Neg. control was set to 100 %

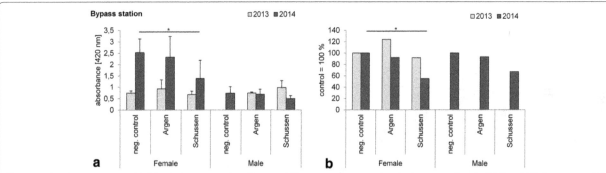

Fig. 7 **a** Absorbance measured in blood plasma samples of 1-year-old brown trout exposed at the bypass stations in 2013 and 2014; means and SD are shown. All samples of a group were analyzed within one semi-quantitative vitellogenin salmonid (Salmoniformes) biomarker ELISA kit (enzyme activity = color intensity is proportional to the concentration of vitellogenin in the sample). *N*-numbers 2013 females: negative control *n* = 4, Argen *n* = 4, Schussen *n* = 3; males: negative control *n* = 0, Argen *n* = 4, Schussen *n* = 4. No significant differences; *p* > 0.05. *N*-numbers 2014 females: negative control *n* = 6, Argen *n* = 6, Schussen *n* = 9; males: negative control *n* = 10, Argen *n* = 13, Schussen *n* = 5. Significant differences with the Tukey–Kramer HSD test: females 2014 neg. control—Schussen p = 0.0231 (*Asterisks* significant differences; *p < 0.05). **b** Values of **a** relative to negative control. Neg. control was set to 100 %. In 2013, no values could be given for males because of absence of males in the neg. control

levels in Fig. 7b. In 2014, VTG levels of females and males were lower at both rivers compared to negative control levels. Fish size did not vary strongly within treatment groups of each year (Fig. 4). Especially females showed no significant differences in their weights in 2014, hence we excluded differences in size as an explanation for differences in VTG levels (see Fig. 4). The fact that the VTG levels in females exposed at the Schussen were significantly lower than the negative control might be explained by the upgrade of the WWTP Langwiese. The additional treatment step at the WWTP Langwiese might have reduced estrogenic activities, and thereby unmasked anti-estrogenic activities which led to reduced VTG levels. Results by Stalter et al. indicated the importance of masking effects to evaluate wastewater [24]. Analyses of the same samples by in vitro yeast assays provided supporting results by showing elevated anti-estrogenicity

and degraded estrogenic activities after the upgrade (not published).

At both bypass stations, brown trout larvae showed no increased VTG values after the upgrade of the WWTP Langwiese compared to the negative control levels (Fig. 8a). On the contrary, the levels are even lower than the negative control levels, which might again be related to unmasked anti-estrogenicity in 2014. These results differ from data we collected prior to the upgrade (Fig. 8, and see also Henneberg et al. [22]). In this previous study, brown trout showed significantly higher VTG levels at the Schussen bypass and at the Argen bypass compared to the negative control after the same exposure time. Estrogen-active compounds were likely causes for the increased VTG levels prior to the upgrade. Due to the fact that we did not observe differences in VTG levels after the upgrade at both bypass stations, we conjecture

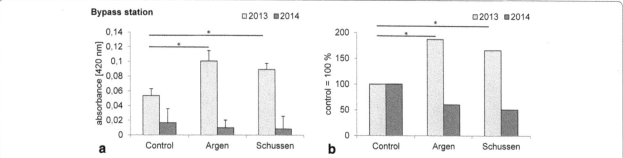

Fig. 8 a Absorbance measured in homogenates of juvenile brown trout exposed for 99 days post-fertilization at the bypass stations in 2014; means and SD are shown. All samples were analyzed within one semi-quantitative vitellogenin salmonid (Salmoniformes) biomarker ELISA kit (enzyme activity = color intensity is proportional to the concentration of vitellogenin in the sample). Each treatment n = 12. No significant differences with the Steel–Dwass test; p > 0.05. For better comparison, previous results from 2013 prior to the upgrade are also shown. These results were already published in PlosOne by Henneberg et al. 2014 [22]. **b** Values of **a** relative to negative control. Neg. control was set to 100 %

that it is mainly annual specific differences that caused these effects, and to a lesser degree the upgrade itself.

In contrast to these results for brown trout, we observed no differences in VTG levels of juvenile and 1-year-old rainbow trout. Previous studies showed that brown trout are more sensitive to environmental stress than rainbow trout [25–27], and our results are in line with this observation. Bjerregaard et al. concluded that the sensitivity of brown trout to estrogens does not differ from the sensitivity of the majority of fish species; first- and second-year brown trout appear to be suitable monitoring organisms to demonstrate estrogenic effects in headwater streams [28]. Hence, our results indicate slight temporary estrogenic effects that might affect feral fish species. However, the differences in VTG levels of brown trout we observed were low, and we conclude that estrogenic effects in the two rivers investigated are generally low.

Organs of the trout we used in the present study were examined in a parallel study to assess their health status before and after the upgrade at the WWTP Langwiese. The results showed that the upgrade led to a better health status of trout and partly also of feral fish species. While this showed that the upgrade reduced toxic effects, the current study showed that estrogenic effects were only slightly reduced.

Conclusion

Overall, our VTG results showed no strong estrogenic effects of WWTP effluents at the Schussen River on trout. After the upgrade of WWTP Langwiese, juvenile and female brown trout showed significantly decreased VTG levels but especially the results for brown trout larvae indicated that annual variation might also play a major role. While rainbow trout showed no significant reduction in VTG levels, we found reduced VTG levels in brown trout, indicating that brown trout might respond more sensitively than rainbow trout.

Furthermore, we did not observe increased VTG levels in males in any experiment. Therefore, we classify the Schussen River as showing only low pollution with estrogens. In particular, neither effluents of the WWTP Langwiese nor effluents of the WWTP Eriskirch caused significantly higher VTG levels in trout, independently of additional wastewater treatment technologies.

Methods
Test organisms

For our investigations we used immature, 1-year-old brown trout (*Salmo trutta* f. *fario*) and rainbow trout (*Oncorhynchus mykiss*) delivered by the fish hatchery Lohmühle, Alpirsbach, Germany. We also obtained freshly fertilized trout eggs from there. For the experiments trout were transported from the hatchery to the exposure sites and directly released in cages or aquaria. The trout which grew up at the fish farm received a mixture of spring water with drinking water quality and stream water which originates in a water protection area (pH 7, nitrate <0.3 mg/L, nitrite <0.0033 mg/L) [29]. All fish were fed with food from the company BioMar, Denmark (INICO Plus for larvae and EFICO alpha for 1-year-old trout) in different particle sizes, depending on fish size. Trout in all our exposure experiments received the same amount of food, except for the negative control (rainbow trout) in 2014, which were sampled directly at the fish farm and for which the amount fed was not under our control.

Exposure experiments at WWTPs and at bypass systems

As a model for a medium-sized WWTP with 40,000 population equivalents we chose the WWTP Eriskirch connected to the Schussen River in the Lake Constance catchment area, South Germany (Fig. 9). At this WWTP, a small-scale model installation was realized in 2013, which included different columns allowing cleaning of

Fig. 9 Overview of sampling sites, bypass stations and examined WWTPs at the Schussen River and Argen River, Lake Constance, South Germany

partial effluent flow by different combinations of ozonation, sand filtration, and granulated activated carbon filter. In 2013 and 2014, 1-year-old rainbow trout were exposed here in aquaria of which one was flown through by the conventional effluent and the second by the additionally treated effluent. In 2013, the additionally treated effluent was proportionately composed of wastewater treated by (1) ozonation + sand filter + granulated activated carbon, and (2) ozonation + granulated activated carbon. In 2014, the composition was changed as follows: (1) ozonation + sand filter; (2) ozonation + granulated activated carbon; and (3) only granulated activated carbon in the ratio 1:1:1. The aquarium with the regular effluent was aerated via a membrane pump to ensure sufficient oxygen concentrations for trout. Daylight was simulated by lamps using timer clocks, and the light/dark photoperiod was adapted to natural daylight. Fish were fed with equal amounts of food by an automatic feeder once a day.

As a model for a large WWTP with 170,000 population equivalents, the WWTP Langwiese was in the focus of our study, also situated at the Schussen River upstream of the WWTP Eriskirch (Fig. 8). At the WWTP Langwiese, an additional powdered activated carbon filter was put into operation after the biological treatment and before the final sand filter in September 2013. At that WWTP,

we exposed 1-year-old rainbow trout in cages (for cage description see [30]) 100 m upstream the WWTP effluent and downstream of it (mixture of 50 % effluent and 50 % Schussen water) in the Schussen River. Trout were fed every second day with a comparable amount of food as the trout at the WWTP Eriskirch received in 2 days. The exposure experiments at the WWTP Langwiese were performed in spring 2013 prior to the upgrade with the powdered activated carbon filter and in spring 2014 after the upgrade.

In addition to the exposure experiments at the WWTPs, we used two bypass stations with 250 L aquaria continuously flown through by fresh river water (0.4 L/s): one setup was located downstream of the WWTP Langwiese at the Schussen River and one at the Argen River as a reference river less influenced by micropollutants [31] (see Fig. 9). Here, fertilized eggs and developing larvae as well as 1-year-old brown trout and rainbow trout were exposed (for a detailed description of the bypass station and trout exposure conditions see [22]).

As a negative control, we kept trout in 250 L aquaria under semi-flow-through conditions in climate chambers at the University of Tübingen. We used filtered tap water and exchanged a third of the water volume once a week. Water was aerated, temperature was kept at 6 °C, a stream pump (Co.: Tunze, Germany) guaranteed a constant

Table 1 *N*-numbers of exposed rainbow trout in 2013 and 2014

Year	Treatment	N-numbers	
		Female	Male
Rainbow trout			
2013	Neg. control	9	1
WWTP Eriskirch	Effluent	6	7
	Additional	8	3
WWTP Langwiese	Upstream WWTP	15	2
	Downstream WWTP	7	4
Bypass	Argen	8	5
	Schussen	4	9
2014	Neg. control	6	13
WWTP Eriskirch	Effluent	5	6
	Additional	4	10
WWTP Langwiese	Upstream WWTP	9	11
	Downstream WWTP	13	8
Bypass	Argen	8	1
	Schussen	6	16

stream, and a filter (Co.: JBL1500e) kept good water conditions. Temperature, ammonium and nitrite concentrations were controlled every other day (ammonium <0.05 mg/L, nitrite <0.01–0.05 mg/L). Light/dark photoperiod was adapted to natural daylight. The semi-static conditions implied that we could not feed fish as much as in the flow-through systems because we had to keep a good water quality. The poor growth of our negative control fish in 2013 was a main reason for us to change the negative control fish in 2014. For that, we sampled 1-year-old trout in 2014 directly at the fish farm where we bought all our trout. To ensure that the development status in all groups was comparable, we sampled the negative control fish at the fish farm shortly before sampling fish at the WWTPs.

To ensure that fish generally react to estrogenic substances by producing VTG, we exposed trout to EE_2 as a positive control. For this, fish were kept at same conditions as negative control fish in 2013, but EE_2 was added in concentrations which ranged from 5 to 20 ng/L. All trout exposed to EE_2 showed extreme higher VTG levels than the negative controls (see Table 4).

Table 2 Significant differences in weights of rainbow trout

Year	Treatment group	p value
Females		
2013	Neg. control 2013—additional treatment WWTP Eriskirch 2013	0.041
	Neg. control 2013—upstream WWTP Langwiese 2013	0.0080
2014	Downstream WWTP Langwiese 2014—bypass Schussen 2014	0.0465
2013 vs 2014	Neg. control 2013—upstream WWTP Langwiese 2014	0.0372
	Neg. control 2013—downstream WWTP Langwiese 2014	0.0081
	Upstream WWTP Langwiese 2013—additional treatment WWTP Eriskirch 2014	0.0092
	Upstream WWTP Langwiese 2013—effluent WWTP Eriskirch 2014	0.0451
	Upstream WWTP Langwiese 2013—neg. control 2014	0.035
	Upstream WWTP Langwiese 2013—bypass Schussen 2014	0.035
	Neg. control 2013—bypass Argen 2014	0.041
	Additional treatment WWTP Eriskirch 2013—upstream WWTP Langwiese 2014	0.0407
	Additional treatment WWTP Eriskirch 2013—downstream WWTP Langwiese 2014	0.0139
Males		
2014	Neg. control 2014—downstream WWTP Langwiese 2014	0.0138
	Additional treatment WWTP Eriskirch 2014—upstream WWTP Langwiese 2014	0.0411
	Additional treatment WWTP Eriskirch 2014—downstream WWTP Langwiese 2014	0.0299
	Bypass Schussen 2014—downstream WWTP Langwiese 2014	0.0076
	Bypass Schussen 2014—upstream WWTP Langwiese 2014	0.0201
2013 vs 2014	Bypass Schussen 2013—bypass Schussen 2014	0.0331

Data were logarithmised to get homoscedastic data and the Steel–Dwass test revealed the following p values

Table 3 *N*-numbers of exposed brown trout in 2013 and 2014

Year	Treatment	N-numbers	
		Female	Male
Brown trout			
2013	Neg. control	4	–
	Argen	4	4
	Schussen	3	4
2014	Neg. control	6	10
	Argen	6	13
	Schussen	9	5

Exposure duration at WWTPs and at bypass systems

Prior to the upgrade at the WWTP Langwiese, we carried out one bypass exposure and one cage exposure experiment in the winter season 2012/2013. After the upgrade at the WWTP Langwiese, one bypass exposure and one cage exposure experiment were performed in the winter season 2013/2014. At the WWTP Eriskirch, we started the first exposure experiment in spring 2013 because the installation of the exposure aquaria was not completed until then. In the second year, 2014 (after the upgrade of the WWTP Langwiese), all exposure experiments started at the same time at all sites. Tables 5 and 6 summarize the time schedule for all exposure experiments, including exposure duration and exposure type.

Ethic statement

This study was carried out in strict accordance with German legislation (animal experiment permit nos. ZO 1/09 and ZP 1/12, District Magistracy of the State of Baden-Württemberg).

Vitellogenin detection
Sampling

One-year-old brown trout and rainbow trout, sampled at each site, were killed with an overdose MS-222 (tricaine mesylate, Sigma-Aldrich, St. Louis, USA). Blood samples were taken immediately from the caudal vein by a sterile syringe, transferred in lithium-heparinized reaction tubes (Co. Sarstedt, Germany), and 4 TIU aprotinin (C. Roth, Germany) per mL blood was added. Samples were centrifuged (4 °C, 10 min, 2500 rpm Eppendorf 5810R) on-site and plasma samples were snap-frozen in liquid nitrogen. Thereafter, plasma aliquots were stored at −80 °C until we determined VTG levels. After taking the blood samples, the length and weight of each fish were measured, gonads were removed for histological examinations and fixed in 2 % glutaraldehyde dissolved in 0.1 M cacodylic acid.

Larvae were killed with an overdose MS-222 (tricaine mesylate, Sigma-Aldrich, St. Louis, USA), and the region between head and pectoral fin from each individual was placed in Eppendorf tubes, snap-frozen, and stored at −80 °C.

All the following steps were undertaken on ice. Homogenates of juvenile trout were prepared by adding homogenization buffer (4 times the sample weight; PBS + 2 TIU aprotinin, C. Roth, Germany), mixing with a plastic pestle, centrifuging (10 min, 4 °C, 20,000×g Eppendorf 5810R) [17] and storing the supernatants at −80 °C.

Table 4 Mean values and SD of exposure experiments with trout using EE$_2$ as positive control

	Females		Males	
	2013	2014	2013	2014
Brown trout				
Mean values (absorbance [420 nm])	326.33	21.77	500.5	4.86
SD	±70.11	±25.09	±114	±3.1
n-number	3	5	2	5
Rainbow trout				
Mean values (VTG [ng/ml])	2,699,183.9	3,812,659.5	3,362,476.1	3,830,203.9
SD	±3,074,723.7	±1,653,878.4		±1,867,237.4
n-number	8	5	1	7
Juvenile trout	**Rainbow trout (VTG [ng/ml])**		**Brown trout (absorbance [420 nm])**	
Mean values	2,030.54		0.0808	
SD	±2811.60		±0.0358	
n-number	8		9	

Table 5 Time schedule for the exposure experiments performed at WWTPs and bypass stations with 1-year-old trout

Start of exposure	End of exposure	Exposure duration (days)	Exposure type	Trout species
Winter season 2012/2013 prior to the upgrade				
15 Nov 2012	24 Jan 2013	70	Laboratory neg. control + EE$_2$ control	Brown and rainbow trout
15 Nov 2012	17 Jan 2013	63	Cage exposure	Rainbow trout
15 Nov 2012	14 Feb 2013	91	Exposure in bypass systems	Brown and rainbow trout
6 Feb 2013	21 Mar 2013	43	Exposure at WWTP Eriskirch	Rainbow trout
Winter season 2013/2014 after the upgrade				
	29 Jan 2014	0	Neg. control from hatchery	Brown and rainbow trout
2 Dec 2013	23 Jan 2014	52	EE$_2$ control	Brown and rainbow trout
2 Dec 2013	4 Feb 2014	64	Cage exposure	Rainbow trout
2 Dec 2013	13 Feb 2014	73	Exposure at WWTP Eriskirch	Rainbow trout
2 Dec 2013	12 Mar 2014	100	Exposure in bypass systems	Brown and rainbow trout

Table 6 Time schedule for exposure experiments performed at the bypass stations with fresh fertilized trout eggs

Start of exposure	End of exposure	Exposure duration	Exposure type	Trout species
Winter season 2012/2013 prior to the upgrade				
07 Dec 2012	20 Mar 2013	103 days	Laboratory neg. control	Rainbow trout
07 Dec 2012	21 Mar 2013	104 days	Exposure in bypass systems	Rainbow trout
Results of exposure experiments using juvenile brown trout are published in Henneberg et al. [22].				
Winter season 2013/2014 after the upgrade				
24 Nov 2013	3 Mar 2014	99 days	Laboratory neg. control	Brown and rainbow trout
24 Nov 2013	4 Mar 2014	100 days	Exposure in bypass systems	Brown and rainbow trout
7 Mar 2014	28 Mar 2014	22 days	EE$_2$ control	Brown and rainbow trout

Vitellogenin ELISA

VTG levels of rainbow trout were measured using the rainbow trout (*Oncorhynchus mykiss*) vitellogenin ELISA kit (V01004402, Biosense Laboratories, Norway). For the analyses of the brown trout samples we used a semi-quantitative kit because the antibody of this kit shows a very good cross-reactivity against brown trout VTG (semi-quantitative vitellogenin Salmonid (*Salmoniformes*) biomarker ELISA kit (V01002402, Biosense Laboratories, Norway)). All steps were performed as described in the protocols. As recommended by the provider of the test kit, a minimum of 1:20 dilution was used and samples were tested in duplicates. The absorbance was measured by a microplate reader (Automated Microplate Reader Elx 8006, Bio-Tek Instruments, INC., USA).

The semi-quantitative ELISA test kit, which is recommended for VTG analyses of salmonids, was used for our brown trout samples. The enzyme activity (absorbance), which is measured by the assay, is proportional to the concentration of VTG in the sample. Purified VTG from Atlantic salmon (*Salmo salar*) was used as a positive control within every assay run. We analyzed all blood samples of females with one 96-well plate (neg. control, Bypass Schussen, Bypass Argen and, EE2 control), all samples of males on the next 96-well plate, etc. Hence, all these samples are comparable within their groups. All steps were performed as described in the protocols by the provider of the test kit.

Statistical analyses

Statistical analyses were performed with JMP 10.0 (SAS Systems, USA). Data were tested for normality using the Shapiro–Wilk W test and for homogeneity of variance with the Levene test. If the data were normally distributed and the variance was homogeneous, the Tukey–Kramer HSD test was conducted. Otherwise, if the data were homoscedastic but not normally distributed, the Steel–Dwass test was used. If the data were normally distributed but not homoscedastic, the Welch's ANOVA was performed.

Authors' contributions

AH: Participated in the design of the study, exposed and sampled trout, carried out immunoassays, performed statistical analyses, prepared manuscript. RT: Designed the study, critically revised manuscript. All authors read and approved the final manuscript.

Acknowledgements

The technical help of S. Krais with laboratory analyses is highly acknowledged. Many thanks are due to M. Weyhmüller for the maintenance of the bypass systems as well as to the staff from the Department of Animal Physiological Ecology (A. Dietrich, M. Di Lellis, D. Maier, K. Peschke, A. and V. Scheil, P. Thellmann, and K. Vincze) for help with sampling.

Funding

We acknowledge support by the Deutsche Forschungsgemeinschaft and Open Access Publishing Fund of Tübingen University. The project SchussenAktivplus is funded by the Federal Ministry for Education and Research (BMBF) and cofounded by the Ministry of Environment Baden-Württemberg. In addition, Jedele & Partner GmbH, Ökonsult GbR, the city of Ravensburg, the AZV Mariatal and the AV Unteres Schussental financially contributed to the project. SchussenAktivplus is connected to the BMBF action plan "Sustainable water management (NaWaM)" and is integrated in the BMBF frame programme "Research for sustainable development FONA". Contract period: 1/2012 to 12/2014, Funding number: 02WRS1281A. All funders had no role in study design, data collection and analysis, decision to publish, or preparation of the manuscript. No competing interests between our study and the commercial funders exist and commercial founders do not alter our adherence to Environmental Sciences Europe policies on sharing data and materials.

Compliance with ethical guidelines

Competing interests

The authors declare that they have no competing interests.

References

1. European Commission. Annex 1. Candidate list of 553 substances. http://ec.europa.eu/environment/chemicals/endocrine/strategy/substances_en.htm. Accessed 30 Mar 2015
2. Jobling S (1998) Natural and anthropogenic environmental oestrogens: the scientific basis for risk assessment. Pure Appl Chem 70(9):1805–1827
3. Sumpter JP (1998) Xenoendocrine disrupters—environmental impacts. Toxicol Lett 102:337–342. doi:10.1016/S0378-4274(98)00328-2
4. Bolong N, Ismail A, Salim MR, Matsuura T (2009) A review of the effects of emerging contaminants in wastewater and options for their removal. Desalination 239(1):229–246. doi:10.1016/j.desal.2008.03.020
5. Purdom CE, Hardiman PA, Bye VVJ, Eno NC, Tyler CR, Sumpter JP (1994) Estrogenic Effects of Effluents from Sewage Treatment Works. Chem Ecol 8(4):275–285. doi:10.1080/02757549408038554
6. Lange R, Hutchinson TH, Croudace CP, Siegmund F, Schweinfurth H, Hampe P et al (2001) Effects of the synthetic estrogen 17 alpha-ethinylestradiol on the life-cycle of the fathead minnow (Pimephales promelas). Environ Toxicol Chem 20(6):1216–1227. doi:10.1002/etc.5620200610
7. Routledge EJ, Sheahan D, Desbrow C, Brighty GC, Waldock M, Sumpter JP (1998) Identification of estrogenic chemicals in STW effluent. 2. In vivo responses in trout and roach. Environ Sci Technol 32(11):1559–1565. doi:10.1021/es970796a
8. Johnson AC, Dumont E, Williams RJ, Oldenkamp R, Cisowska I, Sumpter JP (2013) Do concentrations of ethinylestradiol, estradiol, and diclofenac in European rivers exceed proposed EU environmental quality standards? Environ Sci Technol 47(21):12297–12304. doi:10.1021/es4030035
9. Johnson AC, Sumpter JP (2015) Improving the quality of wastewater to tackle trace organic contaminants: think before you act! Environ Sci Technol. doi:10.1021/acs.est.5b00916
10. Triebskorn R, Amler K, Blaha L, Gallert C, Giebner S, Güde H et al (2013) SchussenAktivplus: reduction of micropollutants and of potentially pathogenic bacteria for further water quality improvement of the river Schussen, a tributary of Lake Constance, Germany. Environ Sci Eur 25(1):1–9. doi:10.1186/2190-4715-25-2
11. Kime DE, Nash JP, Scott AP (1999) Vitellogenesis as a biomarker of reproductive disruption by xenobiotics. Aquaculture 177(1–4):345–352. doi:10.1016/S0044-8486(99)00097-6
12. Ackermann GE, Schwaiger J, Negele RD, Fent K (2002) Effects of long-term nonylphenol exposure on gonadal development and biomarkers of estrogenicity in juvenile rainbow trout (Oncorhynchus mykiss). Aquat Toxicol 60(3–4):203–221. doi:10.1016/S0166-445X(02)00003-6
13. Tyler CR, van Aerle R, Hutchinson TH, Maddix S, Trip H (1999) An in vivo testing system for endocrine disruptors in fish early life stages using induction of vitellogenin. Environ Toxicol Chem 18(2):337–347. doi:10.1002/etc.5620180234
14. Sumpter JP, Jobling S (1995) Vitellogenesis as a biomarker for estrogenic contamination of the aquatic environment. Environ Health Perspect 103(Suppl 7):173
15. OECD. Test No. 229: Fish Short Term Reproduction Assay. OECD Publishing
16. OECD. Test No. 230: 21-day Fish Assay. OECD Publishing
17. Stalter D, Magdeburg A, Weil M, Knacker T, Oehlmann J (2010) Toxication or detoxication? In vivo toxicity assessment of ozonation as advanced wastewater treatment with the rainbow trout. Water Res 44(2):439–448. doi:10.1016/j.watres.2009.07.025
18. Bila D, Montalvao AF, Azevedo DdA, Dezotti M (2007) Estrogenic activity removal of 17β-estradiol by ozonation and identification of by-products. Chemosphere 69(5):736–746
19. Scheurer M, Heß S, Lüddeke F, Sacher F, Güde H, Löffler H et al (2015) Removal of micropollutants, facultative pathogenic and antibiotic resistant bacteria in a full-scale retention soil filter receiving combined sewer overflow. Environ Sci Process Impacts 17(1):186–196. doi:10.1039/c4em00494a
20. An L, Hu J, Yang M (2008) Evaluation of estrogenicity of sewage effluent and reclaimed water using vitellogenin as a biomarker. Environ Toxicol Chem 27(1):154–158. doi:10.1897/07-096.1
21. Copeland PA, Sumpter JP, Walker TK, Croft M (1986) Vitellogenin levels in male and female rainbow trout (Salmo gairdneri richardson) at various stages of the reproductive cycle. Comp Biochem Physiol Part B Comp Biochem 83(2):487–493. doi:10.1016/0305-0491(86)90400-1
22. Henneberg A, Bender K, Blaha L, Giebner S, Kuch B, Köhler H-R et al (2014) Are in vitro methods for the detection of endocrine potentials in the aquatic environment predictive for in vivo effects? Outcomes of the Projects SchussenAktiv and SchussenAktivplus in the Lake Constance Area, Germany. PLoS One 9(6):e98307. doi:10.1371/journal.pone.0098307
23. Bjerregaard LB, Madsen AH, Korsgaard B, Bjerregaard P (2006) Gonad histology and vitellogenin concentrations in brown trout (Salmo trutta) from Danish streams impacted by sewage effluent. Ecotoxicology 15(3):315–327. doi:10.1007/s10646-006-0061-9
24. Stalter D, Magdeburg A, Wagner M, Oehlmann J (2011) Ozonation and activated carbon treatment of sewage effluents: removal of endocrine activity and cytotoxicity. Water Res 45(3):1015–1024. doi:10.1016/j.watres.2010.10.008

25. Schmidt H, Bernet D, Wahli T, Meier W, Burkhardt-Holm P (1999) Active biomonitoring with brown trout and rainbow trout in diluted sewage plant effluents. J Fish Biol 54(3):585–596. doi:10.1111/j.1095-8649.1999. tb00637.x

26. Pickering A, Pottinger T, Carragher J (1989) Differences in the sensitivity of brown trout, Salmo trutta L., and rainbow trout, Salmo gairdneri Richardson, to physiological doses of cortisol. J Fish Biol 34(5):757–768. doi:10.1111/j.1095-8649.1989.tb03355.x

27. Schneeberger HU (1995) Abklärungen zum Gesundheitszustand von Regenbogenforelle (Oncorhynchus mykiss), Bachforelle (Salmo trutta fario) und Groppe (Cottus gobio) im Liechtensteiner-, Werdenberger- und Rheintaler-Binnenkanal [Inaugural Dissertation]: University of Bern, Bern

28. Bjerregaard P, Hansen PR, Larsen KJ, Erratico C, Korsgaard B, Holbech H (2008) Vitellogenin as a biomarker for estrogenic effects in brown trout, Salmo trutta: laboratory and field investigations. Environ Toxicol Chem 27(11):2387–2396. doi:10.1897/08-148.1

29. Schindler J (2015) Besatzforellen. http://www.forellenzucht-lohmuehle. de/besatzforellen.html. Accessed 11 Aug 2015

30. Vincze K, Scheil V, Kuch B, Köhler HR, Triebskorn R (2015) Impact of wastewater on fish health: a case study at the Neckar River (Southern Germany) using biomarkers in caged brown trout as assessment tools. Environ Sci Pollut Res. doi:10.1007/s11356-015-4398-6

31. Triebskorn R, Hetzenauer H (2012) Micropollutants in three tributaries of Lake Constance, Argen, Schussen and Seefelder Aach: a literature review. Environ Sci Eur 24(1):1–24. doi:10.1186/2190-4715-24-8

9

Evaluation of evidence that the organophosphorus insecticide chlorpyrifos is a potential persistent organic pollutant (POP) or persistent, bioaccumulative, and toxic (PBT)

John P Giesy[1], Keith R Solomon[2*], Don Mackay[3] and Julie Anderson[4]

Abstract

A number of chemicals, including several organochlorine pesticides, have been identified as persistent organic pollutants (POPs). Here, the properties of chlorpyrifos (CPY; CAS No. 2921-88-2) and its active metabolite, chlorpyrifos oxon (CPYO; CAS No. 5598-15-2), are assessed relative to criteria for classification of compounds as persistent, bioaccumulative, and toxic substances (PBTs). The manufacture and use of POPs are regulated at the global level by the Stockholm Convention (SC) and the UN-ECE POP Protocol. Properties that result in a chemical being classified as a POP, along with long-range transport (LRT), while understood in a generic way, often vary among jurisdictions. Under the SC, POPs are identified by a combination of bulk (intensive) properties, including persistence and biomagnification, and an extensive property, hazard. While it is known that CPY is inherently hazardous, what is important is the aggregate potential for exposure in various environmental matrices. Instead of classifying chemicals as PBT based solely on a few simple, numeric criteria, it is suggested that an overall weight of evidence (WoE) approach, which can also consider the unique properties of the substance, be applied. While CPY and its transformation products are not currently being evaluated as POPs under the SC, CPY is widely used globally and some have suggested that its properties should be evaluated in the context of the SC, especially in locations remote from application. In Europe, all pesticides are being evaluated for properties that contribute to persistence, bioaccumulation, and toxicity under the aegis of EC Regulation No. 1107/2009: 'Concerning the Placing of Plant Protection Products on the Market.' The properties that contribute to the P, LRT, B, and T of CPY were reviewed, and a WoE approach that included an evaluation of the strength of the evidence and the relevance of the data to the classification of CPY and CPYO as POPs or PBTs was applied. While toxic under the simple classification system used in EC Regulation No. 1107/2009, based on its intensive properties and results of monitoring and simulation modeling, it was concluded that there is no justification for classifying CPY or its metabolite, CPYO, as a POP or PBT.

Keywords: Stockholm Convention; EC Regulation No. 1107/2009; Chlorpyrifos oxon; Long-range transport

Background

A number of chemicals, including several organochlorine pesticides, have been identified as persistent organic pollutants (POPs). The POPs were first brought to the attention of the general public by Rachel Carson in her book *Silent Spring* [1]. In that now famous book, she pointed out that a number of chemicals, including the pesticide dichlorodiphenyltrichloroethane (DDT) and its transformation products, dichlorodiphenyldichloroethylene (DDE) and dichlorodiphenyldichloroethane (DDD), were not only persistent but also biomagnified in food chains, caused adverse effects in non-target organisms, such as birds, and underwent long-range transport (LRT) to more remote and pristine areas, such as the Arctic and Antarctic. At about the same time, it was recognized that a number of other organochlorine pesticides and the industrial chemical polychlorinated biphenyls (PCBs) also had properties consistent with them being POPs. Since that time, these and

* Correspondence: ksolomon@uoguelph.ca
[2]Centre for Toxicology, School of Environmental Sciences, University of Guelph, Guelph, ON N1G 2 W1, Canada
Full list of author information is available at the end of the article

additional chemicals have been identified as POPs and the manufacture and use of these substances are regulated at the global level by the Stockholm Convention (SC) [2] and the UN-ECE POP Protocol [3]. While many of the chemicals classified as POPs have been organochlorines, some such as those that contain the terminal degradation product, perfluorooctanesulfonate (PFOS) are not. As in many regulatory systems, the SC uses the precautionary approach; however, this includes detailed scientific review by the POPs Review Committee, where there is an opportunity to consider the inherent properties of the chemical under review.

While understood in a generic way, the properties that are used to derive criteria for classification of a chemical as a POP with LRT, or a PBT, are used differently among jurisdictions [4,5]. Under the global aegis of the SC, POPs are identified by a combination of intensive properties (independent of concentration), including persistence, biomagnification, and chemical and physical properties that result in harmful interactions with biological systems, and extensive properties (dependent on concentration), including toxicity, hazard, and risk. In addition to the SC [2], several additional frameworks have been developed to assess chemicals based on the properties of persistence, bioaccumulation, and toxicity (P, B, and T). Some of these frameworks are international, such as the Convention for the Protection of the Marine Environment of the North-East Atlantic [6]. Others are regional, such as the EU legislation Registration, Evaluation, Authorisation and Restriction of Chemicals (REACH [7]), with a focus on chemicals in commerce, and EC Regulation No. 1107/2009 [8], which is focused on pesticides. National frameworks include, for example, the Toxic Substances Management Policy [9], the Toxics Release Inventory Reporting [10], and the Chemicals Management Plan in Canada [11].

Classifying chemicals as POPs or having the properties of PBTs is used to assist industries in making decisions about the development of chemicals and governments in priority setting and regulation of these chemicals. The concepts of persistence, bioaccumulation, and toxicity are commonly used in the scientific literature, as is the internationally used concept of a POP. PBT, as a term, appears to have originated in policies of the Japanese government in the 1970s, even though the term did not appear in the peer-reviewed scientific literature until the 1990s [5]. This term and underlying concepts are being used increasingly by policy makers in regulatory decisions. Unfortunately, inconsistent definitions and criteria for classifying chemicals as being PBT vary among jurisdictions and have been changing over time. Furthermore, this very simplistic method of classification does not take into account the unique properties of chemicals or the environments to which they are released [12]. These shortcomings are exacerbated by both poor quality of data and, in some cases, little or a complete

lack of data, such as was the case for perfluoroundecanoic acid [5].

There are a number of uncertainties in these approaches that require interpretation of metrics such as persistence in various media, bioaccumulation, and toxicity. Since everything can be toxic, the critical issue is not the inherent toxicological properties of a chemical, which is its potency, but the concentration to which it can accumulate into various matrices of the environment. Ultimately, interpretation of the potential for harm that can be caused by a chemical of concern (COC) is duration and intensity of exposure that determines the severity and rate of damage. Injury occurs when the rate of damage exceeds the rate of elimination and/or repair. So the concept of toxicity needs to be considered not in abstract or absolute terms, but relative to exposures. Of the three principal parameters used to classify chemicals, toxicity is the least well described and interpretable. Adverse effects are only observable when the concentration (exposure) of a substance exceeds the threshold for effects for a sufficient duration. Because of its intensive properties, a chemical, such as the organophosphorus pesticide chlorpyrifos (CPY), can have relatively great potency to cause adverse effects, but if the concentrations in various matrices do not exceed thresholds for adverse effects, there is no adverse effect. Risk is defined as the likelihood for exceedence of a threshold (used here in the inclusive sense) and is always expressed as a probability. This has been known for some time, as attributed to Antoine Arnauld in a monastic text in 1662: 'If, therefore, the fear of an evil ought to be proportionate, not only to its magnitude, but also to its probability...' (page 368 in [13]). Several properties drive the probability of exposure but the most important parameter is persistence. If a COC is sufficiently persistent, then there is always the potential for accumulation and toxicity. Even if a compound degrades relatively rapidly, if it is released continuously or organisms are exposed for a sufficient duration, it might be present in sufficient quantities to exert toxicity. Such substances have been termed 'pseudo persistent.' So here the classification of CPY as a POP is considered not only relative to specific, absolute 'trigger' values for the classifying parameters but also relative to what is likely to occur in the environment. That is, while it is known that CPY is inherently hazardous, what is the aggregate potential for exposure in various environmental matrices [14,15]? Instead of classifying chemicals as PBT based solely on a few simple, numeric criteria, a transparent weight of evidence (WoE [16]) approach, which can also consider the unique properties of COCs, should be applied [5,17]. This approach allows all relevant scientific data to be considered on a case-by-case basis, but the process of classification requires more description of the process and expert evaluation of the results of the multiple lines of evidence.

While CPY and its transformation products are not currently being formally evaluated as POPs under the SC, it has undergone simplified screening as an alternative to endosulfan. This screening suggested that CPY might meet all Annex D criteria (be a POP) but there are only equivocal or insufficient data [18]. CPY is widely used globally and some have suggested that its properties should be evaluated in the context of POPs [19]. In Europe, all pesticides (excluding biocides used to control bacteria and fungi) are being evaluated for properties that contribute to persistence, bioaccumulation, and toxicity (PBT) under the aegis of EC Regulation No. 1107/2009: 'Concerning the Placing of Plant Protection Products on the Market' [8]. Under EC Regulation No. 1107/2009, products identified as POPs under the SC are not allowed to be used.

Properties that contribute to P, LRT, B, and T of CPY have recently been reviewed [15,20]. These reports were part of a series devoted to assessing risks to environments associated with the use of CPY in the United States of America (USA). This report builds on these previous reports but refines the assessment with a WoE approach that includes an evaluation of the strength of evidence and relevance of the data to classification of CPY as a POP or PBT. There is general recognition that industry has a responsibility to evaluate its commercial products, especially chemicals such as pesticides, for their possible environmental impacts. One approach for doing this is to use POP and PBT criteria as a basis for quantitative evaluation of properties, even when there is little likelihood that the chemical will be considered or declared to meet these criteria. In short, established POPs and PBTs are used as 'benchmarks' against which the chemical in question can be compared. It is partly in this spirit that this evaluation was undertaken.

Chemicals can be assessed and classified as PBTs under several auspices with varying sets of guidelines. While there is some guidance on how the classification should be done [21], none of these processes are inherently assessments of risk. That is, they do not consider

probabilities of exceeding threshold concentrations for defined effects in the environment. At best, these processes are an evaluation of measured or predicted parameters that relate to persistence in various media and the potential to bioconcentrate or biomagnify.

Since there were no predefined criteria for identification of POPs, they have been developed over time by various individuals and/organizations from empirical observations of a number of chemicals that were observed to be persistent, biomagnified, and transported over long distances. Thus, the chemical, physical, biological, and environmental properties of the so-called 'dirty dozen' [22,23] were used as the basis for the trigger values for persistence (P), bioaccumulation (B), toxicity (T), and propensity for long-range transport (LRT) that are currently used under the SC (Table 1). As has been pointed out elsewhere [17,24], there are no consistently applied criteria for classification of B other than the bioconcentration factor (BCF) in EC Regulation No. 1107/2009. Although the bioaccumulation factor (BAF) is also used in the SC, other criteria for B, such as the biomagnification factor (BMF) and trophic magnification factor (TMF), have not been used explicitly, even though they are better descriptors because they incorporate the potential for dietary uptake and biotransformation in the aggregate measure of accumulation. Similarly, under the SC, toxicity is simply stated as 'significant adverse... effects' or 'high toxicity' with no indication of what 'significant' or 'high' means.

Criteria for classification of pesticides or other chemicals as PBT under EC Regulation No. 1107/2009 or the program for REACH, which entered into force on 1 June 2007 (Table 2), are similar to those used for POPs (Table 1), but LRT is omitted and the triggers for P and B are more stringent. As has been pointed out elsewhere, criteria used to classify POPs and PBTs are single values [17] and the classification process, particularly for pesticides under EC Regulation No. 1107/2009, does not consider additional data on intensive properties as well as environmental fate and toxicity that are available for pesticides. Since REACH does not have jurisdiction over pesticides, such as CPY,

Table 1 Criteria for categorization of compounds as POPs and LRT substances under the SC and UNECE

Persistent (P)	Bioaccumulative (B)	Toxic (T)	Potential for long-range transport (LRT)
Water: DT_{50}[1] > 2 months	BCF or BAF >5,000 or Log K_{OW} >5	No specific criteria other than 'significant adverse human health and/or environmental effects' (in Article 8, 7 (a))	Air: DT_{50} > 2 days. Monitoring or modeling data that shows long-range transport via air, water, or biota
Sediment: DT_{50} > 6 months	High bioaccumulation in other species, high toxicity or ecotoxicity		
Soil: DT_{50} > 6 months			
Other evidence of persistence	Monitoring data in biota indicating that the bioaccumulation potential is sufficient to justify its consideration within the SC		Concentrations of potential concern detected in remote locations

From [2,3], [1]DT_{50}; note that the SC uses the term half-life but does not state whether this is is for dissipation or for degradation (transformation).

Table 2 Criteria for the categorization of compounds as PBT under REACH or EC Regulation No. 1107/2009

Persistent (P)	Bioaccumulative (B)	Toxic (T)
Marine water: $t\frac{1}{2}$ >60 days	BCF >2,000 in aquatic species	Chronic NOEC <0.01 mg/L or is a carcinogen, mutagen, or toxic for reproduction, or other evidence of toxicity
Fresh water $t\frac{1}{2}$ >40 days		
Marine sediment: $t\frac{1}{2}$ >180 days		
Freshwater sediment: $t\frac{1}{2}$ >120 days		
Soil: $t\frac{1}{2}$ >120 days		

From [7,8].

classification of chemicals in commerce under this legal instrument is not further discussed, except for purposes of comparison.

Assessment of chemicals to determine if they should be classified as POPs under the SC is a lengthy process involving nomination of candidate substances by a party or group of parties, review of data, and final recommendations from a review committee (the POPs RC) [2]. This process is open, but there is no definitive framework for classification and criteria are sometimes inconsistently applied [24]. After a COC is classified as a POP, it is added to Annex A (elimination), B (exemptions), or C (unintentional) of the SC. Since, under the SC, the UN does not have regulatory jurisdiction over the parties (signatory nations), ratification of classification and any subsequent phase-out and/or banning of the manufacture and use of the POPs are undertaken individually by the parties. In fact, the USA, which is a major player in the manufacture and use of chemicals, is not a signatory of the SC. Phase-out can take several years because time is provided for users to find substitutes and, in some cases, such as DDT, specific exemptions for continued availability may be granted for acceptable purposes such as for the protection of human health.

Within the European Union (EU), criteria for assessment of plant protection products (PPPs) for PBT or POP properties are given in EC Regulation No. 1107/2009 and assessments of individual COCs are conducted by Rapporteur Member States (RMS) of the EU in much the same way as registration of new active ingredients. There is no explicit framework or guidance for classification other than a draft document from the EU Directorate General for Health and Consumer Affairs (DG SANCO) [21], and, unlike REACH [7], there is no guidance for how studies are to be evaluated or how the relevance of the data in these studies is to be assessed. REACH recommends the use of a WoE approach for assessing data on chemicals in commerce but does not describe how this is to be done. EC Regulation No. 1107/2009 does not mention WoE at all. Under EC Regulation No. 1107/2009, if a PPP is classified as P, B, and/or T, exceeding trigger values for all three criteria ultimately results in a ban of the use of the product in the EU. Exceeding two of the criteria

results in the PPP being listed for substitution with alternative pesticides that do not exceed established trigger values.

Since pesticides are designed to be toxic to at least some groups of organisms, the criterion for assessment of toxicity is likely to capture all pesticides. Therefore, classification of PPPs as PBTs under this scheme is primarily driven by the P and B. The trigger for classification as T (Table 2) is 'Chronic NOEC <0.01 mg/L or is a carcinogen, mutagen, or toxic for reproduction, or other evidence of toxicity.' The NOEC trigger is strictly for aquatic organisms, which will bias classification of insecticides as T because they are usually equally or more toxic to crustaceans than they are to insects. Few PPPs are deliberately applied to water, so fate and movement in the environment are important drivers of concentrations in water, yet these factors are not considered in classification. Finally, there is no consideration of toxicity for terrestrial species, despite the fact that it is to this environmental compartment that most pesticides are routinely applied.

Problem formulation

Registration and re-registration of pesticides in most jurisdictions require a large number of expensive and demanding studies under both laboratory and field conditions on fates and effects of pesticides in the environment as a whole. As illustrated in Figure 1, assessments of risk used during registration of pesticides are focused on protection of non-target organisms that enter and use the treated areas as habitat or that might be affected if the pesticide moves off the target agroecosystem. The assessment of risk conducted during registration includes characterization of bioaccumulation and metabolism in key species and toxicity to a range of species. In the process of decision-making, toxic potency to non-target organisms is considered and combined with exposures inside and outside of the agroecosystem to assess the acceptability of risks from the use of the pesticide.

Risk, which is the relationship between toxicity and exposure, is not considered in the probabilistic sense in the classification of chemicals as POPs. The review process under the SC is designed to 'evaluate whether a chemical is likely, as a result of its long-range environmental

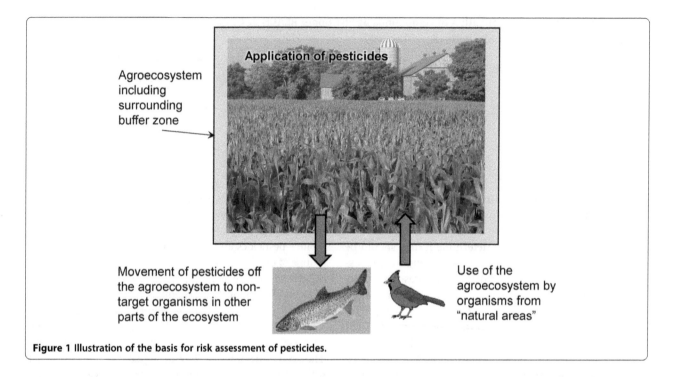

Figure 1 Illustration of the basis for risk assessment of pesticides.

transport, to lead to significant adverse human health and/or environmental effects, such that global action is warranted' [Annex E in 2], whereby the process is based on a deterministic hazard quotient (Annex E (b)). Under EC Regulation No. 1107/2009, binary criteria are used to categorize substances by comparing the properties of the compound to simple threshold or trigger values. This simplistic approach is appropriate for lower tier screening or priority setting, but it is not appropriate as a final step in decision-making.

Goals for protection, sometimes referred to as 'assessment endpoints', are usually either identified explicitly or implicitly in regulations. In terms of humans and the environment, the goals of EC Regulation No. 1107/2009 are '...to ensure a high level of protection of both human and animal health and the environment and at the same time to safeguard the competitiveness of Community agriculture.' [8]. In the absence of more specific goals, it is logically assumed that the concern is for the general environment, not for a particular local scenario. For classification of COCs as POPs under the SC, this is a global concern. POPs identified under the SC are not permitted for use in the EU, so the environment of concern under EC Regulation No. 1107/2009 is that within the EU, which is the jurisdiction of regulation. With this in mind, characterization of P for the regional as well as the global environment was accomplished by refinement of the generalized assessment presented previously [15].

Properties of chlorpyrifos
The physical and chemical properties of CPY have been summarized relative to assessment of risk to the

environment of the use of this product in agriculture in the USA [15,20,25] and are thus not repeated here. The focus of the following sections is on characterizing the P, B, and T properties of CPY in relation to criteria for classification under the SC and EC Regulation No. 1107/2009. Since Annex II 3.3 of EC Regulation No. 1107/2009 specifically includes metabolites, they were included in the assessment.

Under environmental conditions, several transformation products of CPY are formed [20] and have been considered in the assessment of risks [26]. CPYO is assessed in this document, but trichloropyridinol (TCPy) has not been identified as a metabolite of toxicological or environmental concern [26,27] and was excluded from consideration here.

Because of similarities in the structure of TCPy to trichlorophenol, from which dioxins and furans are known to be formed, the possibility of this occurring with CPY was considered. Dibenzo dioxins and furans were not detected (limit of detection (LOD) 0.006 to 0.0008 ng/g) in formulations of CPY [28]. A recent study reported the formation of 2,3,7,8-tetrachloro-1,4-dioxino-[2,3-b:5,6-b'] dipyridine (TCDD-Py), an analog of 2,3,7,8-tetrachlorodibenzo-p-dioxin (TCDD), when pure (2 mg) CPY was pyrolyzed in sealed ampules at 380°C, but not 300°C or 340°C, for 15 min in the presence of 10 mL of air [29]. Greater amounts of TCDD-Py ($\approx$100-fold) were formed from pure TCP under all of the above conditions. TCDD-Py is unstable under the conditions of synthesis of CPY from TCP [30], suggesting that, even if it is formed, it will not become a contaminant in the commercial product. In a study of effects of combustion on the fate of TCP, only TCP and CO_2 were identified in smoke from cigarettes containing residues (900 ng/cigarette) of ^{14}C-labeled TCP. Detection limits for hexane-extractable non-polar compounds (2% of total radioactivity applied) such as TCDD-Py were not provided [31].

Dibenzo-p-dioxins have been observed in formulations of chlorinated pesticides, such as 2,4-D, exposed to sunlight [32]. A study of photodegradation of ^{14}C ring-labeled CPY in buffered and natural waters treated with 0.5 and 1 mg CPY/L did not reveal the presence of polar compounds except CPY and dichloropyridinyl phosphoro-thioate esters and 96% of the degradates formed were polar compounds [33]. Dioxins such as TCDD are rapidly photolyzed by sunlight in the presence of a hydrogen donor with a half-life of the order of hours [34]. Thus, if TCDD-Py was formed in sunlight, it might be expected to be photolabile and non-persistent in the environment.

A search of the literature failed to reveal the isolation and identification of TCDD-Py in the environment, either because it is not formed in detectable amounts, because it is rapidly degraded, or because it has not been analyzed for. The only papers that reported on its formation and/or biological activity [29,35] did not conduct analyses of environmental samples. They also did not test whether TCDD-Py was formed by photolysis from CPY or TCP.

TCDD-Py is only moderately toxic to rats. It has an oral median lethal dosage (LD50) of 300 mg/kg body mass (bm) in rats (strain unspecified), about four orders of magnitude less toxic than TCDD, which has an LD50 of 0.022 to 0.045 mg/kg bm [36]. Tests in female Sprague–Dawley rats exhibited loss of body mass but no lethality or gross pathological findings at an even greater acute oral dose of 600 mg/kg bm [37]. The same study reported no evidence of chloracne on the ears of NZ white rabbits treated 18 times with a solution containing 50 mg TCDD-Py/L [37].

On the basis of this evidence, we conclude that TCDD-Py is either not formed from CPY or TCP under normal conditions of use or the amounts formed are so small that it has escaped notice in the analyses of bioaccumulative substances in environmental samples. In addition, the relatively low toxicity of TCDD-Py indicates that, even if formed in the environment, it presents little risk to humans or the environment. Thus, TCDD-Py was not included in the following assessment.

Analysis plan

Since there was little guidance in categorizing POPs and PBTs [with the exception of 21], WoE was used to select the most appropriate data for inclusion in the assessment. WoE is a phrase that is widely misused in the literature [16] and has been applied to a number of procedures for assessment of risk. Here, WoE is used as a quantitative procedure for evaluating the strength of studies, based on how they were conducted, and the relevance of the data from the studies to characterization of the COC, CPY, as a

POP or PBT chemical. Strength of studies was evaluated by a numerical scoring system (see the 'Quality assurance' section). Relevance was also assessed, particularly in the case of persistence, where studies were conducted at very large rates of application, such as for control of termites, which are inconsistent with current uses, and in the case of bioconcentration, where studies were conducted at exposures greater than the maximum solubility of the CPY in water. All of the available data were evaluated (see Additional file 1), and then, on the basis of strengths of the studies, those studies that provided the most robust data were selected for inclusion in the assessment of the PBT properties of CPY. Studies conducted under non-relevant conditions were then excluded to provide the most robust and relevant data for the characterization. This procedure is different from the assessment conducted by Mackay et al. [15] where all data were used, regardless of their strength or relevance.

Because extreme (worst-case) values observed in specific conditions are not representative of all situations, mean values were used for comparison to the criteria for classification, a process which has been recommended in the literature [17] and the draft guidance of SANCO [21]. Because most of the processes related to P or B at environmentally relevant concentrations are driven by first-order or pseudo-first-order kinetics and thus are lognormally distributed, geometric mean values are the most appropriate for comparing triggers for classification and were used in this assessment.

Since persistence of CPY in the environment is dependent on its unique properties as well as the properties of and conditions in the surrounding environment, in the context of global persistence, all acceptable values for assessment of CPY as a POP were combined [15]. Because of the regional focus of EC Regulation No. 1107/2009, data for persistence were segregated into studies from the EU and from other regions. These data were analyzed separately.

In addition to consideration of characteristics related to P, B, and T, other lines of evidence were also used as a means of corroboration of the more simple criteria for classification. As in the SC, one line of evidence was based on reports of bioaccumulation of CPY in organisms in the field. Concentrations in aquatic systems in the USA have been thoroughly reviewed [25] and were not included in the assessment presented here other than in the context of assessing exceedences of criteria for toxicity and their rapid response to changes in pattern of use.

Sources of data

An exhaustive review of the literature was conducted in support of an ecological risk assessment of CPY [14, et seq.], and results were compiled into an electronic database, which formed the basis for the information used in this report. The extensive search of the available literature was conducted by using Web of Knowledge®, a database with access to a number of other digital collections and databases. In addition, searches were conducted with Google Scholar using keyword string searches to access other available peer-reviewed resources. Recent reviews of the literature on the properties of CPY [15,20] provided an overview of pertinent studies. Also included in the reviewed body of work were relevant unpublished studies from Dow AgroSciences and its affiliates, which were provided to the authors in their original forms.

From the results of the extensive literature search, those studies judged to be directly related to persistence and/or bioaccumulation of CPY were retained for further assessment. These included studies on persistence in sediment, soil, and water, performed under laboratory or field conditions, and studies from any international jurisdiction. In total, 41 papers or reports on bioaccumulation of CPY in biota and 90 papers or reports on persistence of CPY in soil, sediment, and/or water were included in the scoring evaluation. Data on toxicity to aquatic organisms were previously screened for quality [38] and were used in this assessment without further characterization.

Quality assurance

Those papers retained from the literature search were subjected to a WoE assessment with the aim of identifying those studies that should form the basis of the final report. Scoring used in our evaluation was based on criteria developed for inclusion of data in the International Uniform Chemical Information Database (IUCLID) [39] and more current updates. The system was augmented by application of numerical scores as has been done previously for selection of toxicity data for CPY [38]. Specific scoring criteria were developed for each type of exposure, including bioaccumulation in sediment-dwelling organisms, aqueous bioconcentration, bioaccumulation from dietary exposure, persistence in soil, persistence in sediment, and persistence in water. These criteria were based on OECD methods for testing [40] under the relevant conditions (OECD Methods 305, 308, 309, and 315). Criteria assessed the strength of the methods used in each study, and a score of 0, 1, or 3 was assigned for each criterion, indicating that the criterion was not met, was attempted but not fully met, or was fully achieved, respectively. The scoring matrices are provided (see Additional file 1: Table S1).

A consensus score was also available as part of the strength of methods as a mechanism to adjust scores for those few studies where an invalidating error was not captured by the standard matrix for evaluation. In addition to assessing the strength of methods, the relevance of the study was also assessed by setting limits for rates of application that would be considered environmentally relevant (based upon current recommended use rates

[20]) and assigning a score of 0, 1, or 3 for relevance. A full score was assigned if the rate of application was equal to or less than the relevant rates of 200 µg/L (accumulation in water), 1,000 µg/L (persistence in water), or 8 mg/kg (accumulation or persistence in soil/sediment), a score of 1 was assigned if the rate was within 25% of the set value (i.e., ≤250 µg/L, 1,250 µg/L, or 10 mg/kg), and a score of 0 was assigned if the rate was more than 25% above the set value for environmental relevance. Criteria for relevance also evaluated whether a description of variance was provided, whether dietary spiking was performed at appropriate concentrations (residues not exceeding 5 µmol/g) and/or whether BAF was calculated kinetically from the depuration rate constant, as appropriate for each type of study.

Briefly, the process of evaluation proceeded as follows: a study that had been identified through review of the literature and screening process was scored according to the appropriate matrix. Scores were recorded, along with a description of the study methods, and a summary of the reasons for the assigned score for each criterion. Where a specific study element or criterion was not explicitly addressed in the paper and could not be readily inferred from the results, it was assumed that the criterion was not met and a reduced score was assigned. If a missing element made the assessment impossible or if it was not possible to score a particular criterion for an individual study, that criterion was omitted from the scoring. This most commonly occurred for the use of a solvent in studies of bioaccumulation, where a solvent was not required, but in the case where it was used, it should be one from an approved list. In cases where no solvent was used, that criterion was omitted from the assessment.

In total, assessments were completed for 44 bioaccumulation studies and 90 persistence studies, and the scores are summarized in Additional file 1: Tables S2 through 5. It was not possible to obtain a copy of one report on bioaccumulation and two papers on persistence that had been identified during the review of the literature, so these studies could not be assessed. After assessment, those studies that received scores of less than 50% for strength of methods were excluded for the purposes of this assessment. In total, 23 studies of bioaccumulation of CPY and 44 studies of persistence of CPY were included in the final assessment.

Main text
Persistence

Persistence was assessed by determining the half-lives for degradation in soils, sediments, and surface waters. Information from both laboratory and field studies was considered. In the case of field studies, some losses might have occurred via volatilization; however, this is a realistic loss process affecting exposure in the field. Thus, these dissipation data are appropriate for inclusion in the assessment as long as the degradation in the compartment into which partitioning occurs, in this case air, is considered as well. Studies conducted in the laboratory at normal ranges of temperature (15°C to 35°C) were included in the assessment because these temperatures were representative of the surface layers of the soil in the regions from which the samples were obtained. Since the relatively large K_{OC} of CPY limits mobility in agricultural soils, surficial temperatures are more representative of the environment in which CPY would be expected to occur. Data from field studies were assumed to have been conducted at realistic temperatures.

Use of a half-life implies first-order kinetics, but most environmental degradation processes are of second order in the concentration of the chemical and the concentration of the reactant, either a chemical or the number of microorganisms. Concentrations of reactants can vary widely; thus, half-lives are expected to vary considerably, especially in field studies. Use of extreme values for assessment purposes can be highly misleading since they likely reflect unusually small reactant concentrations. The geometric mean was used in this assessment as a more rigorous approach and is recommended for assessment of PBT [21].

Soil

In soils, CPY can be degraded by hydrolysis and also microbial transformation. Half-lives for dissipation from soils via all pathways ranged from 1.1 to 1,576 days (see Additional file 1: Table S2). There were several outliers that were excluded from the assessment because the application rates were large (i.e., 1,000 mg/kg for control of termites) [41] and were phased out in 2000. Lesser rates of application

(10 and 100 mg/kg) were included in the assessment. Studies of CPY in soils were divided into those conducted under European conditions and those conducted with soils from other areas of the world. Studies were also stratified by whether they were conducted under laboratory or field conditions.

For European soils [42,43] and non-European soils (Figure 2) [41,44-50] studied under laboratory conditions, the geometric means were 73 and 21 days, respectively. These geometric means do not exceed the criterion for classification as a POP under the SC (180 days) or as persistent under EC Regulation No. 1107/2009 (120 days). The geometric mean half-life for all laboratory-based data for soils without exclusion of studies was 32 days [15], also less than the criteria for POPs or PBT.

Geometric means of half-lives derived from field studies in European soils [51-57] and non-European locations [44,58-66] (Figure 3) were 20 and 13 days, respectively. None of the geometric means for dissipation of CPY in soils under field conditions exceeded the threshold to be classified as being persistent in soils under the SC or EC Regulation No. 1107/2009. The geometric mean half-life for all field-based data for soils without exclusion of studies was 22 days [15], also less than the criteria for POPs or PBT.

Given the relatively short half-lives observed in soil, CPY is very unlikely to accumulate in soils as a result of repeated use in agriculture. Thus, treated soils will not act as a reservoir for other matrices such as water. This is consistent with rapid decreases in concentrations of CPY in surface water after changes in patterns of use (see the 'Measured concentrations in surface waters' section).

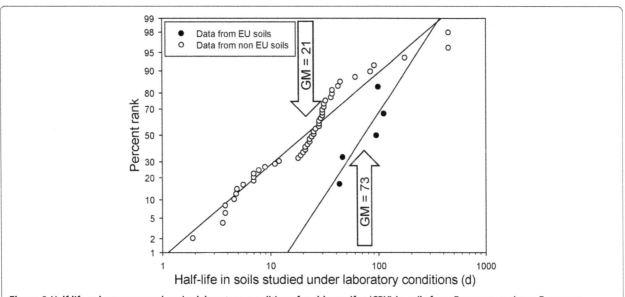

Figure 2 Half-life values measured under laboratory conditions for chlorpyrifos (CPY) in soils from European and non-European locations. Geometric means are indicated by vertical arrows.

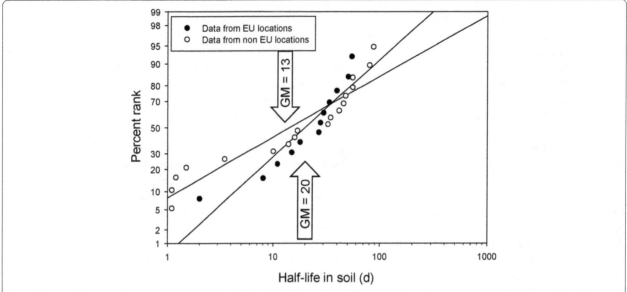

Figure 3 Half-life values measured under field conditions for chlorpyrifos (CPY) in soils from European and non-European locations. Geometric means are indicated by vertical arrows.

Sediment

In sediments, under laboratory and field conditions, the half-life of CPY ranged from 1 to 223 days (see Additional file 1: Table S3). The geometric mean of all studies was 29 days. When only those studies that scored in the top 50% of studies were included, the geometric mean was 25 days. When studies conducted with European and non-European sediments were considered separately (Figure 4), the geometric means were 40 and 19 days, respectively. None of the geometric means exceeded the threshold for classification of CPY in sediments as being

persistent under EC Regulation No. 1107/2009. Even the maximum values of 200 and 230 days were only slightly greater than the 180 days necessary to classify CPY as being persistent in sediment. The geometric mean half-life for all studies (without exclusions) was 38 days [15], which is also less than the criteria for POPs or PBT (180 and 120 days, respectively).

Water

In water, CPY can be degraded abiotically by aqueous hydrolysis, photolysis, and microbial transformation.

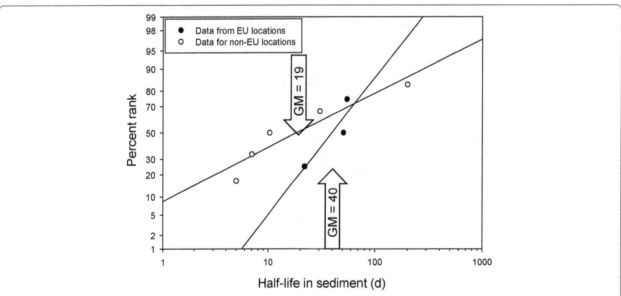

Figure 4 Half-life values measured under laboratory conditions for chlorpyrifos (CPY) in sediments from European and non-European locations. Geometric means are indicated by vertical arrows.

The following sections summarize data on persistence in water.

Decomposition in aquatic systems In water, hydrolysis is one of the primary mechanisms of degradation of CPY. Aqueous hydrolysis of CPY is inversely proportional to pH [67] (see Additional file 1: Figure S1). In aquatic systems at 25°C, half-lives of 73, 72, and 16 days were measured at pH 5, 7, and 9, respectively [summarized in 67]. In distilled water, in the absence of light and microorganisms, the half-life ranged from as little as 0.01 to 210 days, depending on pH [68-70] (see Additional file 1: Table S4D). Data for half-life measured in distilled were not included in the assessment as this matrix is not environmentally realistic. At pH >6 to <10, the half-life of CPY ranged from 4.5 to 142 days. When the pH was greater than 10, the half-life was as short as 0.01 day. In this assessment, hydrolysis in natural water was considered realistic and was included, regardless of pH.

Half-lives of 22 to 51 days have been reported from metabolism studies conducted in aerobic aquatic systems [71,72]. In the presence of natural sunlight, in sterile pH 7 phosphate-buffered solution, the half-life was 30 days [33]. Thus, dissipation attributable to photolysis was not much different from that attributable to hydrolysis alone. In their simulations of the dynamics of CPY in aquatic systems, the US EPA [27] used an aqueous hydrolysis half-life of 81 days at pH 7.0. In surface waters measured in the laboratory, half-lives of CPY ranged from 1.29 to 126 days [69,70,73,74] (see Additional file 1: Table S4B).

In an analysis of half-lives in water-only studies with WoE scores greater than 50%, the geometric mean half-life for all waters tested in the laboratory was 6 days. The geometric mean half-life for all laboratory-based data without exclusion of studies was 21 days [15]. When studies conducted with waters from European locations (Figure 5 and Additional file 1: Table S4A) [75,76] and non-European locations (Figure 5 and Additional file 1: Table S4B) [69,70,73,74] were considered separately, the geometric means were 2.2 and 11 days, respectively. There were no field data from the EU, but the geometric mean of field data from non-EU locations [77,78] was 6 days (see Additional file 1: Table S4C). None of these geometric mean values exceeds the criterion for persistence (60 days) in water for classification as a POP under the SC or as persistent (40 days) under EC Regulation No. 1107/2009.

Overall evaluation of persistence
The geometric means of half-lives of CPY in soils, sediments, and surface waters were less than the thresholds for classification of a compound as being persistent in soils, sediments, and water under the SC or EC Regulation No. 1107/2009. These conclusions are the same as those reached in an assessment of all the data on persistence in the earlier assessment by Mackay et al. [15].

No studies on persistence CPYO have been reported in the literature, and it has not been detected as a metabolite in studies on dissipation of CPY in soils [15]. This, along with its short half-life in water (4.7 days at pH 7), supports the conclusion of Mackay et al. [15] that CPYO is less persistent than the parent compound, CPY, and therefore does not trigger the criterion for persistence under the SC and EC Regulation No. 1107/2009.

Bioaccumulation

The major focus of assessing bioaccumulation under the SC (POPs) and EC Regulation No. 1107/2009 (PBT) is on concentration from the matrix into the organism as a BCF. Data for BCF (and BAF and biota-sediment accumulation factor (BSAF)) from the literature were selected based on the quality of the study and relevance of the exposure concentration (see Additional file 1: Table S5). These data are a subset of a larger data set from Mackay et al. [15]. BCF data for fish were separated from other aquatic organisms. For an amphibian, invertebrates, and plants, BCF, BAF, and BSAF were taken as equivalent for the purposes of analysis and values are presented graphically (Figure 6). As for toxicity values (below), all data were combined in the analysis.

Empirical values for BCFs for fish ranged from 0.6 to 5,100 [79-94] (see Additional file 1: Table S5A). From the distribution, the smallest value was clearly an outlier and was omitted from the calculation of the geometric mean and the regression. The geometric mean value for BCF for fish was 853, less than the criterion for EC Regulation No. 1107/2009 (2,000) and the SC (5,000).

BCFs are often estimated from the octanol-water partition coefficient (K_{OW}), especially when there is a lack of empirical data. In the case of CPY, K_{OW} is approximately 100,000 [15]. Thus, a fish containing 5% lipid would be expected to have a BCF of approximately 5,000, which is at the extreme of empirical observations. This suggests that CPY is subject to loss processes from fish in addition to respiratory loss, the obvious process being metabolic biotransformation, but growth dilution and egestion may also apply.

The BCF (and BAF and BSAF) for invertebrates and plants (see Additional file 1: Table S5B) ranged from 3.4 to 5,700, with a geometric mean of 204. For plants, the greatest value was reported for duckweed (*Lemna minor*, 5,700) with water lettuce (*Pistia stratiotes*, 3,000), the third largest and both greater than algae (*Oedogonium cardiacuin*, 72). The larger values for the two macrophytes might be more reflective of adsorption to the surface of the plant than uptake into the plant. The BCF values in the invertebrates ranged from 3.4 to 691, both in mollusks. The value for the only amphibian (*Ambystoma mexicanum*) in the data set

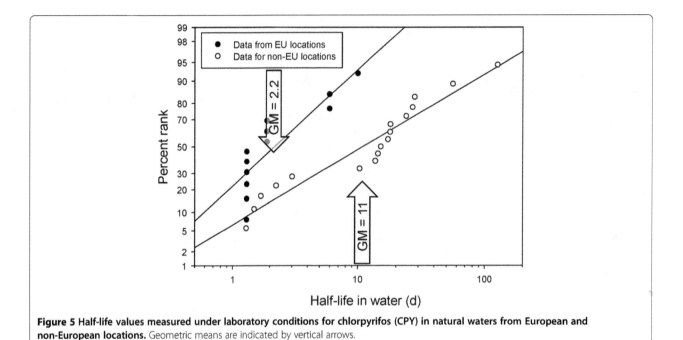

Figure 5 Half-life values measured under laboratory conditions for chlorpyrifos (CPY) in natural waters from European and non-European locations. Geometric means are indicated by vertical arrows.

was 3,632, which is in the same range as those reported for fishes. The geometric mean value for invertebrates and plants (204) was less than the criterion for EC Regulation No. 1107/2009 (2,000) and the SC (5,000).

No data on biomagnification of the toxicologically relevant metabolite of CPY, CPYO, were found in the literature. Although CPYO is formed in the atmosphere, it is reactive and has a shorter half-life in water (4.7 days at pH 7) than the parent (geometric mean of 2.2 to 11 days) [15]. Despite extensive sampling of surface waters in areas

of more intensive use, CPYO has never been detected above the LOD [25]. Lack of observed bioconcentration or biomagnification of CPYO is completely consistent with the greater reactivity of the molecule and its greater solubility and smaller K_{OW} than CPY [15]. All the evidence suggests that CPYO does not biomagnify.

Concentrations in aquatic biota

Concentrations of chemicals such as CPY in organisms collected in the field are another line of evidence of

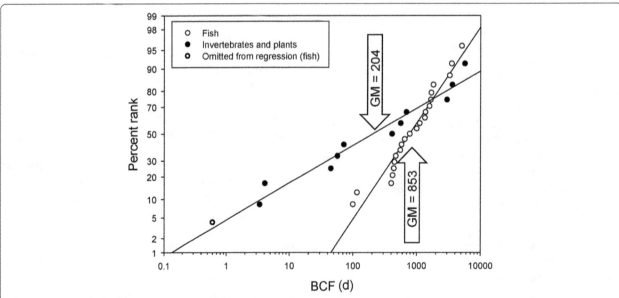

Figure 6 Graphical presentation of BCF (BAF and BSAF) values for chlorpyrifos (CPY) in aquatic organisms. Geometric means are indicated by vertical arrows.

bioconcentration and biomagnification. In an extensive survey of chemicals in fish from lakes and reservoirs of the USA [95], residues of CPY were not detected. The method detection limit was 59 µg/kg, and 486 samples from predatory fish and 395 samples from bottom-dwelling fish were analyzed. Concentrations of CPY in samples of fish from lakes in Western National Parks of the USA were all <1 µg/kg (wet mass; wet. ms.) [96]. Taken together, this is further evidence of lack of significant bioconcentration or biomagnification of CPY into fishes or biomagnification in the aquatic food chain.

Accumulation in terrestrial organisms

Apart from the detections in plants in montane areas of California discussed in Mackay et al. [15], there are reports of detections of small concentrations in plants from more remote areas such as in the panhandle of Alaska [96] (2.4 ng/g lipid mass (l. ms.) in needles of conifers, see the 'Measured concentrations near areas of use' section). Chlorpyrifos was detected in lichen (0.073 ± 0.23 ng/g l. ms.), mushrooms (0.78 ± 0.82 ng/g l. ms.), green plants (0.24 ± 0.47 ng/g l. ms.), caribou muscle (0.57 ± 0.19 ng/g l. ms.), and wolf liver (<0.10 ng/g l. ms.) in Nunavut [97].

Trophic magnification

A characteristic of POPs is biomagnification or trophic transfer in food chains, and this provides another line of evidence to assess CPY. Only one study was found on movement of CPY in food webs. Several current-use pesticides were measured in a terrestrial food web in the Canadian Arctic [97], and CPY was detected frequently in moss and mushrooms (83% to 86%) but at lesser frequency in lichens, willow, and grass (44% to 50%). Concentrations ranged from 0.87 ng/g l. ms. in moss to 0.07 ng/g l. ms. in lichen. Concentrations in caribou muscle and total body burden provided BMFs of 7.8 and 5.1 compared to lichen; however, the BMF from caribou to wolf was <0.10 (based on concentrations < MDL). Trophic magnification factors for the lichen-caribou-wolf food chain were all <1 for muscle, liver, and total body burden. That small concentrations of CPY were found in the Arctic is indicative of some long-range transport, but the lack of BMF between caribou and wolf and TMFs of <1 are indicators of no significant trophic magnification of CPY in the food chain. These additional

lines of evidence support the laboratory and microcosm data, which indicate that CPY does not trigger the criterion for bioaccumulation under the SC or EC Regulation No. 1107/2009.

Toxicity

Toxicity of CPY to aquatic organisms has been reviewed in detail previously in Giddings et al. [38], to birds in Moore et al. [98], and to the honeybee in Cutler et al. [99]. Rather than repeat this information here, the reader is referred to these papers and their relevant supplemental information. The following sections summarize the toxicity data and discuss this in relation to the classification criteria for POPs and PBTs.

Acute toxicity in aquatic organisms

Because CPY dissipates and degrades rapidly in aquatic environments and is only present for short durations (≤4 days), data on acute toxicity were selected as the most appropriate for assessment of risks in aquatic systems [38]. There were numerous published studies on laboratory toxicity tests for aquatic organisms. These data were screened for quality and a subset of the higher quality and most relevant studies were used in the assessment [38]. Toxicity values were analyzed as distributions (species sensitivity distributions (SSDs)) using SSD Master Version 3.0 software [100] and 5th centiles (HC5 concentrations) used to characterize toxicity of CPY to major taxa (Table 3).

These acute toxicity values were not separated by type of medium (saltwater or freshwater) or origin of the species (tropical, temperate, Palearctic, or Nearctic). Analysis of the large amount of toxicity data available for CPY has shown that there are no significant differences in sensitivity between these groups [101]. Thus, these data are appropriate for classification in the global context (POPs) and in the regional context (PBT).

Toxicity in aquatic meso- and microcosms

Several studies of effects of CPY have been conducted in aquatic meso- and microcosms (cosms) and were reviewed in Giddings et al. [38]. These studies were conducted in various jurisdictions and climatic zones, including Europe (Netherlands and Mediterranean locations), the US Midwest, Australia, and Thailand. Half of the 16 studies reported no-observed-adverse-effect concentrations (NOAEC$_{eco}$) values

Table 3 Toxicity values for CPY in aquatic organisms

Taxon	5th centile (µg/L)	95% CI (µg/L)	Comments
Crustaceans	0.034	0.022 to 0.051	Based on 23 species with a range of 0.035 to 457 µg CPY/L
Insects	0.087	0.057 to 0.133	Based on 17 species with a range of 0.05 to >300 µg CPY/L
Fish	0.812	0.507 to 1.298	Based on 25 species with a range of 0.53 to >806 µg CPY/L
Amphibians	Too few species for a SSD, range was 19 to a questionable value of 5,174 µg CPY/L for three species		
Plants	Too few species for a SSD, range was 138 to 2,000 µg CPY/L		

as 'less than' values. For the eight studies in cosms where NOAEC$_{eco}$s were available, all were ≥0.1 µg/L and the geometric mean was 0.14 µg/L. The NOAEC$_{eco}$ is based on observation of short-term effects in some sensitive organisms, from which there is rapid recovery. For a pesticide, such as CPY, which degrades relatively rapidly in the environment, this is an appropriate measure of a threshold for toxicity under realistic environmental conditions. Because studies in cosms incorporate toxicity of organisms from the region, as well as processes related to fate that may be influenced by local conditions such as climate and hydro-geo-chemistry, there may be regional differences in responses. This was not the case for CPY; the NOAEC$_{eco}$ values were the same regardless of location of the study. This not only is consistent with lack of region-specific toxicity tests but also suggests that the fate processes that can influence exposures in aquatic systems are not different between regions. Thus, it was not necessary to separate the studies in cosms for purposes of classification of POPs and PBTs.

Toxicity to terrestrial organisms

Because CPY is used as a pesticide to protect crops from damage by arthropods, it is obviously toxic to terrestrial stages of insects. This is a benefit of use and is not considered an adverse effect. However, toxicity to valued arthropods can be considered an adverse effect and, in the case of the honeybee, was characterized in a risk assessment of CPY [99]. CPY is toxic to the honeybee by direct contact (topical toxicity) with the spray and also via the oral route. The former route of exposure is only relevant when bees are present during or shortly after spraying and is mitigated by restrictions on the label (see the 'Reports of toxicity under current conditions of use' section below). Topical 24- to 48-h LD50 values for formulated CPY range from 0.024 to 0.54 (geometric mean =0.123) µg a.i./bee and 0.059 to 0.115 (geometric mean =0.082) µg a.i./bee for the technical product. Oral 24- to 48-h LD50s ranged from 0.114 µg a.i./bee for the technical to 0.11 to 1.1 (geometric mean =0.36) µg a.i./bee for the formulated material [99]. Significant toxicity to honeybees has only been associated with direct exposure during spraying and/or during foraging for nectar and/or pollen in recently treated fields (0 to 3 days post spray). Toxicity has not been reported to be caused by CPY outside the foraging range of the bees, and residues in

samples of brood comb have not been casually linked to colony collapse disorder [102]. There is no evidence to suggest that small concentrations measured outside the areas of use are toxic to bees or other beneficial insects. As the honeybee is found in the EU, North America, and other parts of the globe, there is no need to consider this species differently across locations. The conclusions regarding toxicity to honeybees thus apply to considerations of POPs and PBTs.

Toxicity to birds has been characterized previously by Moore et al. [98]. Because of rapid dissipation of CPY in the environment and in animals, acute toxicity data were considered most relevant for assessing risks. Acute LD50s ranged from 8.55 to 92 (geometric mean =30.5) mg CPY/kg bm in 14 species of birds. Few chronic toxicity data were available, but values for the NOEC and LOEL in the mallard duck exposed via diet for 28 days were 3 and 18.7 mg CPY/kg bm/day, respectively [98]. Risks of CPY to birds foraging in treated fields were considered *de minimis* for most species, except sensitive species foraging in crops with large application rates (e.g., citrus). This conclusion was consistent with the lack of observed mortality of birds in field studies conducted in North America and the EU. Mammals are less sensitive to CPY than birds. Acute LC50 values for laboratory test species ranged from 62 to 2,000 mg CPY/kg bm [103]. Several assessments of risk have concluded that birds are more sensitive and more likely to be exposed and are protective of risks from CPY in wild and domestic mammals and that risks to these organisms are *de minimis* [98,104]. Given *de minimis* or very small risks from exposures in areas of use, concentrations of CPY reported from semi- and remote locations present even lesser risks to birds or mammals.

Acute and chronic dietary toxicity values for CPYO have been measured in birds (Table 4). Although data were few, toxicity values were similar to those for CPY, suggesting, as would be expected from the mechanism of action, that CPYO has similar toxicity to the parent CPY.

Chronic toxicity in aquatic organisms

Although there are no known situations where exposures of aquatic organisms to CPY are long-term, some toxicity tests, such as mesocosm studies, have used repeated exposures with no hydraulic flushing to assess the equivalent of repeated exposures. The most sensitive NOEC reported

Table 4 Acute and dietary toxicity values for CPYO in birds

Species	Observation or feeding time (days)	Toxicity value (mg/kg)	95% CI	Reference
Bobwhite quail	7 observations	LD50 = 8.8 bm	7.2 to 10.7	[105]
Zebra finch	7 observations	LD50 ≥ 30 bm	NA	[106]
Mallard duck	5 exposure, 8 observations	LC50 = 523 mg/kg diet	363 to 796	[107]
Bobwhite quail	5 exposure, 8 observations	LC50 = 225 mg/kg diet	173 to 292	[108]

NA, not applicable.

for an aquatic organism was from one of these studies: 0.005 µg CPY/L for *Simocephalus vetulus* in a mesocosm experiment [75]. This value is relevant to assessment of CPY as a PBT chemical.

Reports of toxicity under current conditions of use

The above conclusions of lack of significant toxicity to aquatic and terrestrial organisms under current conditions of use in the USA is supported by the very few reports on fish, invertebrate, bee, and bird kills reported in the last 12 years [38,98,99]. Where these few incidents have occurred, most have been the result of accidents or deliberate misuse.

Toxicity in relation to classification as a POP

The criterion for toxicity for classification of POPs is 'significant adverse effects' without a clear definition of 'significant' or the location of the effects. We have interpreted that to mean that the use of CPY results in unacceptable risks in areas outside but not directly adjacent to the area of application (i.e., edge of field). As a pesticide, risks to target organisms in the agricultural field are accepted, but risks to non-target organisms, especially outside the areas of application, are considered undesirable.

None of the data on toxicity of CPY or CPYO to non-target organisms suggests that there are significant adverse effects in the environment outside of the areas of use [15,38,98,99]. Even in areas of use, risks to birds and mammals are small or *de minimis*. The data on toxicity of CPY and CPYO to birds, mammals, and aquatic organisms determined under laboratory conditions is robust. These data are complemented by studies in aquatic cosms, which are more representative of exposures in natural environments, showing similar patterns of toxicity and including species that have not been tested in the laboratory under guideline protocols. There are some uncertainties. Not all species have been tested and many groups of marine species have not been tested at all; however, this is not unique and applies to pesticides other than CPY and to chemicals in general.

Considering all of the data on toxicity, we conclude that CPY and CPYO do not exceed the POPs criterion of 'significant adverse effects' (Table 1) for toxicity to organisms in the environment.

Toxicity in relation to classification as a PBT

The criterion for classification of pesticides as toxic under EC Regulation No. 1107/2009 is 'Chronic NOEC <0.01 mg/L (10 µg/L) or is a carcinogen, mutagen, or toxic for reproduction, or other evidence of toxicity' (Table 2). As has been discussed before [17], the criterion refers only to aquatic organisms and terrestrial organisms are not considered. The NOEC for *S. vetulus* (0.005 µg CPY/L) is less than the criterion, so CPY would be classified as T. In addition, the acute toxicity values for CPY for many crustaceans and insects, and even some fish, were <10 µg CPY/L [38]. Given that CPY is an insecticide and that crustaceans and insects are the most sensitive taxa [38], this is not unexpected. However, several additional factors that place toxicity in perspective must be considered. CPY is not applied directly to water, so exposures in this environment are indirect and small [25]. Since CPY is not persistent in water or other environmental compartments, chronic toxicity values are not environmentally realistic or appropriate for classification of toxicity. There is robust evidence to show that CPY is not sufficiently persistent in any environmental compartments to justify durations of exposure associated with chronic toxicity. Thus, it would have been inappropriate to compare concentrations in remote regions to those associated with chronic effects of CPY. No chronic toxicity data for CPYO were available; however, it has a relatively short half-life in water [15] and has not been detected in surface waters, even in areas of high use [25].

Carcinogenicity, mutagenicity, or reproductive toxicity of CPY were not assessed in this evaluation, but have been assessed in recent reviews by the US EPA as part of the re-registration process. Based on current use patterns, CPY was not identified as a mutagen, carcinogen, reproductive toxicant, or immunotoxic agent [26]. The very small concentrations reported in semi- and remote areas do not represent a risk to humans through drinking water or via the food chain.

Discussion
Atmospheric transport

The potential for LRT is considered in both water and air. Since the half-life of CPY in water does not exceed the criterion for persistence in water (see the 'Water' section), it is unlikely that LRT in water would be a significant issue. Thus, the potential for LRT of CPY and CPYO in the atmosphere was assessed in detail. The criterion for LRT in air under the United Nations Economic Commission for Europe [3] is that the half-life is ≥2 days (Table 1) or that monitoring or modeling data demonstrates long-range transport. Since masses of air containing volatilized CPY can move, a static determination of the half-life in air is not instructive. The issue is: can CPY persist long enough to move significant distances from where it is released and deposit into soils and water at concentrations sufficient to cause adverse effects? Evidence that CPY is subject to LRT is provided in reports of concentrations in air and other media at locations remote from sites where CPY is applied in agriculture [15].

The assessment reported by Mackay et al. [15] used a combination of analyses, including measured concentrations at locations distant from sources, in conjunction with mass balance modeling. Predictions of atmospheric transport were made by the use of simple mass balance models such as

TAPL3 and the OECD Tool [15,109,110]. These models have been used in regulatory contexts and characterize LRT as a characteristic travel distance (CTD), which is defined as the distance that approximately two thirds of the originally released mass of CPY or CPYO is transported from the source before it is deposited or transformed. Detailed assessments of the properties of CPY [15,20,111] and its fate in the environment and potential risks [14] have been published previously [67,112]. CTDs of several pesticides, including CPY, have been estimated [113]. Results of these modeling exercises have suggested a CTD of 280 to 300 km for CPY if it is assumed that the atmospheric half-life is 12 h, the narrow range being the direct result of close similarities between the model equations. As is discussed below, this estimate of CTD reflects an unrealistically long atmospheric half-life.

Predicted concentrations in the environment

The assessment of LRT presented here went beyond determination of CTD and the related characteristic travel time (CTT) and also included consideration of estimates of concentrations of CPY and CPYO in other environmental media such as rain, snow, and terrestrial phases, as well as in the atmosphere at more remote locations, including higher altitudes [15]. A relatively simple mass balance model was developed and used to predict concentrations in various media at various distances from sources where CPY was applied in agriculture, which could be compared to measured concentrations of CPY in air and other media. Results of the model can then serve as a semi-quantitative predictive framework that is consistent with observations.

As an example of dissipation of a parcel of air containing 100 ng CPY/m^3, which is typical of concentrations 1 km from application sources, a model was developed to assess the concentration as it is conveyed downwind [15]. The mass would be decreased as a result of transformation processes, primarily reaction with •OH radicals, deposition, and dilution by dispersion. Oxidation results in formation of primarily CPYO. By using the TAPL3 simulation of a relatively large environmental area and a half-life in air of 3.0 h and conservative (longer duration) half-lives in other media and assuming an emission rate to air of 1,000 kg/h, the resulting mass in air is 4,328 kg, the residence time in air and the CTT is 4.3 h, and the corresponding rate constant for total loss is 0.231/h. The CTD of approximately 62 km is the product of 4.3 h and the wind velocity of 14.4 km/h. The rate of transformation is 993 kg/h, and the net losses by deposition to water, vegetation, and soil total about 7 kg CPY/h, which corresponds to a rate constant of 0.0016/h and is less than 1% of the rate of degradation. The critical determinant of potential for LRT is the rate of transformation from reactions with •OH radicals in air. If the half-life is increased by an

arbitrary factor of 4 to 12 h, as was assumed in [113], the CTD increases to 244 km [15].

Results of simulation models predicted concentrations and partial pressures or fugacities (expressed in units of nPa) at several distances from application of CPY in typical agricultural uses. A simple but approximate approach to estimate concentrations of CPY at distances from sources is to use a dispersion model to estimate concentrations at ground level from a ground-level source using standard air dispersion parameters [114]. Near the area of application, such as at a distance of 1 km and assuming a 0.1-h air transit time, air concentrations ($C_{1 km}$) were assigned a value of 100 ng CPY/m^3 (approximately 700 nPa). Concentrations of CPY are primarily controlled by rates of evaporation and dispersion rather than reactions with •OH. At a distance of 120 km and a transit time of 8.4 h, which is equivalent to two CTDs, 84% of the volatilized CPY would have been transformed and the concentration of CPY in air would be 0.022 ng CPY/m^3 (0.16 nPa). At steady state, rain water would have a concentration of 0.1 ng CPY/L and snow a concentration of 1.5 ng CPY/L. If a very conservative half-life of 12 h for CPY were assumed, the fraction of CPY transformed would be only 38% and thus greater concentrations of CPY would persist for longer distances. At a distance of 300 km and a transit time of about 20 h, which is equivalent to approximately five CTDs, 1.0% of the initial mass of CPY would remain because the CPY would have been subjected to nearly seven half-lives. Concentrations at this distance from the source would likely be 0.0003 ng CPY/m^3 (0.002 nPa) or less. Concentrations of 0.003 ng CPYO/m^3 would be expected. Thus, at this distance from the source, CPYO would be the primary product present, at a concentration which is near the typical limit of quantitation. Rain, if at equilibrium with air, would be expected to contain a concentration of 0.001 ng CPY/L and snow 0.02 ng CPY/L. Given an assumed half-life of 3 h and the time to be transported this distance, it is unlikely that, under normal conditions, significant quantities could travel more than 300 km. Observations of detectable amounts of CPY at greater distances, such as 1,000 km [115], suggest that, at least under certain meteorological conditions as may apply at high latitudes or times of low solar radiation and less production of •OH radicals, the half-life is longer than was assumed in this analysis. The significant conclusion is that partial pressures, fugacities, and concentrations in air at distances of 100 s of km are expected to be reduced by a factor of a million or more from those within a km from sources.

Measured concentrations near areas of use

While the vapor pressure of CPY is considered to be moderate, CPY can be measured in the air during and after application. In the 12 h following application of the liquid formulation to the surface, approximately 10% to

20% of the applied material volatilizes, but variability is expected diurnally, with temperature, rainfall, and soil moisture content. Sorption then 'immobilizes' the CPY and subsequent volatilization is slower, with a rate of approximately 1% per day that decreases steadily to perhaps 0.1% per day in the subsequent weeks [15]. Concentrations in air that exceed 20 ng CPY/m^3 have been observed near sources of application in agriculture [15]. Concentrations of CPY in air immediately above a potato field in the Netherlands at noon in midsummer ranged from 14,550 to 7,930 ng/m^3 at 1 and 1.9 m above the crop 2 h after application [116]. These decreased to a range of 2,950 to 1.84 ng/m^3 after 8 h and to 26 to 15 ng/m^3 in the 6 days following application. Concentrations of CPY in air following an application of 4.5 kg/ha to turf were in the range of 1,000 to 20,000 ng/m^3 [117]. This might be a 'worst case' in terms of concentrations and represents approximately 10% of the saturation concentration in air, i.e., the vapor pressure/RT, where RT is the gas constant-absolute temperature group. Concentrations of approximately 100 ng CPY/m^3 are regarded as typical of areas immediately downwind (approximately 1 km) of application sites [15].

Measured concentrations and deposition in semi-remote locations

Chlorpyrifos and CPYO have been detected in the environment [15]. Concentrations in the range of 0.01 to 10 ng CPY/m^3 that have been reported at distances of up to 100 km from sources are considered to be regional. Concentrations less than 0.01 ng CPY/m^3 have been observed in more remote areas. Approximately 70% of the data for concentrations in air were in the range of 0.01 to 1.0 ng CPY/m^3. For rain, the greatest frequency (40%) was in the range 1 to 10 ng CPY/L. Concentrations of CPY in snow exhibited similar patterns, but with more concentrations in the range 0.01 to 0.1 ng CPY/L [15].

Apart from the detections in plants in montane areas of California discussed in Mackay et al. [15], there are reports of detections of small concentrations from more remote areas, such as the panhandle of Alaska [96]. Concentrations of CPY in lichen were < MDL (1 ng/g l. ms.) and mean concentrations as great as 2.4 ng/g l. ms. in needles of conifers in Denali National Park, Wrangell-St. Elias National Park and Preserve, Glacier Bay National Park, Katmai National Park and Preserve, the Stikine-LeConte Wilderness, and the Tongass National Forest in samples collected between 2002 and 2007. The amounts of CPY measured were small in comparison to those reported at locations closer to regions of release [15] and are not suggestive of the transport of toxicologically significant amounts of CPY. It is thus not surprising that small but detectable concentrations can be found in remote locations such as Svalbard [113,115]. The largest concentration in a remote location was found in ice corresponding to the 1980s from Svalbard. While that

concentration was 16 ng CPY/L, concentrations measured more recently are generally <1 ng CPY/L. Residues of CPY and CPYO were absent in the surface section of the core, representing 1990 to 1998 [115], despite this likely being the period of greatest global use. A survey of concentrations of CPY in a north–south transect of lakes in Canada reported the presence of residues of CPY [113]. Greater concentrations were reported in lakes with agricultural inputs (mean =0.00065 µg/L). Concentrations and frequency of detection decreased with increasing latitude, with mean concentrations of 0.00082, <0.00002, and 0.00027 µg/L for remote mid-latitude, subarctic, and arctic lakes, respectively. These were grab samples and the temporal profile of exposures are not known; however, all concentrations are several orders of magnitude less than the HC5 for crustaceans (0.034 µg/L, Table 3) or the $NOAEC_{eco}$ of ≥0.1 µg/L (see the 'Toxicity in aquatic meso- and microcosms' section) for repeated exposures in microcosms.

Measured concentrations in surface waters

Chlorpyrifos (but not its toxicologically significant product of transformation, CPYO) has been detected in surface waters, particularly in areas of intensive use [25]. In several regions of the USA, these concentrations have decreased since the late 1990s and early 2000s [118-120], most probably as a result of changes in patterns of use [25]. Thus, rather than an upward trend in concentrations, the frequency of detection and the concentrations measured in surface waters have declined. This is not indicative of persistence in the environment.

Conclusions

While both CPY and CPYO are classified as "toxic", based on the assessment of persistence and bioaccumulation, all the lines of evidence suggest that neither would be classified as persistent or bioaccumulative under the SC or EC Regulation No. 1107/2009. Based on the analysis of LRT, neither CPY nor its most toxic transformation product, CPYO, would be transported at sufficiently great concentrations to cause adverse effects in humans or the environment in remote areas. Based on the simple criterion for toxicity in EC Regulation No. 1107/2009, CPY (and by extension, CPYO) would be classified as toxic; however, when a more refined assessment of 'risk' is considered instead of 'hazard', it does not present unacceptable risks to humans or organisms in the environment. Based on the wording of the SC, CPY and CPYO do not present a significant adverse risk to humans and the environment. These conclusions are based on the selection of higher quality data but are similar to those reached by inclusion of all the data [15].

Abbreviations

B: bioaccumulative; BAF: bioaccumulation factor; BCF: bioconcentration factor; bm: body mass; BMF: biomagnification factor; BSAF: biota-sediment accumulation factor; COC: chemical of concern (substance of concern); CPY: chlorpyrifos; CPYO: chlorpyrifos oxon; CTD: characteristic travel distance; CTT: characteristic travel time; K_{OC}: water-soil partition coefficient corrected for the amount of organic carbon in the soil; K_{OW}: octanol-water partition coefficient; l. ms.: lipid mass; LC50: lethal concentration for 50% of test individuals; LRT: long-range transport; NOEC: no-observed-effect concentration; nPa: nanoPascals; P: persistent; PBT: persistent, bioaccumulative, and toxic; POP: persistent organic pollutant; REACH: Registration, Evaluation, Authorisation and Restriction of Chemicals; SC: Stockholm Convention; T: toxic; $t\frac{1}{2}$: half-life; TCDD: 2,3,7,8-tetrachlorodibenzo-p-dioxin; TCDD-Py: 2,3,7,8-tetrachloro-1,4-dioxino-[2,3-b:5,6-b'] dipyridine; TCP: trichlorophenol; TCPy: trichloropyridinol; TMF: trophic magnification factor; UNEP: United Nations Environment Programme.

Competing interests

The authors declare that they have no competing interests.

Authors' contributions

The authors JPG, KRS, and DM contributed directly to the concepts, analyses, development of conclusions, and writing of the paper. JA provided technical assistance and assessment of the strength and relevance of the studies, and KRS formatted and prepared the paper for submission. All authors read and approved the final manuscript.

Acknowledgements

Funding for this assessment was provided by Dow AgroSciences, LLP, USA. The opinions expressed in this paper are those of the authors alone. All of the references are available from publishers or from the authors except for those reports which are considered to contain confidential business information. These reports have been provided to the appropriate regulatory agencies for use in their reviews and deliberations relative to chlorpyrifos. If readers wish to obtain specific information contained in these reports, requests will be passed on to the registrant on a case-by-case basis. JPG was supported by the Canada Research Chairs Program, a Visiting Distinguished Professorship in the Department of Biology and Chemistry and State Key Laboratory in Marine Pollution, City University of Hong Kong, the 2012 'High Level Foreign Experts' (#GDW20123200120) program, funded by the State Administration of Foreign Experts Affairs, the P.R. China to Nanjing University, and the Einstein Professor Program of the Chinese Academy of Sciences.

Author details

[1]Department of Veterinary Biomedical Sciences and Toxicology Centre, University of Saskatchewan, Saskatoon, SK S7B 5B3, Canada. [2]Centre for Toxicology, School of Environmental Sciences, University of Guelph, Guelph, ON N1G 2 W1, Canada. [3]Centre for Environmental Modelling and Chemistry, Trent University, Peterborough, ON K9J 7B8, Canada. [4]Stantec, 603-386 Broadway Ave, Winnipeg, MB R3C 3R6, Canada.

References

1.   Carson R: Silent Spring. Boston: Houghton Mifflin; 1962.
2.   United Nations Environmental Programme: Stockholm Convention on Persistent Organic Pollutants (as amended in 2009 and 2011). Geneva: Secretariat of the Stockholm Convention; 2009:64. http://chm.pops.int/TheConvention/Overview/TextoftheConvention/tabid/2232/Default.aspx.
3.   United Nations Economic Commission for Europe: Protocol to the 1979 Convention on Long-Range Transboundary Air Pollution on Persistent Organic Pollutants. Geneva: United Nations Economic Commission for Europe; 1998:49. http://www.unece.org/env/lrtap/full%20text/1998.POPs.e.pdf.
4.   Moermond C, Janssen M, de Knecht J, Montforts M, Peijnenburg W, Zweers P, Sijm D: PBT assessment using the revised Annex XIII of REACH: a comparison with other regulatory frameworks. Integr Environ Assess Manag 2011, 8:359–371.
5.   Consensus Panel: Scientific and Policy Analysis of Persistent, Bioaccumulative, and Toxic Chemicals: a Comparison of Practices in Asia, Europe, and North America. The Report of a Consensus Panel. Bloomington: School of Public and Environmental Affairs, Indiana University; 2013:98. http://www.indiana.edu/~spea/faculty/pdf/scientific_policy_analysis_of_persistent_bioaccumulative_and_toxic_chemicals_PBT_.pdf.
6.   OSPAR: Convention for the Protection of the Marine Environment of the North-East Atlantic. London: OSPAR Commission; 1992:33. http://www.ospar.org/html_documents/ospar/html/ospar_convention_e_updated_text_2007.pdf.
7.   European Community: Regulation (EU) No 253/2011 of 15 March 2011 amending Regulation (EC) No 1907/2006 of the European Parliament and of the Council on the Registration, Evaluation, Authorisation and Restriction of Chemicals (REACH) as regards Annex XIII. Off J Europ Commun 2011, 54:7–12.
8.   European Community: Regulation (EC) No 1107/2009 of the European Parliament and of the Council of 21 October 2009 concerning the placing of plant protection products on the market and repealing Council Directives 79/117/EEC and 91/414/EEC (91/414/EEC). Off J Europ Commun 2009, 52:1–50.
9.   Environment Canada: Toxic Substances Management Policy - Persistence and Bioaccumulation Criteria. Final Report of the Ad Hoc Science Group on Criteria. Ottawa: Environment Canada; 1995:26.
10. USEPA: Persistent bioaccumulative toxic (PBT) chemicals; final rule. 40 CFR Part 372. Fed Reg 1999, 64:58665–587535.
11. Chemicals Management Plan. In [http://www.chemicalsubstanceschimiques.gc.ca/plan/index-eng.php]
12. Siloxane D5 Board of Review: Report of the Board of Review for Decamethylcyclopentasiloxane (D5). Ottawa: Environment Canada; 2011:83. http://www.ec.gc.ca/lcpe-cepa/default.asp?lang=En&n=515887B7-1.
13. Arnauld A: The Port Royal Logic. Edinburgh: James Gordon & Hamilton Adams and Co; 1861:1662. translation from French by T.S. Bayes published in 1861.
14. Giesy JP, Solomon KR, Mackay D, Giddings JM, Williams WM, Moore DRJ, Purdy J, Cutler GC: Ecological risk assessment for chlorpyrifos in terrestrial and aquatic systems in the United States – overview and conclusions. Rev Environ Contam Toxicol 2014, 231:1–12.
15. Mackay D, Giesy JP, Solomon KR: Fate in the environment and long-range atmospheric transport of the organophosphorus insecticide, chlorpyrifos and its oxon. Rev Environ Contam Toxicol 2014, 231:35–76.
16. Weed DL: Weight of evidence: a review of concepts and methods. Risk Anal 2005, 25:1545–1557.
17. Solomon KR, Matthies M, Vighi M: Assessment of PBTs in the EU: a critical assessment of the proposed evaluation scheme with reference to plant protection products. Environ Sci EU 2013, 25:1–17.
18. United Nations Environmental Programme: Assessment of Alternatives to Endosulfan. Geneva: Secretariat of the Stockholm Convention; 2010:4.
19. Watts M: Chlorpyrifos as a Possible Global POP. Oakland: Pesticide Action Network North America; 2012:34. www.pan-europe.info/News/PR/121009_Chlorpyrifos_as_POP_final.pdf.
20. Solomon KR, Williams WM, Mackay D, Purdy J, Giddings JM, Giesy JP: Properties and uses of chlorpyrifos in the United States. Rev Environ Contam Toxicol 2014, 231:13–34.
21. SANCO: DG SANCO Working Document on "Evidence Needed to Identify POP, PBT and vPvB Properties for Pesticides". Brussels: European Commission Health and Consumers Directorate-General; 2012:5.
22. Ritter L, Solomon KR, Forget J, Stemeroff M, O'Leary C: Persistent Organic Pollutants. Report for the Intergovernmental Forum on Chemical Safety. Guelph: Canadian Network of Toxicology Centres; 1995:44. http://www.chem.unep.ch/pops/ritter/en/ritteren.pdf.
23. Ritter L, Solomon KR, Forget J, Stemeroff M, O'Leary C: A Review of Selected Persistent Organic Pollutants. DDT-Aldrin-Dieldrin-Endrin-Chlordane-Heptachlor-Hexachlorobenzene-Mirex-Toxaphene-Polychlorinated Biphenyls-Dioxins and Furans. Review Prepared for IPCS. Geneva: International Programme on Chemical Safety (IPCS) of the United Nations; 1995:149. http://www.who.int/ipcs/assessment/en/pcs_95_39_2004_05_13.pdf.
24. Solomon KR, Dohmen P, Fairbrother A, Marchand M, McCarty L: Use of (eco)toxicity data as screening criteria for the identification and

classification of PBT/POP compounds. *Integr Environ Assess Manag* 2009, **5**:680–696.

25. Williams WM, Giddings JM, Purdy J, Solomon KR, Giesy JP: **Exposures of aquatic organisms to the organophosphorus insecticide, chlorpyrifos resulting from use in the United States.** *Rev Environ Contam Toxicol* 2014, **231**:77–118.

26. USEPA: *Chlorpyrifos. Preliminary Human Health Risk Assessment for Registration Review.* Washington, DC: United States Environmental Protection Agency, Office of Chemical and Pollution Prevention; 2011:159. http://www. regulations.gov/#!documentDetail;D=EPA-HQ-OPP-2008-0850-0025.

27. USEPA: *Revised Chlorpyrifos Preliminary Registration Review Drinking Water Assessment.* Washington, DC: United States Environmental Protection Agency, Office of Chemical Safety and Pollution Prevention; 2011:272. http://www.epa.gov/oppsrrd1/registration_review/chlorpyrifos/EPA-HQ-OPP-2008-0850-DRAFT-0025%5B1%5D.pdf.

28. Holt E, Weber R, Stevenson G, Gaus C: **Polychlorinated dibenzo-p-dioxins and dibenzofurans (PCDD/Fs) impurities in pesticides: a neglected source of contemporary relevance.** *Environ Sci Technol* 2010, **44**:5409–5415.

29. Sakiyama T, Weber R, Behnisch P, Nakano T: **Formation of the pyridine-analogue of 2,3,7,8-TCDD by thermal treatment of chlorpyrifos, chlorpyrifos-methyl and their major degradation product 3,5,6-tri-chloro-2-pyridinol.** *Organohalogen Compd* 2012, **74**:1441–1444.

30. Krumel KL, Wollowitz S, Beyer J, Hummel R: *Dursban. Insecticide Process Research: the Synthesis and Characterization of Possible Trace Impurities in Chlorpyrifos.* Midland: Dow Chemical Company; 1987:50.

31. McCall PJ, Levan LW: *A Study of Pyrolysis of 3,5,6-Trichloro-2-pyridinol in Cigarette Tobacco.* Midland: Dow Chemical Company; 1987:45.

32. Holt E, Weber R, Stevenson G, Gaus C: **Formation of dioxins during exposure of pesticide formulations to sunlight.** *Chemosphere* 2012, **88**:364–370.

33. Batzer FR, Fontaine DD, White FH: *Aqueous Photolysis of Chlorpyrifos.* Midland: DowElanco; 1990:189.

34. Crosby DG, Wong AS: **Environmental degradation of 2,3,7,8-tetrachloro-p-dioxin (TCDD).** *Science* 1977, **195**:1337–1338.

35. Hanno K, Oda S, Nakano T, Mitani H: **Preliminary assessment of thermal treatments of chlorpyrifos and 3,5,6-trichloro −2-pyridinol (a potential precursor of the pyridine analogue of 2,3,7,8-T4CDD) using the early developmental stage embryos of medaka (*Oryzias latipes*).** *J Water Environ Technol* 2013, **11**:319–329.

36. Weis CD: **Synthesis of 1,4-dioxino[2,3-b:5,6-b']dipyridine.** *J Heterocycl Chem* 1976, **13**:145–147.

37. Henck JW, Kociba RJ: *2,3,7,8-Tetrachloror-1,4-dioxino-[2,3-b:5,6-b']dipyridine: Acute Oral Toxicity and Chloracnegenic Potential.* Midland: Dow Chemical Company; 1980:6.

38. Giddings JM, Williams WM, Solomon KR, Giesy JP: **Risks to aquatic organisms from the use of chlorpyrifos in the United States.** *Rev Environ Contam Toxicol* 2014, **231**:119–162.

39. Klimisch H-J, Andreae M, Tillmann U: **A systematic approach for evaluating the quality of experimental toxicological and ecotoxicological data.** *Reg Toxicol Pharmacol* 1997, **25**:1–5.

40. OECD: *OECD Guidelines for the Testing of Chemicals, Section 3, Degradation and Accumulation.* Paris: OECD; 2011:321. http://www.oecd-ilibrary.org/environment/oecd-guidelines-for-the-testing-of-chemicals-section-3-degradation-and-accumulation_2074577x;jsessionid=1xfkgbcx5erx7x-oecd-live-02.

41. Racke K, Fontaine DD, Yoder RN, Miller JR: **Chlorpyrifos degradation in soil at termiticidal application rates.** *Pestic Sci* 1994, **42**:43–51.

42. de Vette HQM, Schoonmade JA: *A Study on the Route and Rate of Aerobic Degradation of ¹⁴C-Chlorpyrifos in Four European Soils.* Indianapolis: Dow AgroSciences; 2001:191.

43. Coppola L, Castillo MP, Monaci E, Vischetti C: **Adaptation of the biobed composition for chlorpyrifos degradation to Southern Europe conditions.** *J Agric Food Chem* 2007, **55**:396–401.

44. McCall PJ, Oliver GR, McKellar RL: *Modeling the Runoff Potential and Behavior of Chlorpyrifos in a Terrestrial-Aquatic Watershed.* Midland: Dow Chemical; 1984:118.

45. Racke KD, Laskowski DA, Schultz MR: **Resistance of chlorpyrifos to enhanced biodegradation in soil.** *J Agric Food Chem* 1990, **38**:1430–1436.

46. Korade DL, Fulekar MH: **Rhizosphere remediation of chlorpyrifos in mycorrhizospheric soil using ryegrass.** *J Haz Matter* 2009, **172**:1344–1350.

47. Kuhr R, Tashiro H: **Distribution and persistence of chlorpyrifos and diazinon applied to turf.** *Bull Environ Contam Toxicol* 1978, **20**:652–656.

48. Sardar D, Kole RK: **Metabolism of chlorpyrifos in relation to its effect on the availability of some plant nutrients in soil.** *Chemosphere* 2005, **61**:1273–1280.

49. Getzin L: **Degradation of chlorpyrifos in soil: influence of autoclaving, soil moisture, and temperature.** *J Econ Entomol* 1981, **74**:158–162.

50. Graebing P, Chib JS: **Soil photolysis in a moisture- and temperature-controlled environment. 2. Insecticides.** *J Agric Food Chem* 2004, **52**:2606–2614.

51. Reeves G, Old J: *The Dissipation of Chlorpyrifos and its Major Metabolite (3,5,6-Trichloro-2-pyridinol) in Soil Following a Single Spring Application of Dursban 4 (EF-1042), Spain-2000.* Indianapolis: Dow Agrosciences; 2002:85.

52. Reeves G, Old J: *The Dissipation of Chlorpyrifos and its Major Metabolite (3,5,6-Trichloro-2-pyridinol) in Soil Following a Single Spring Application of Dursban 4 (EF-1042), UK-2000.* Indianapolis: Dow Agrosciences; 2002:83.

53. Reeves G, Old J, Reeves G, Old J: *The Dissipation of Chlorpyrifos and its Major Metabolite (3,5,6-Trichloro-2-pyridinol) in Soil Following a Single Spring Application of Dursban 4 (EF-1042), France-2000.* Indianapolis: Dow Agrosciences; 2002:73.

54. Rouchaud J, Metsue M, Gustin F, van de Steene F, Pelerents C, Benoit F, Ceustermans N, Gillet J, Vanparys L: **Soil and plant biodegradation of chlorpyrifos in fields of cauliflower and Brussels sprouts crops.** *Toxicol Environ Chem* 1989, **23**:215–226.

55. Reeves G, Old J: *The Dissipation of Chlorpyrifos and its Major Metabolite (3,5,6-Trichloro-2-pyridinol) in Soil Following a Single Spring Application of Dursban 4 (EF-1042), Greece-2000.* Indianapolis: Dow Agrosciences; 2002:81.

56. Koshab A, Nicholson A, Berryman T: *The Dissipation of Chlorpyrifos and Fate of its Major Metabolite, 3,5,6-Trichloropyridin-2-ol, in Soil Following Application of Dursban Fluessing (EF 747) to Bare Soil, Germany-1992.* Wantage: DowElanco Europe, Letcombe Laboratory; 1994:41.

57. Koshab A, Nicholson A, Kinzel P, Draper R: *The Dissipation of Chlorpyrifos and Fate of its Major Metabolite, 3,5,6-Trichloropyridin-2-ol, in Two Soil Types Following Application of Dursban Fluessing (EF 747) to Bare Soil, Germany-1991.* Wantage: DowElanco Europe, Letcombe Laboratory; 1993:34.

58. Laabs V, Amelung W, Pinto A, Altstaedt A, Zech W: **Leaching and degradation of corn and soybean pesticides in an Oxisol of the Brazilian Cerrados.** *Chemosphere* 2000, **41**:1441–1449.

59. Chai LK, Mohd-Tahir N, Bruun Hansen HC: **Dissipation of acephate, chlorpyrifos, cypermethrin and their metabolites in a humid-tropical vegetable production system.** *Pest Manag Sci* 2009, **65**:189–196.

60. Putnam RA, Nelson JO, Clark JM: **The persistence and degradation of chlorothalonil and chlorpyrifos in a cranberry bog.** *J Agric Food Chem* 2003, **51**:170–176.

61. Chapman RA, Harris CR: **Persistence of chlorpyrifos in a mineral and an organic soil.** *J Environ Sci Hlth B* 1980, **15**:39–46.

62. Szeto S, Mackenzie J, Vernon R: **Comparative persistence of chlorpyrifos in a mineral soil after granular and drench applications.** *J Environ Sci Hlth B* 1988, **23**:541–557.

63. Davis A, Kuhr R: **Dissipation of chlorpyrifos from muck soil and onions.** *J Econ Entomol* 1976, **69**:665–666.

64. Fontaine D, Wetters JH, Weseloh JW, Stockdale GD, Young JR, Swanson ME: *Field dissipation and leaching of chlorpyrifos.* Midland: Dow Chemical USA; 1987:116.

65. Pike K, Getzin L: **Persistence and movement of chlorpyrifos in sprinkler-irrigated soil.** *J Econ Entomol* 1981, **74**:385–388.

66. Getzin LW: **Factors influencing the persistence and effectiveness of chlorpyrifos in soil.** *J Econ Entomol* 1985, **78**:412–418.

67. Racke KD: **Environmental fate of chlorpyrifos.** *Rev Environ Contam Toxicol* 1993, **131**:1–151.

68. Macalady DL, Wolfe NL: **New perspectives on the hydrolytic degradation of the organophosphorothioate insecticide chlorpyrifos.** *J Agric Food Chem* 1983, **31**:1139–1147.

69. Meikle RW, Youngson CR: **The hydrolysis rate of chlorpyrifos, O-O-diethylO-(3,5,6-trichloro-2-pyridyl) phosphorothioate, and its dimethyl analog, chlorpyrifos-methyl, in dilute aqueous solution.** *Arch Environ Contam Toxicol* 1978, **7**:13–22.

70. Sharom M, Miles J, Harris C, McEwen F: **Persistence of 12 insecticides in water.** *Water Res* 1980, **14**:1089–1093.

71. Kennard LM: *Aerobic Aquatic Degradation of Chlorpyrifos in a Flow-through System.* Indianapolis: DowElanco; 1996:97.

72. Reeves GL, Mackie JA: *The Aerobic Degradation of ¹⁴C-Chlorpyrifos in Natural Waters and Associated Sediments.* Indianapolis: Dow Agrosciences; 1993:107.

73. Pablo F, Krassoi F, Jones P, Colville A, Hose G, Lim R: **Comparison of the fate and toxicity of chlorpyrifos—laboratory versus a coastal mesocosm system.** *Ecotoxicol Environ Safety* 2008, 71:219–229.

74. Liu B, McConnell L, Torrents A: **Hydrolysis of chlorpyrifos in natural waters of the Chesapeake Bay.** *Chemosphere* 2001, 44:1315–1323.

75. Daam MA, Van den Brink PJ: **Effects of chlorpyrifos, carbendazim, and linuron on the ecology of a small indoor aquatic microcosm.** *Arch Environ Contam Toxicol* 2007, 53:22–35.

76. van Wijngaarden RP, Brock TC, Douglas MT: **Effects of chlorpyrifos in freshwater model ecosystems: the influence of experimental conditions on ecotoxicological thresholds.** *Pest Manag Sci* 2005, 61:923–935.

77. Laabs V, Wehrhan A, Pinto A, Dores E, Amelung W: **Pesticide fate in tropical wetlands of Brazil: an aquatic microcosm study under semi-field conditions.** *Chemosphere* 2007, 67:975–989.

78. Nhan DD, Carvalho FP, Nam BQ: **Fate of ¹⁴C-chlorpyrifos in the tropical estuarine environment.** *Environ Technol* 2002, 23:1229–1234.

79. Banni M, Jebali J, Guerbej H, Dondero F, Boussetta H, Viarengo A: **Mixture toxicity assessment of nickel and chlorpyrifos in the sea bass *Dicentrarchus labrax*.** *Arch Environ Contam Toxicol* 2011, 60:124–131.

80. Eaton J, Arthur J, Hermanutz R, Kiefer R, Mueller L, Anderson R, Erickson R, Nordling B, Rogers J, Pritchard H: **Biological effects of continuous and intermittent dosing of outdoor experimental streams with chlorpyrifos.** In *Aquatic Toxicology and Hazard Assessment: Eighth Symposium ASTM STP.* Philadelphia: American Society for Testing and Materials; 1985:85–118.

81. Thomas CN, Mansingh A: **Bioaccumulation, elimination, and tissue distribution of chlorpyrifos by red hybrid tilapia in fresh and brackish waters.** *Environ Technol* 2002, 23:1313–1323.

82. Douglas MT, Bell IB: *The Bioaccumulation and Depuration of Chlorpyrifos.* Kings Lynn: DowElanco Europe; 1991:10.

83. Goodman L, Hansen D, Cripe G, Middaugh D, Moore J: **A new early life-stage toxicity test using the California grunion (*Leuresthes tenuis*) and results with chlorpyrifos.** *Ecotoxicol Environ Safety* 1985, 10:12–21.

84. Goodman LR, Hansen DJ, Middaugh DP, Cripe GM, Moore JC: **Method for early life-stage toxicity tests using three atherinid fishes and results with chlorpyrifos.** In *Aquatic Toxicology and Hazard Assessment: Seventh Symposium, ASTM STP 854.* Edited by Cardwell RD, Purdy R, Bahner RC. Philadelphia: American Society for Testing and Materials; 1985:145–154.

85. Hedlund R: *Bioconcentration of Chlorpyrifos by Mosquito Fish in a Flowing System.* Midland: Dow Chemical Company; 1973:17.

86. Tsuda T, Shigeru A, Mihoko K, Toshie F: **Accumulation and excretion of pesticides used in golf courses by carp (*Cyprinus carpio*) and willow shiner (*Gnathopogon caerulescens*).** *Comp Biochem Physiol C Comp Pharmacol* 1992, 101:63–66.

87. Hansen DJ, Goodman LR, Cripe GM, Macauley SF: **Early life-stage toxicity test methods for gulf toadfish (*Opsanus beta*) and results using chlorpyrifos.** *Ecotoxicol Environ Safety* 1986, 11:15–22.

88. Macek KJ, Walsh DF, Hogan JW, Holz DD: **Toxicity of the insecticide Dursban(R) to fish and aquatic invertebrates in ponds.** *Trans Am Fish Soc* 1972, 101:420–427.

89. Murphy P, Lutenske N: *Bioconcentration of Chlorpyrifos in Rainbow Trout (Salmo gairdneri Richardson).* Indianapolis: DowElanco; 1986:49.

90. Deneer J: **Uptake and elimination of chlorpyrifos in the guppy at sublethal and lethal aqueous concentrations.** *Chemosphere* 1993, 26:1607–1616.

91. Jarvinen AW, Nordling BR, Henry ME: **Chronic toxicity of Dursban (chlorpyrifos) to the fathead minnow (*Pimephales promelas*) and the resultant acetylcholinesterase inhibition.** *Ecotoxicol Environ Safety* 1983, 7:423–434.

92. Welling W, De Vries J: **Bioconcentration kinetics of the organophosphorus insecticide chlorpyrifos in guppies (*Poecilia reticulata*).** *Ecotoxicol Environ Safety* 1992, 23:64–75.

93. Cripe GM, Hansen DJ, Macauley SF, Forester J: **Effects of diet quantity on sheepshead minnows (*Cyprinodon variegatus*) during early life-stage exposures to chlorpyrifos.** In *Aquatic Toxicology and Environmental Fate: Ninth Symposium ASTM Spec Tech Publ.* Philadelphia: American Society for Testing and Materials; 1986:450–460.

94. El-Amrani S, Pena-Abaurrea M, Sanz-Landaluze J, Ramos L, Guinea J, Camara C: **Bioconcentration of pesticides in zebrafish eleutheroembryos (*Danio rerio*).** *Sci Tot Environ* 2012, 425:184–190.

95. USEPA: *The National Study of Chemical Residues in Lake Fish Tissue.* Washington, DC: Environmental Protection Agency, Office of Water; 2009:242.

96. Landers DH, Simonich SL, Jaffe DA, Geiser LH, Campbell DH, Schwindt AR, Schreck CB, Kent ML, Hafner WD, Taylor HE, Hageman KJ, Usenko S, Ackerman LK, Schrlau JE, Rose NL, Blett TF, Erway MM: *The Fate, Transport, and Ecological Impacts of Airborne Contaminants in Western National Parks.* Corvallis: U.S. Environmental Protection Agency, Office of Research and Development, NHEERL, Western Ecology Division; 2008:350. water.usgs.gov/nrp/proj.bib/.../2008/landers_simonich_etal_2008a.pdf.

97. Morris AD, Muir DCG, Solomon KR, Teixeira C, Duric M, Wang X: **Bioaccumulation and trophodynamics of current use pesticides and endosulfan in the vegetation-caribou-wolf food chain in the Canadian Arctic.** *Environ Toxicol Chem* 2014, 33:1956–1966.

98. Moore DRJ, Teed RS, Greer C, Solomon KR, Giesy JP: **Refined avian risk assessment for chlorpyrifos in the United States.** *Rev Environ Contam Toxicol* 2014, 231:163–217.

99. Cutler GC, Purdy J, Giesy JP, Solomon KR: **Risk to pollinators from the use of chlorpyrifos in North America.** *Rev Environ Contam Toxicol* 2014, 231:219–265.

100. CCME: *Determination of Hazardous Concentrations with Species Sensitivity Distributions, SSD Master.* Ottawa: Canadian Council of Ministers of the Environment; 2013:38.

101. Maltby L, Blake NN, Brock TCM, van den Brink PJ: **Insecticide species sensitivity distributions: the importance of test species selection and relevance to aquatic ecosystems.** *Environ Toxicol Chem* 2005, 24:379–388.

102. Wu JY, Anelli CM, Sheppard WS: **Sub-lethal effects of pesticide residues in brood comb on worker honey bee (*Apis mellifera*) development and longevity.** *PloS ONE* 2011, 6:e14720.

103. Barron MG, Woodburn KB: **Ecotoxicology of chlorpyrifos.** *Rev Environ Contam Toxicol* 1995, 144:1–93.

104. Solomon KR, Giesy JP, Kendall RJ, Best LB, Coats JR, Dixon KR, Hooper MJ, Kenaga EE, McMurry ST: **Chlorpyrifos: ecotoxicological risk assessment for birds and mammals in corn agroecosystems.** *Human Ecol Risk Assess* 2001, 7:497–632.

105. Hubbard PM, Beavers JB: *Chlorpyrifos-Oxon: an Acute Oral Toxicity Study with the Northern Bobwhite.* Indianapolis: Dow AgroScience; 2010:55.

106. Hubbard PM, Beavers JB: *Chlorpyrifos-Oxon: an Acute Oral Toxicity Study with the Zebra Finch (Poephila guttata).* Indianapolis: Dow AgroScience; 2010:71.

107. Hubbard PM, Martin KH, Beavers JB: *Chlorpyrifos-Oxon: a Dietary LC50 Study with the Mallard.* Indianapolis: Dow AgroScience; 2010:93.

108. Hubbard PM, Martin KH, Beavers JB: *Chlorpyrifos-Oxon: a Dietary LC50 Study with the Northern Bobwhite.* Indianapolis: Dow AgroScience; 2010:89.

109. Wegmann F, Cavin L, MacLeod M, Scheringer M, Hungerbuhler K: **The OECD software tool for screening chemicals for persistence and long-range transport potential.** *Environ Model Software* 2009, 24:228–237.

110. Beyer A, Mackay D, Matthies M, Wania F, Webster E: **Assessing long-range transport potential of persistent organic pollutants.** *Environ Sci Technol* 2000, 34:699–703.

111. Gebremariam SY, Beutel MW, Yonge DR, Flury M, Harsh JB: **Adsorption and desorption of chlorpyrifos to soils and sediments.** *Rev Environ Contam Toxicol* 2012, 215:123–175.

112. Giesy JP, Solomon KR, Coates JR, Dixon KR, Giddings JM, Kenaga EE: **Chlorpyrifos: ecological risk assessment in North American aquatic environments.** *Rev Environ Contam Toxicol* 1999, 160:1–129.

113. Muir DC, Teixeira C, Wania F: **Empirical and modeling evidence of regional atmospheric transport of current-use pesticides.** *Environ Toxicol Chem* 2004, 23:2421–2432.

114. Turner DB: *Workbook of Atmospheric Dispersion Estimates.* Boca Raton: Lewis Publ. CRC Press; 1994.

115. Hermanson MH, Isaksson E, Teixeira C, Muir DC, Compher KM, Li YF, Igarashi M, Kamiyama K: **Current-use and legacy pesticide history in the Austfonna Ice Cap, Svalbard, Norway.** *Environ Sci Technol* 2005, 39:8163–8169.

116. Leistra M, Smelt JH, Weststrate JH, van den Berg F, Aalderink R: **Volatilization of the pesticides chlorpyrifos and fenpropimorph from a potato crop.** *Environ Sci Technol* 2005, 40:96–102.

117. Vaccaro JR: **Risks associated with exposure to chlorpyrifos formulation components.** In *Pesticides in Urban Environments: Fate and Significance, ACS*

Symposium Series, Volume 522. Edited by Racke KD, Leslie AR. Washington, DC: American Chemical Society; 1993:197–396.

118. Sullivan DJ, Vecchia AV, Lorenz DL, Gilliom RJ, Martin JD: *Trends in Pesticide Concentrations in Corn-Belt Streams, 1996–2006.* Reston: U.S. Geological Survey; 2009:75.

119. Ryberg KR, Vecchia AV, Martin JD, Gilliom RJ: *Trends in Pesticide Concentrations in Urban Streams in the United States, 1992–2008.* Reston: U.S. Geological Survey; 2010:101.

120. Johnson HM, Domagalski JL, Saleh DK: **Trends in pesticide concentrations in streams of the western United States, 1993–2005.** *J Am Water Res Assoc* 2011, **47**:265–286.

10

Contaminant levels in the European eel (*Anguilla anguilla*) in North Rhine-Westphalian rivers

Barbara Guhl[*], Franz-Josef Stürenberg and Gerhard Santora

Abstract

Background: Populations of the European eel (*Anguilla anguilla*) are declining rapidly and are now considered below safe biological limits. High pollution levels are one of the possible reasons for this decline. Contaminant levels are also of concern with regard to human consumption. This study examined the contamination levels of eels from the North Rhine-Westphalian catchment area of the Rhine and from adjacent rivers. A total of 119 eels from 13 sampling sites were analysed for polychlorinated dibenzo-p-dioxins and furans (PCDD/PCDFs), polychlorinated biphenyls (PCBs), polybrominated diphenyl ethers (PBDEs), mercury, perfluorooctane sulfonate (PFOS) and hexachlorobenzene (HCB).

Results: North Rhine-Westphalian eels had very high levels of contaminants comparable to eels from other European water bodies which are strongly influenced by anthropogenic activities. Mean values for PCDD/PCDFs ranged between 0.5 and 5.4 pg WHO_{2005} toxicity equivalents (TEQ)/g, for PCDD/PCDF + dl-PCBs between 6.3 and 44.7 pg WHO_{2005} TEQ/g, for indicator PCBs between 165 and 1,630 ng/g wet weight (ww), for 6 PBDEs between 9.2 and 242 µg/kg ww, for mercury between 0.069 and 0.314 mg/kg ww, for PFOS between 8.3 and 49 µg/kg ww and for HCB between 3.4 and 50 µg/kg ww. For certain sampling sites, high contamination contents of the eels could be attributed to local sources. Congener patterns for PCBs and PBDEs were described, and biota to suspended matter ratios were calculated.

Conclusions: Pollution levels in eels from North Rhine-Westphalia are declining with regard to some contaminants but are still very high. Due to the high contaminant contents, eels from the rivers investigated are not suitable for human consumption. Furthermore, the concentrations of endocrine disrupting contaminants such as PCBs, PCDD/PCDFs and PBDEs in the eels are in a range which might have deleterious effects on the eel populations.

Keywords: Biota monitoring; Contaminant levels; Dioxins; PCB; HCB; PBDE; PFOS; Mercury; European eel (*Anguilla anguilla*)

Background

The European eel (*Anguilla anguilla* L.) used to be an abundant species but since the beginnings of the 1980s, populations have been declining sharply [1]. Amongst the possible reasons for this decline are over exploitation, habitat loss [2], migration barriers [3], the introduction of pathogens such as *Anguillicoloides crassus* [4] and adverse effects of contaminants [5,6].

Due to their high body lipid content and due to ecological features such as a bottom dwelling way of life, longevity and a predatory feeding mode eels are particularly prone to bioaccumulate lipophilic contaminants.

Polychlorinated biphenyls (PCBs) and polychlorinated dibenzo-p-dioxins and furans (PCDD/PCDFs) especially have been suspected to impair aquatic organisms due to their endocrine disrupting mode of action [7]. Eels are thought to starve during their 6 months journey to their spawning grounds in the Sargasso Sea. The lipid reserves of the body are released, and the contaminants stored in the lipid tissue are put into circulation again [8]. Thus, during the migration, PCBs and PCDD/PCDFs may reach harmful concentrations in the blood [9]. Polybrominated diphenyl ethers (PBDEs) are another group of pollutants which are also potential endocrine disruptors. *In vitro* tests have demonstrated various agonistic and antagonistic activities of PBDEs with respect to the aryl hydrocarbon receptor and androgen receptor and to a lesser extend to

* Correspondence: barbara.guhl@lanuv.nrw.de
North Rhine-Westphalian State Agency for Nature, Environment and Consumer Protection, Leibnizstr 10, 45659 Recklinghausen, Germany

the estrogen and progesterone receptors [10-12]. They are additionally suspected to effect neurobehavioral development [13]. Other contaminants which are objects of scientific investigations due to recent legal regulation are perfluorooctane sulfonate (PFOS), hexachlorobenzene (HCB) and mercury. For PFOS, numerous effects are reported, e.g. hepatic damage, disturbance of DNA metabolism [14] and adverse effects on protein expression [15]. Hoff et al. [16] found evidence that HCB has an impact on various blood parameters. Like other metal pollutants, mercury has been demonstrated to impair general fitness as assessed by Fulton's condition factor and to display immunotoxic effects [17,18].

Contaminants in biota have been subject of EU Regulation due to their potential adverse effects on predatory species (secondary poisoning) and on human health. EU Directive 2013/39/EU sets environmental quality standards (EQS) for 11 priority substances in biota, amongst others, for the above-mentioned parameters mercury; HCB; PCDD/PCDFs; dioxin-like PCBs (dl-PCBS); the sum of the six indicator PCBs 28, 52, 101, 138, 153 and 180; the sum of the BDE congeners 28, 47, 99, 100, 153 and 154 and PFOS, which have to be met by 2018.

Contaminants in eels have been monitored extensively with a particular emphasis on PCDD/PCDFs and PCBs. For example, low toxicity equivalents (TEQ) of PCDD/PCDFs + dl-PCBs, mostly not exceeding the EQS_{biota} of 6.5 ng WHO_{2005}-TEQ/kg set by EU Directive 2013/39/EU, were reported from the Loire [7], and high concentrations, in the range of 20 to nearly 100 ng WHO_{2005}-TEQ/kg, were detected in eels from contaminated sites such as the Elbe near the port of Hamburg [19] and from certain Belgian rivers [5]. To protect the human health, EU Regulation 1259/2011 sets a maximum level of 300 ng/g ww for the indicator PCBs. In a review, Tapie et al. [20] presented data for the concentration of the six indicator PCBs + PCB 118 which ranged from <10 ng/g ww in Irish waters to well over 1,000 ng/g ww in the Dutch Haringvliet and in the Rhone. About half of the studies considered in the review reported concentrations exceeding the EU EQS.

Generally, PBDE concentrations monitored in fish were two to four orders of magnitude higher than the new biota EQS_{biota} of 0.0085 µg/kg ww. Tapie et al. [20] also compiled data about BDE-47 which is usually three-fourths of the sum of 6 PBDE. The data were reported for different fish species and ranged from <0.001 µg/kg ww in near-natural Irish waters to 25 µg/kg ww in the Netherlands. Even higher values from Dutch waters were reported from Pujolar et al. [21] who measured a mean concentration of 92 µg/kg ww (sum of ten PBDEs) in female silver eels from the highly polluted Dessel Schoten Canal. Sühring et al. [22] analysed nine PBDEs in silver eels from the upper part of the Rhine and found a mean concentration

of 21.3 µg/kg. Silver and yellow eels from the Elbe contained an average of 8.3 and 8.9 µg/kg, respectively.

HCB concentrations detected in eel were mostly below the EU EQS_{biota} of 10 µg/kg, e.g. [23] but also under exceptional circumstances up to concentrations of 192 µg/kg [24]. Mercury was well above the EQS_{biota} of 20 µg/kg ww in all monitoring studies but usually below the limit for human consumption of 1 mg/kg ww set by the EU Commission Regulation 1881/2006. Noël et al. [25] compiled data on Hg concentrations in eels from European waters which ranged between 10 and 800 µg Hg/kg ww.

EU Directive 2013/39/EU sets an EQS_{biota} for PFOS of 9.1 µg/kg ww. In a review, Hloušková et al. [26] reported values for the sum of perfluoroalkyl substances (PFASs) in fish (PFOS normally comprises at least 90% of the PFASs present) in a range of 0.2 to up to 1,963 µg/kg ww. The highest values were reported from polluted Czech rivers. Data on PFOS concentrations in eel are scarce. Only data on PFOS contents in the liver have been published. Roland et al. [15] determined mean liver concentrations of 31.1 µg/kg in a relatively unpolluted canal and 230.1 and 329.9 µg/kg for two polluted sites. Hoff et al. [16] measured mean hepatic PFOS concentrations of 1,387 µg/kg ww in eels from four Belgian water bodies which are heavily polluted and related to the high concentrations to fluorochemical production units in the vicinity of the sampling sites.

The objectives of this study were the description of contamination patterns in eels from the North Rhine-Westphalian Rhine region with respect to the EU priority substances and the evaluation of the potential risk to the eel populations as well as to human health via consumption of eels.

Results and discussion

Figure 1 depicts the location of the sampling sites in North Rhine-Westphalia (for further details see Methods). Table 1 sums up the information on the biometric data of the eels. On average eels were between 7 and 14 years old and differed accordingly in size and weight. Eels from the Sieg were relatively small (mean length 47 cm, mean weight 203 g). The oldest and heaviest eels were collected from the Niers (mean age 14 years, mean weight 861 g). Despite these variations, the mean lipid content of all sampling sites did not differ significantly (α =0.1). For most sites, the condition factor varied between 0.17 and 0.22 which is a common range for healthy eels. Only eels from the Ruhr had an exceptionally high condition factor of 0.28 which can be attributed to the fact that the sample included three very old (>15 years) and heavy eels. As additional information, it should be pointed out that more than half of all specimens had skin lesions, hematoma or other external signs of impaired physical health.

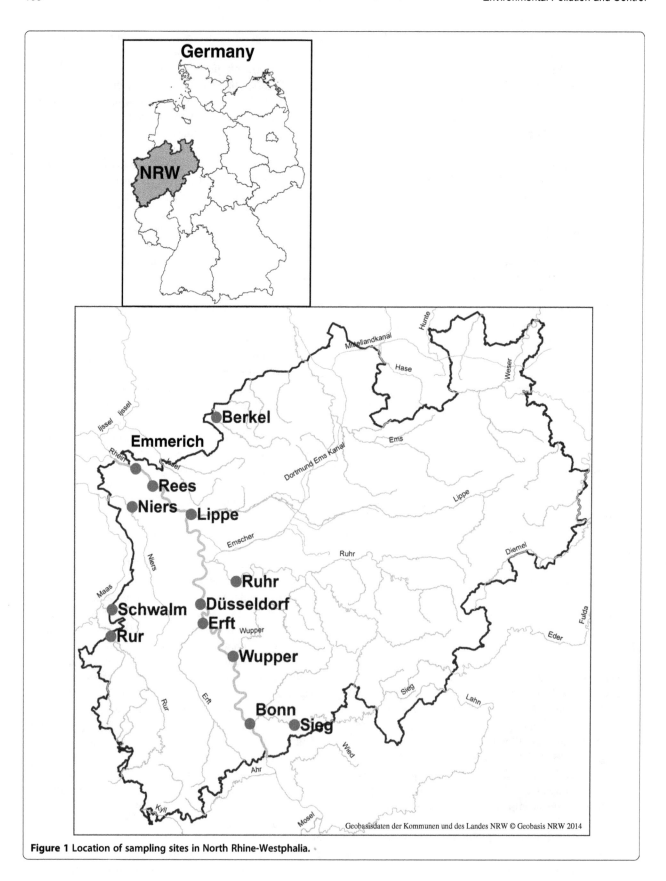

Figure 1 Location of sampling sites in North Rhine-Westphalia.

Table 1 Arithmetic means and standard deviations for biometric parameters

Sampling site	n	Age (years)	Length (cm)	Weight (g)	Lipid content (%)	Condition factor
Rhein Bonn	9	9.5 ± 2.8	59 ± 12	407 ± 230	22.9 ± 9.7	0.18 ± 0.02
Rhein Düsseldorf	11	7.1 ± 1.8	47 ± 13	250 ± 169	13.9 ± 8.8	0.22 ± 0.08
Rhein Rees	9	8.7 ± 2.1	62 ± 10	529 ± 245	22.7 ± 5.8	0.20 ± 0.02
Rhein Emmerich	11	n.d.[a]	60 ± 10	405 ± 194	26.7 ± 11.2	0.18 ± 0.02
Sieg	10	8.6 ± 2.4	47 ± 7	203 ± 92	13.8 ± 8.0	0.19 ± 0.03
Wupper	10	7.6 ± 2.8	49 ± 11	278 ± 165	18.9 ± 8.8	0.21 ± 0.03
Ruhr	9	13.0 ± 4.0	56 ± 12	632 ± 604	26.5 ± 14.1	0.28 ± 0.17
Erft	10	8.8 ± 3.2	47 ± 8	230 ± 127	23.4 ± 12.6	0.20 ± 0.03
Lippe	9	8.3 ± 2.8	52 ± 10	280 ± 171	21.6 ± 11.8	0.17 ± 0.03
Niers	5	14.0 ± 4.6	734 ± 14	861 ± 395	29.1 ± 6.3	0.20 ± 0.02
Schwalm	8	13.4 ± 2.1	67 ± 8	597 ± 194	20.8 ± 7.8	0.20 ± 0.03
Rur	10	12.2 ± 2.3	55 ± 6	316 ± 116	25.4 ± 9.6	0.19 ± 0.04
Berkel	8	9.0 ± 4.3	54 ± 13	375 ± 252	20.6 ± 13.4	0.19 ± 0.03

[a]n.d. =not determined.

Results for the sum of PCDDs and PCDFs; for indicator PCBs, HCB, PFOS, PBDEs and mercury are presented in Table 2. In the literature, contaminant concentrations are presented per wet weight (ww), dry weight or lipid normalised (lw). The main focus of this study was on a compliance check for priority substances according to EU Directive 2013/39/EU and on the investigation of a possible risk for human health. Therefore, contaminant concentrations are presented as arithmetic means per wet weight. Where necessary, for comparison with published data, results are converted to lipid normalised values.

PCDDs, PCDFs and indicator PCBs

Mean concentrations for the sum of PCDDs and PCDFs are generally below the EQS of Directive 2013/39/EU of 3.5 pg WHO_{2005} PCDD/PCDF TEQ/g. Exceptions are the values for the Rhine sites Rees and Emmerich where mean concentrations reached 4.7 and 5.4 pg WHO_{2005} PCDD/PCDF TEQ/g, respectively (significantly different from all other sampling sites). In contrast, mean concentrations of the sum of PCDD/PCDF + dl-PCB were nearly always above the EQS of 6.5 pg WHO_{2005} PCDD/PCDF +dl-PCB TEQ/g and also above the maximum level of 10 pg WHO_{2005} PCDD/PCDF + dl-PCB TEQ/g in EU Regulation

Table 2 Arithmetic means and standard deviations for chemical parameters

	n[a]	Σ PCDD/PCDF (pg WHO_{2005} TEQ/g)	Σ PCDD/PCDF + dl-PCB (pg WHO_{2005} TEQ/g)	Indicator PCB (ng/g)	HCB (µg/kg)	Σ PBDE (µg/kg)	PFOS (µg/kg)	Hg (mg/kg)
Rhein Bonn	9/9	1.8 ± 0.7	17.0 ± 5.9	253 ± 81	23.2 ± 14.7	18.4 ± 5.6	22.9 ± 9.8	0.210 ± 0.116
Rhein Düsseldorf	7/11	1.2 ± 0.7	19.4 ± 7.9	382 ± 152	27.8 ± 11.9	20.7 ± 9.9	27.5 ± 23.6	0.165 ± 0.057
Rhein Rees	9/9	4.7 ± 1.8	39.4 ± 12.9	698 ± 269	35.7 ± 12.8	74.6 ± 30.7	13.7 ± 9.9	0.273 ± 0.058
Rhein Emmerich	11/11	5.4 ± 2.3	44.7 ± 17.8	982 ± 302	50.2 ± 25.2	76.2 ± 28.2	35.8 ± 17.9	0.195 ± 0.059
Sieg	9/10	0.5 ± 0.3	6.3 ± 2.8	165 ± 60	3.4 ± 1.7	9.2 ± 4.0	13.4 ± 2.9	0.069 ± 0.021
Wupper	9/10	0.8 ± 0.3	14.7 ± 3.0	369 ± 88	18.5 ± 6.1	14.3 ± 3.4	30.6 ± 16.9	0.314 ± 0.131
Ruhr	8/9	2.2 ± 1.2	32.2 ± 16.6	800 ± 318	17.1 ± 6.5	23.1 ± 12.8	32.0 ± 14.3	0.077 ± 0.024
Erft	9/10	1.5 ± 0.8	14.5 ± 6.2	301 ± 93	11.6 ± 7.1	13.6 ± 6.6	14.6 ± 7.1	0.075 ± 0.070
Lippe	9/9	1.1 ± 0.5	25.2 ± 22.4	507 ± 544	19.3 ± 8.2	241.9 ± 143.7	15.6 ± 6.7	0.146 ± 0.061
Niers	5/5	2.2 ± 0.5	28.6 ± 6.0	643 ± 197	11.6 ± 2.2	14.1 ± 4.4	49.0 ± 12.5	0.080 ± 0.018
Schwalm	8/8	1.00 ± 0.4	12.3 ± 4.0	216 ± 75	8.8 ± 3.1	8.5 ± 2.5	16.9 ± 7.0	0.106 ± 0.039
Rur	10/10	2.1 ± 1.2	39.6 ± 26.4	1,630 ± 762	12.1 ± 4.8	53.4 ± 30.4	42.3 ± 32.5	0.124 ± 0.030
Berkel	7/8	1.2 ± 1.3	15.8 ± 10.2	357 ± 377	6.2 ± 4.3	20.8 ± 22.6	8.3 ± 3.0	0.145 ± 0.067

[a]The first figure represents the number of data for HCB and PFOS, and the second figure represents the number of data for PCDD/PCDFs, dl-PCBs, indicator PCBs, PBDEs and Hg.

1259/2011. Highest mean TEQ concentrations were determined for eels from Emmerich (44.7 pg WHO_{2005} PCDD/PCDF+dl-PCB TEQ/g), Rees (39.4 pg WHO_{2005} PCDD/PCDF+dl-PCB TEQ/g), the Rur (39.6 pg PCDD/PCDF+dl-PCB WHO_{2005} TEQ/g) and the Ruhr (32.2 pg WHO_{2005} PCDD/PCDF+dl-PCB TEQ/g). Only eels from the Sieg had a mean concentration below the EQS (6.3 pg WHO_{2005} PCDD/PCDF+dl-PCB TEQ/g). These low concentrations cannot be attributed to the relatively low lipid contents of the Sieg eels as lipid normalised values are still low. The Rhine sites Rees and Emmerich and the Rur differed significantly from most other sites, but not from the Wupper, Ruhr, Niers and Lippe.

Similar TEQ concentrations were reported from Belgian waters [5] and from the Elbe in Germany [19]. Dioxin and dl-PCB levels were lower in the Loire (mean concentration 4.4 pg WHO_{2005} PCDD/PCDF + dl-PCB TEQ/g, [7]) and in Irish rivers (0.49 to 4.9 pg WHO_{1998} PCDD/PCDF + dl-PCB TEQ/g [27]).

TEQ concentrations for dl-PCBs were 8 to 16 times higher than TEQ concentrations for PCDD/PCDFs. The highest ratio was found for the Wupper with a mean value of 18.5. In eels from European waters, dl-PCB contamination is usually of higher importance than PCDD/PCDF contamination. High concentrations of dl-PCBs were e.g. also found in eels from Flemish waters [5], from the catchment area of the Elbe [19] and the Tiber in Rome [28]. In contrast, McHugh et al. [27] reported dl-PCB concentrations between 0.17 and 1.24 pg WHO_{1998} dl-PCB TEQ from relatively clean Irish rivers and therefore ratios between dl-PCBs and PCDD/PCDF of 0.1 to 4.3. Knutzen et al. [29] found PCDD/PCDF concentrations between 5.0 and 22.9 pg WHO_{1998} PCDD/PCDF TEQ and dl-PCB concentrations between 1.4 and 3.9 pg WHO_{1998} dl-PCB TEQ in Norwegian fjords which had been contaminated by discharges of waste water from a magnesium factory in the past.

All values recorded were well above the value of 4 pg PCDD/PCDF + dl-PCB WHO_{1998} TEQ/g which was determined by Palstra et al. [30] as a threshold value for adverse effects on the reproductive system of eels.

For indicator PCBs, the mean concentrations exceeded the maximum level for human consumption of 300 ng/g at most sampling sites. Exceptions were the eels from the Schwalm and the Sieg with 215 and 165 ng/g, respectively. Values in the Rhine eels ranged from 253 ng/g in Bonn to 982 ng/g in Emmerich. They are well below the concentrations found in Rhine eels in 1995 and 1998/2000 by Heinisch et al. [31]. All results in their paper are presented only as graphs and therefore approximations. Values for eels in Bonn were over 500 ng/g in 1995 and even over 600 ng/g in 1998/2000. Concentrations in Emmerich were about 1,500 in 1995 and 1,200 ng/g in 1998/2000, respectively. This comparison with our dataset documents a decrease of PCB pollution reflecting the general reduction of PCBs present in the atmosphere as well as a lower availability of PCBs in the Rhine after changes in production processes in major industrial sites in North Rhine-Westphalia.

By far the highest PCB concentrations were determined for eels from the Rur (mean concentrations of 1,630 ng/g). The catchment area of the Rur has a mining history. According to Detzel et al. [32], hydraulic fuel used in the mining industry has been a major source for PCB release into the environment. Additionally, PCB concentrations in the Rur were influenced by the emissions of paper industry upstream the sampling site. The high PCB concentrations in eels from the Rur were also described by de Boer et al. [33] who investigated the contamination of Dutch eels over a period of 30 years. For the Dutch part of the Rur, they reported concentrations of the sum of indicator PCBs + PCB 118, which decreased between 1982 and 2006 from 44,372 to 7,087 ng/g lw and therefore considerably more than in the Rhine (mean concentrations of 1,630 ng/g ww for Rur eels from this study correspond to 6,668 ng/g lw). Generally, PCB concentrations in eels from North Rhine-Westphalian waters were in the range of values found in other European waters impacted by industry and dense human population. Tapie et al. [20] determined the sum of indicator PCBs + PCB 118 in eels from the Gironde values ranging between 1,141 and 5,746 ng/g lw. High concentrations were as well reported from eels in a Belgian canal which reached up to 8,700 ng/g ww [34] and from Flemish waters, where eels reach concentrations between 78 and 4,812 ng/g lw [35].

Concentrations of PCDD/PCDFs, indicator PCBs and dl-PCBs were highly positively correlated with lipid content ($p < 0.001$) and also with age (PCDD/PCDFs $p = 0.02$, indicator PCB $p = 0.004$, dl-PCBs $p = 0.003$). But while PCDD/PCDFs and dl-PCB were positively correlated with length ($p < 0.001$ and $p = 0.001$, respectively), indicator PCBs were not. Similarly, indicator PCBs were not correlated with weight (PCDD/PCDFs $p = 0.012$, dl-PCBs $p = 0.003$).

Dioxins, PCBs and other contaminants with endocrine disrupting properties may contribute to the decline of the eel populations [5]. Several studies tried to relate the contaminant burden of eels to enzymatic activity and transcriptomic responses. Pujolar et al. [21] found a correlation for the up-regulation of the transcription of genes associated with detoxification and a down-regulation of the transcription of genes associated with the oxidative phosphorylation pathway in eels with a PCB burden comparable to many North Rhine-Westphalian sites. Kammann et al. [36] investigated the ethoxyresorufin-O-deethylase (EROD) activity in eels from the Elbe and found that the EROD activity corresponded to PAH metabolite concentration in eel bile and to pollution levels from the sampling sites.

HCB

Mean HCB concentrations exceeded the EQS of 10 µg/kg ww at most sampling sites. At the Sieg, all eels met the EQS (mean value 3.4 µg/kg) and at the Berkel and Schwalm, more than 50% of all eels sampled complied with the EQS (mean values 6.2 and 8.8 µg/kg, respectively). Along the Rhine, mean HCB concentrations increased from 23.2 µg/kg at Bonn to 50.2 µg/kg at Emmerich (27.8 µg/kg at Düsseldorf and 35.7 µg/kg at Rees). Differences between Bonn and Emmerich were significant. Also, HCB concentrations differed significantly between the Rhine sites and the other rivers. The increasing HCB concentrations in eels along the Rhine can at least be partly attributed to former production processes of the chemical industry based in North Rhine-Westphalia. Although HCB concentrations were comparatively high in the Rhine, eels sampled in 2009 demonstrate a decrease of HCB over time which had already been detected in data compiled by Heinisch et al. [37] on HCB concentrations in Rhine eels from 1990/1991, 1995 and 1998/2000. As mentioned above, results in their paper are presented only as graphs and therefore approximations. They show a decrease from approximately 180 µg HCB/kg in 1990 to 125 µg/kg in 2000 for eels from the Rhine site Emmerich. In our study, HCB concentrations in eels from this site were further reduced by about 60% to 50 µg/kg. The reduction of the HCB contaminant levels in eels is also due to the fact that production of HCB in the chemical industry in the upper part of the Rhine catchment area ceased in the 1990s and that the use of HCB as a fungicide was banned in Germany in 1981.

Comparable values to the HCB concentrations in the Rhine or even higher concentrations have been determined in eels from Dutch waters. De Boer et al. [33] reported mean HCB concentrations of 22 to 243 µg/kg lw. The latter value originated from eels from the Dutch part of the Rhine from the year 2004. The corresponding mean concentrations from this study for Emmerich, south of the Dutch border, of 178 µg/kg lw and for Rees, further south of the border, 155 µg/kg lw are in good agreement with their results. At other European waters, HCB concentrations in eels were far lower, e.g. in Italian lagoons (1 to 4 µg/kg lw [6]) or Scottish waters where HCB levels often were below the quantification limit [23].

As expected from its lipophilic nature, HCB was positively correlated with lipid content ($p < 0.001$). It was also positively correlated with length ($p = 0.006$), but not with the other biometric parameters.

Sum of PBDE

The PBDE concentrations always exceeded the EQS of 0.0085 µg/kg ww for the substances established by EU Directive 2013/39/EU by a factor of 10^3 to 10^5 (range of mean concentrations 9.2 to 241.9 µg/kg). Highest concentrations were found in eels from the Lippe, where an industrial complex with mainly chemical industry has been established for a long time. Means for the sampling sites Lippe (241.9 µg/kg), Rees (74.6 µg/kg) and Emmerich (76.2 µg/kg) - which are the Rhine sampling sites downstream of the confluence of the Lippe - differed significantly from all other sampling sites. Sühring et al. [22] investigated BDE-47, BED-66, BDE-99, BDE-100, BDE-153, BDE-154 and BDE-183 in yellow and silver eels from the Elbe and in silver eels from the upper part of the Rhine and found mean values of 8.9 and 8.3 µg/kg for yellow eels and silver eels of the Elbe and 21.3 µg/kg for the Rhine. PBDE concentrations in the range of 10 to 100 µg/kg ww (sum of BDE-28, BDE-47, BDE-49, BDE-66, BDE-85, BDE-99, BDE-100, BDE-153, BDE-154, BDE-183 and BDE-209) have been reported from a number of Belgian waters [38], with a maximum value of 5,811 µg/kg lw (for comparison, Lippe 242 µg/kg ww =1,242 µg/kg lw). In eels from 60 Flemish sites sums for 14 BDE congeners (mainly congeners 47, 99 and 100) ranged between 12 and 1,400 µg/kg with a mean of 110 µg/kg [35]. PBDE concentrations from Irish waters were lower and ranged from 1.0 to 7.1 µg/kg ww for the sum of 11 PBDE congeners [27].

In this study, the sum of 6 PBDEs was significantly positively correlated with the lipid content ($p = 0.006$) and weakly negatively correlated with the condition factor ($p = 0.05$) but not with any other of the biometric parameters.

There is still few evidence for possible effects of PBDEs at environmental concentrations, but Tomy et al. [39] found evidence for lower plasma levels of thyroxine in Lake Trout with BDE concentrations similar to those found in eel from the Lippe and the Rhine sites Rees and Emmerich. Kierkegaard et al. [40] investigated the effects of BDE-209 on Rainbow Trout over a period of 120 days. They detected significantly reduced numbers of leucocytes and lymphocytes and an increase of liver weight. But they used high doses resulting in a 5 to 25 times higher body burden than the sum of PBDEs measured in the eels from North Rhine-Westphalia.

PFOS

Also, the PFOS concentrations often exceeded the EQS of 9.1 µg/kg ww with mean concentrations between 8.3 (Berkel) and 49.0 µg/kg (Niers). High concentrations were also found in the Rur (42.3 µg/kg). Mean values from the Niers and the Rur differed significantly from most other rivers but not from the Rhine eels and those from the Wupper. Only five eels were collected from the Niers, so the mean value for this site should be interpreted with caution. Nevertheless, the high PFOS concentration could be taken as an indication of the high proportion of municipal waste water present in the Niers. In North Rhine-Westphalia, the main source for PFOS is the plating industry. Until 2008, PFOS was widely used as fire extinguishing agent [41]. PFOS is also often present in consumer goods. As mentioned above, the Rur has been affected by paper industry emissions. PFOS has been used in the production process of special papers which explains the high PFOS concentrations in eels from the Rur. For eels, only data on PFOS contents in the liver have been published which are of limited value for the interpretation of tissue concentrations. Roland et al. [15] determined mean liver concentrations between 31 and 390 µg/kg from eels collected from Belgian waters. Hoff et al. [16] measured mean hepatic PFOS concentrations of 1,387 µg/kg ww in eels from four heavily polluted Belgian water bodies and related them to the high concentrations to fluorochemical production units in the vicinity of the sampling sites. In this study, they found a strongly significant correlation between liver PFOS concentrations and serum alanine aminotransferase activity which they interpreted as indication for induction of hepatic damage. In North Rhine-Westphalia, a survey on PFOS in the muscle tissues of several fish species, with eel being one of them, was conducted [41]. The median concentration of all fishes from surveillance monitoring was 7.1 µg/kg, and the median concentration from impacted sites was 24.4 µg/kg. Hloušková et al. [26] investigated PFT levels in a mixed sample consisting mainly of bream (*Abramis brama*), European chub (*Squalius cephalus*) and Roach (*Rutilis rutilus*) and found a median concentration of 5.7 µg/kg. The results of both studies indicate that the PFOS concentrations measured in eels during this study are comparatively high.

In this study, PFOS did not show any significant correlation with any of the biometric parameters.

Mercury

For mercury concentrations, in accordance with results from other European waters, all eels from North Rhine-Westphalia had concentrations below the EU limit for mercury in fish for human consumption (1.0 mg/kg ww) but above the EQS of 20 µg/kg ww for secondary poisoning. Highest concentrations were measured in eels from the Wupper (mean concentration 0.314 mg/kg) and the Rhine (Rees 0.273 mg/kg, Emmerich 0.195 mg/kg, Bonn 0.210 mg/kg and Düsseldorf 0.165 mg/kg) and lowest concentrations for eels from the Sieg (0.069 mg/kg) The mean concentrations from the Wupper eels and from the Rhine eels at Rees differed significantly from most other sampling sites but not from each other. Eels in two Spanish rivers had comparable mean mercury concentrations between 0.155 and 0.533 mg/kg [42], and in Portuguese lagoons, eels contained between 0.55 and 0.285 mg Hg/kg [43]. Noël et al. [25] reviewed further data from the studies in other countries including Poland, the Czech Republic, Germany, UK, Slovenia and Hungary and reported a range of values between 10 and 800 µg/kg. In their study of French rivers, they found mean concentrations of 0.199 mg/kg and reported a strong positive correlation between Hg concentration and the condition factor ($p < 0.0001$) but no correlation between Hg levels and body weight or length. In this study, mercury concentrations were significantly positively correlated with age ($p = 0.029$). Also, there was a weak positive correlation with length ($p = 0.049$) and weight ($p = 0.055$) but not with the condition factor. After a review of experimental data on Hg effects on fish, Sandheinrich and Wiener [44] estimated that the threshold value for negative effects of Hg on fish is between 0.3 and 0.7 mg Hg/kg in the whole body homogenates which are at least about a quarter lower than the concentrations in the filet. Therefore, a deleterious effect of the Hg concentrations measured in the eels from this study is unlikely.

PCB congener patterns

Based on the eel data for PCB contamination, the Rhine sites Bonn and Emmerich and the Rur were chosen for a comparison between PCB congener patterns in eel and in the suspended matter from corresponding sampling sites. In the Rhine and the Rur, PCB 153 comprised 40% to 43% of the sum of indicator PCBs in eel, PCB 138 accounted for 23% to 28% and PCB 180 for 12% to 16%. PCB 101 and PCB 52 had a slightly higher proportion in eels from Emmerich (16% and 8%, respectively) compared to eels from Bonn and from the Rur (7% to 9% and 3% to 5%). PCB 28 did not exceed 1% (Figure 2). The pattern of PCBs in suspended matter varied slightly with PCB 138 being the most dominant congener in the Rhine while in the Rur, PCB 153 had a slightly higher percentage than PCB 138. PCB 28 had mean percentages of 8% to 9% in suspended matter and therefore higher concentrations than in eels. According to Tapie et al. [20], PCB 153 is not regarded as being metabolisable and therefore an indicator for bioaccumulation. Müller et al. [45] examined PCB concentrations in eels and in sediment samples in the urban water bodies of the city of Berlin, Germany. They found similar PCB patterns in eels and in the sediment with PCB 138 and 153 dominating in eels and in the sediment while PCB 28 was more prominent in sediment samples compared to eel tissue. On the whole, PCB patterns in suspended matter and in biota found in this study were similar to PCB patterns in eels recorded from many European waters and also from North American water bodies reflecting the commercial PCB mixtures used in the past. Examples are data for eels from the Loire [7], from Scottish water bodies [23] and also for the American eel *Anguilla rostrata* from the Hudson River [46].

The bioconcentration ratio between contaminant content of suspended matter and eel tissue was calculated

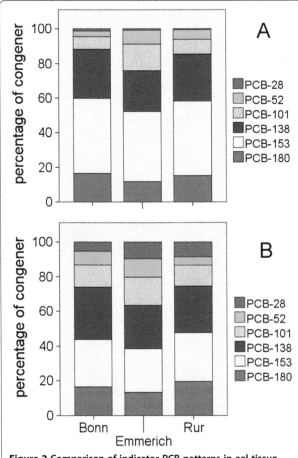

Figure 2 Comparison of indicator PCB patterns in eel tissue and in suspended matter. Congener patterns are presented for the Rhine at Bonn and Emmerich and the Rur **(A)** congener pattern in eel and **(B)** congener pattern in suspended matter.

between PCB concentration of suspended matter expressed as dry weight and PCB concentration of eel expressed as dry weight and as lipid weight. The dry weight ratio for the Rhine at Bonn was 39 (76 for lw), at Emmerich 66 (116) and in the Rur 54 (92). Harrad and Smith [47] reported concentration ratios between sediment (normalised for the carbon content of the sediment) and eel tissue for individual PCB congeners in the range of <1 and 10. The ratios found in this study are far higher which might be explained by the 10 to 10^2 higher PCB concentrations in North Rhine-Westphalian rivers compared to the River Severn investigated by Harrad and Smith [47]. As mentioned above, Müller et al. [45] also determined PCB concentrations in eels and in sediment samples from corresponding sampling sites. They expressed all eel data per wet weight (ww) and found a ratio of approximately 5. The water bodies examined had comparable PCB levels to the suspended matter of the Rur. In this study, the dry weight of the eels investigated was approximately 30% of the wet weight. Based on this estimate, eels from Berlin waters had an eel to

sediment ratio of approximately 15 which is still below the values found in this study.

Considering the 12 dl-PCB congeners, PCB 118 was clearly dominating with 50% to 67% of the sum of dl-PCBs. PCB 105, 156 and 167 were present with 5% to 15%. All other PCB congeners were present in negligible concentrations (Figure 3). This congener pattern is quite common in eels from European waters and has also been reported, e. g. from Belgium [5] and Portugal [48].

PBDE congener patterns

PBDE patterns in eel and in suspended matter were compared for the Rhine sites Düsseldorf and Emmerich and the Lippe (Figure 4). Lippe eels had very high PBDE concentrations, and as a consequence, eels at the Rhine sites below the confluence of the Lippe had an elevated PBDE content. Therefore, the sampling sites at Düsseldorf upstream of the confluence of the Lippe and at Emmerich downstream of the Lippe were chosen to track a possible influence of the Lippe contamination on the Rhine. The comparison was limited to BDE-28, BDE-47, BDE-99, BDE-100, BDE-153, BDE-154 and BDE-209 as these were the only BDE congeners determined in suspended matter.

At all sites, BDE-47 was the dominant congener with 60% to 62% at the two sites at the Rhine and 74% in the Lippe. BDE-100 was present with 28% to 30% in the Rhine and 17% in the Lippe. The other congeners were present in minor concentrations. A similar congener pattern with BDE-47 dominating and BDE-100 as the second most important congener was reported from the eel populations from the Scheldt [38], from the Elbe [49] and by Sühring et al. [50] for eel populations from several European rivers. In contrast, Sühring et al. [50] found equal concentrations of BDE-47 and BDE-100 in yellow eels from North America. They related this difference to the continuous exposition of American eels to technical penta- and octa-PBDE mixtures which had been banned in Europe in 2004. In contrast to the biota PBDE pattern, the Rhine suspended matter comprised almost exclusively of BDE-209 (95% to 97%), reflecting the continuous use of deca-BDE as fire protection agent in consumer goods. In the suspended matter of the Lippe, BDE-209 accounted for only 25%, and BDE-99 and BDE-47 were present with 35% and 24%, respectively. This conspicuous PBDE pattern, which can be related to the influence of local chemical industry, did not lead to a change in the PBDE pattern in eels. Roosens et al. [51] investigated PBDE patterns in eels and in the sediment from various locations in Flemish water bodies. They also found high BDE-209 concentrations in the sediment and only rarely found elevated BDE-209 concentrations in eels. They attributed this discrepancy to a poor uptake of BDE-209 from the sediment, a low solubility in water and a high degradation in the fish. The debromination of BDE-209 to lower brominated congeners

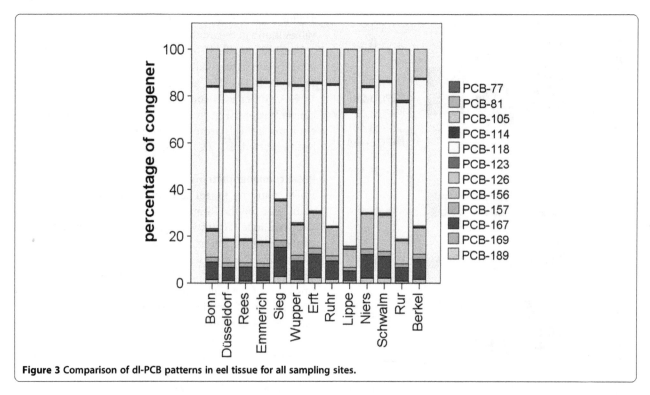

Figure 3 Comparison of dl-PCB patterns in eel tissue for all sampling sites.

has been described by several authors, e. g. [40,52-54] and seems to be a major source of penta-PBDEs in fish.

Ratios between PBDE concentrations in eel tissue and in suspended matter ranged between 0.6 (Düsseldorf) and 2.6 (Lippe) for eel dry weight and 1.5 (Düsseldorf) and 4.9 (Lippe) for eel lipid weight. These ratios are considerably lower than the ratios calculated for PCBs despite the fact that the sum of PBDEs was significantly positively correlated with the lipid content. The low ratios can be attributed to the poor transfer of BDE-209 congener from the suspended matter/sediment to the eels. This is in agreement with the higher eel to suspended matter ratios in the Lippe. According to Mariottini et al. [55], BDE-47 is considered to have the highest bioconcentration factor of PBDE congeners. In the Lippe suspended matter, BDE-47 was present in approximately the same concentrations as BDE-209.

Figure 5 shows the pattern of BDE-28, BDE-47, BDE-49, BDE-66, BDE-71, BDE-99, BDE-100, BDE-153, BDE-154 and BDE-209 in eel tissue for all sampling sites. The pattern resembles the profile depicted in Figure 4 although additional BDE congeners were included in the calculation. At all sites, BDE-47 dominated the BDE pattern, with BDE-100 being the second most important congener. Together, these two congeners accounted for 85% to 94% of the BDEs measured.

Conclusions

Eels from North Rhine-Westphalian waters showed a high level of contamination. Over the last decades, PCB

and HCB concentrations have decreased in the Rhine and its major tributaries and subsequently in eels and other fish but were still present in concentrations well above the EQS of the EU. In 2012, as a consequence of the high PCB contamination, the North Rhine-Westphalian government issued a recommendation not to consume wild eels from North Rhine-Westphalian waters [56].

Recent research has accumulated evidence that pollution may contribute to the decline of eel populations. For PCDD/PCDFs, PCBs and PBDEs, there are investigations indicating effects on the immune system, the reproductive system and the endocrine system of the fish. Due to the complex life cycle of eels and due to the large number of other simultaneously acting environmental factors, unequivocal evidence for the direct impact of pollutants on population dynamics is difficult. But it might be taken as an indication that the decrease in recruitment in the populations of the European eel during the last 30 years coincided with a strong intensification of agriculture and with the industrial production of a plethora of new substances.

Eels are prone to accumulate contaminants due to their high body lipid content, their longevity and their benthic way of life. Hence, the contamination levels of the eels in North Rhine-Westphalian waters reflected the contamination of the water bodies investigated. The influence from ambient contamination levels was particularly striking for eels from the four sampling sites at the Rhine as their contamination concentrations reflected the increasing pollution levels for most substances along the

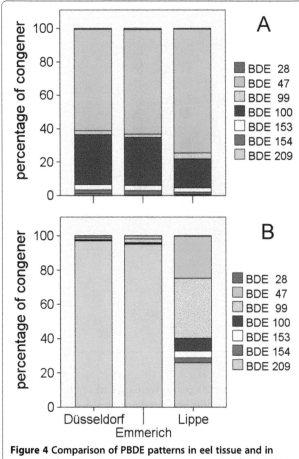

Figure 4 Comparison of PBDE patterns in eel tissue and in suspended matter. Congener patterns are presented for the Rhine at Bonn and Emmerich and the Rur (BDE-28, BDE-47, BDE-99, BDE-100, BDE-153, BDE-154 and BDE-209) **(A)** congener pattern in eel and **(B)** congener pattern in suspended matter.

river course. For PBDEs, the rise of mean concentrations between Düsseldorf and Rees is a result of the load of the River Lippe, which reaches the Rhine between the two sampling sites. PFOS and mercury concentrations in eels did not increase along the North Rhine-Westphalian stretch of the Rhine. Contamination patterns at certain sampling sites could be traced back to local point sources as was shown for the conspicuous PCB pollution of the Rur.

Methods
Sampling of eels
In order to set up eel management plans in 2009 and 2010, the North Rhine-Westphalian government commissioned the Rhineland Fishery Association to carry out a survey of eel populations in ten major rivers. Eel populations were monitored using point abundance sampling with electrofishing equipment from a boat (electrofishing equipment EFKO FEG 8000, Leutkirch, Germany 8 KW,

up to 600 V, direct current, 60 points with a distance of 3 m between points). The length of each eel caught was measured before the eels were released again. As part of the survey, a subsample of 124 eels was collected from 13 sites in 10 rivers (Sieg, Wupper, Ruhr, Lippe, Erft, Rur, Niers, Schwalm, Berkel and four sites - Bonn, Düsseldorf, Rees and Emmerich - at the Rhine, Figure 1) for biometrical measurements and for inspection of the health status. Sites were generally situated in the downstream regions of the rivers, apart from the sites at the Rhine which covered the North Rhine-Westphalian stretch of the river. Usually, ten to eleven eels were taken, but as there were only small numbers of eels at the sampling sites at the Niers, the Schwalm and the Berkel five eels were taken at the Niers and eight eels at the Schwalm and the Berkel. The eels collected roughly represented the size-class distribution of the population at the sampling site. The eels were killed immediately by over narcotisation with 300 mg Benzocain/l (Merck, Darmstadt, Germany), and transported frozen to the lab of the North Rhine-Westphalian State Agency for Nature, Environment and Consumer Protection (LANUV NRW). The sampling and killing was carried out by a licensed fish biologist. In the lab, the length, weight and sex were determined, and the condition factor was calculated (condition factor C_F = weight (g) × 100/ length (cm)^3). Heads were dissected and sent to the Thünen Institute of Fisheries Ecology, Hamburg, Germany, where the age was determined using otolith preparation following ICES [57]. The developmental stage was determined using the eye index according to Durif et al. [58]. The eels were with very few exceptions in stages 1 to 3. Eels from the Schwalm and from the lower Rhine were predominantly in stage 3 which is according to Durif et al. [58], the stage just before the onset of metamorphosis. At all other sampling sites, stages 1 to 3 were present in about equal numbers. It was decided to take tissue samples of the yellow eels for the analysis of contaminants. Five eels which were judged to be silver eels according to their eye index were excluded from further analysis (one each from the Ruhr, the Erft and the Rhine sites Bonn, Düsseldorf and Rees). Therefore, 119 eels altogether were chemically analysed. The left filet of each specimen was sampled, including the skin (at least 100 g; from a large specimen, the middle part of the filet was chosen), and stored frozen.

Analysis of eels
In 2011, the tissue samples were analysed for PCDDs, PCDFs, PCBs, PBDEs, HCB, mercury, perfluorooctanoic acid (PFOA) and PFOS by Eurofins GfA Lab Service (Hamburg, Germany). Prior to further treatment for analysis of organic contaminants, eel samples were lyophilized and homogenised. For dry matter determination, lyophilised subsamples were subjected to a drying procedure at 103°C for at least 4 h. Dry matter content was calculated

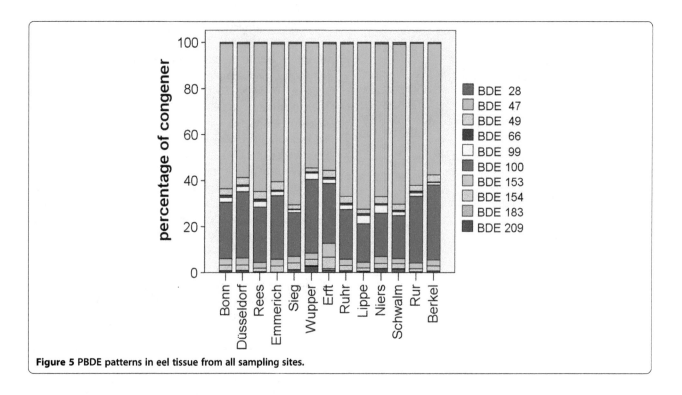

Figure 5 PBDE patterns in eel tissue from all sampling sites.

taking into account the dry matter content after lyophilisation as well as the dry matter after the drying procedure.

PCDD/PCDFs, dl-PCBs and indicator PCBs

In brief, samples were analysed for 17 2,3,7,8-substituted congeners of PCDDs and (PCDFs, 12 dl-PCBs and the 6 indicator PCBs 28, 52, 101, 138, 153 and 180 using high-resolution gas chromatography and high-resolution mass spectrometry (HRGC/HRMS) analogue to a method described by Neugebauer et al. [59]. For each native PCDD/PCDF and PCB congener to be determined, the corresponding $^{13}C_{12}$- internal standard was added to the samples before the extraction procedure. After spiking, the samples were Soxhlet-extracted overnight with appropriate organic solvents for ultratrace-analyses (e.g. toluene). Lipid determination was performed gravimetrically in these crude extracts as extractable lipids. Subsequently, these extracts were subjected to a cleanup procedure using a multi-column chromatography system (involving carbon-on-glass fibre or carbon-on-celite for PCDD/PCDF and PCB). Further $^{13}C_{12}$-labelled internal standards were added to the extracts for the determination of the recovery of the internal standards added before. Analyses were performed by HRGC/HRMS (Waters Autospec, Eschborn, Germany or DFS mass spectrometers, Thermo Fisher Scientific, Bremen, Germany). For each substance, two isotope masses were measured. The quantification was carried out by means of isotope dilution analysis with the use of internal and external standards. In addition to values for individual congeners, calculations of the TEQ according to

the WHO-system 2005 [60] were carried out by taking into account the quantification limit for non-quantifiable compounds (upperbound procedure).

PBDEs

Samples were analysed for 24 congeners of PBDEs (including BDE-17, BDE-28, BDE-47, BDE-49, BDE-66, BDE-71, BDE-77, BDE-85, BDE-99, BDE-100, BDE-119, BDE-126, BDE-138, BDE-153, BDE-154, BDE-156, BDE-183, BDE-184, BDE-191, BDE-196, BDE-197, BDE-206, BDE-207 and BDE-209) using gas chromatography and mass spectrometry (GC/MS) analogue to a method by Päpke et al. [61].

A mixture of $^{13}C_{12}$-labelled internal standards was added to the samples before the extraction procedure. After spiking, the samples were extracted by means of Soxhlet extraction using a mixture of appropriate polar and non-polar solvents for ultratrace-analyses (e.g. hexane/acetone). After extraction, a cleanup procedure was performed using concentrated sulfuric acid additionally to column chromatography involving activated silica gel resp. alumina. Further $^{13}C_{12}$-labelled internal standards were added to the extracts for the determination of the recovery of the internal standards added before. PBDEs were analysed by means of GC/MS. For each substance, two isotope masses were measured. The quantification was carried out by means of isotope dilution analysis with the use of internal and external standards.

As many of the BDEs analysed were always or almost always below the limit of quantification, only the results for BDE-28, BDE-47, BDE-49, BDE-66, BDE-99, BDE-100,

BDE-153, BDE-154, BDE-183 and BDE-209 are presented in this publication.

PFOS/PFOA

A mixture of $^{13}C_{12}$-labelled internal standards was added to the samples before the extraction procedure. After spiking, the samples were extracted by means of ultrasonic extraction with appropriate polar solvents (e.g. methanol) for ultratrace-analyses (e.g. nanograde). After extraction, the cleanup was carried out involving carbon black. A further $^{13}C_{12}$-labelled internal standard was added to the extracts for the determination of the recovery of the internal standards added before. The analysis was performed using liquid chromatography and tandem mass spectrometry (LC/MS-MS). For each substance, two isotope masses were measured. The quantification was carried out with the use of internal and external standards. As values for PFOA rarely exceeded the limit of quantification, only the results for PFOS are presented in this publication.

HCB

Before extraction, the eel samples were dried by thorough mixing with anhydrous sodium sulfate and homogenised. A mixture of $^{13}C_{12}$-labelled internal standards was added to the samples before the extraction procedure. After spiking, the samples were extracted with a mixture of appropriate polar and non-polar solvents for ultratrace-analyses (e.g. hexane/acetone). The extraction was followed by a cleanup procedure using a column system (involving florisil resp. alumina). Further $^{13}C_{12}$-labelled internal standards were added to the extract for the determination of the recovery of the internal standards added before. The measurement was taken by means of HRGC/HRMS. For each substance, two isotope masses were measured. The quantification was carried with the use of internal and external standards.

Mercury (Hg)

Eel homogenate was digested by means of microwave using nitric acid. After reduction of mercury compounds by Tin(II)-chloride, the total mercury content was determined by means of cold vapor atomic absorption spectrometry (CV-AAS) following DIN EN 13806.

Sampling of suspended matter

Sampling and analysis of suspended matter was carried out by LANUV NRW as part of a separate surveillance monitoring program.

Suspended matter was collected using a flow through pump (Carl Padberg Cepa Z 61, Lahr, Germany) which was deposited in the water approximately 0.5 m below the surface. The time necessary to collect sufficient material was estimated by measuring the turbidity of the water (WTW IQ Sensor Net, System 184 with sensor with VISOTurb 700/Q, Weilheim, Germany). Usually, the pump collected material over 21 to 24 h with a flow rate of 1,000 to 1,100 l/hr. The material was weighed in the field and transported refrigerated to the lab where it was freeze-dried, ground by a grinding mill equipped with a zirconium dioxide mortar and pestles (KM1, Retsch, Haan, Germany) and sieved for the 63-µm fraction.

Data for suspended matter were available as annual means of usually between two and five individual samples per site. As sampling sites were sampled in different years, values for suspended matter were calculated as means for all annual means available between 2005 and 2012 (3 to 8 values).

Analysis of suspended matter

PBDEs

PBDEs in suspended matter were identified using $^{13}C_{12}$ standards purchased from LGC Standards, Wesel, Germany. PBDEs were analysed according to DIN EN ISO 22032.

In brief, PBDEs were extracted with toluene from the freeze-dried <63-µm fraction of the suspended matter using Soxhlet extraction. Subsequently, the crude extracts were purified using multi-layer column chromatography with silica gel. Activated alumina was used as a second purification step. Subject to the grade of purification, these steps were repeated. The purified extracts were analysed by gas chromatography and mass spectrometry (GCMS Thermo Finnigan 2000, Thermo Fisher Scientific, Bremen, Germany).

PCDD/PCDFs and PCBs

Quantification of PCDD/PCDFs and PCBs in suspended matter was performed using isotope dilution analysis according to EN 1948 part 2, 3 and 4. The $^{13}C_{12}$ standards were purchased from Cambridge Isotopes Laboratories (Andover, USA).

In brief, the freeze-dried <63-µm fraction of the suspended matter was extracted with toluene using Soxhlet extraction. Prior to extraction, the samples were spiked with $^{13}C_{12}$-quantification standard mixtures. The crude sample extracts were subjected to a cleanup procedure consisting of a solid-phase multi-layer column chromatography with silica modified with 44% H_2SO_4, 33% NaOH or 10% $AgNO_3$. Furthermore, PCBs and PCDD/PCDFs were separated using basic alumina. Subsequently, the eluates were concentrated via rotary evaporation and nitrogen flow to a final volume of approximately 10 to 20 µl (PCDD/PCDF) or 100 µl (PCB). After addition of recovery standards, PCDD/PCDFs and dl-PCB were analysed using high-resolution gas spectrometry (model 6890 Series 2, Agilent, Santa Clara, USA) and mass spectrometry (model DFS, Thermo Fisher Scientific, Bremen,

Germany). Indicator PCBs were analysed using gas chromatography (model 6890 N, Agilent, Santa Clara, USA) and low-resolution mass spectrometry (model 5973 N, Agilent, Santa Clara, USA). A detailed description of the analytical method can be found in Klees et al. [62].

Quality assurance
Analysis of eels
Quality control for the analysis of PCDD/PCDFs and PCBs was carried out in accordance with EU Regulation 1883/2006. According to these requirements, the limit of quantification (LOQ) of an individual congener was the concentration of an analyte in the extract of a sample which produces an instrumental response at two different ions to be monitored with a signal/noise (S/N) ratio of 3:1 for the less sensitive signal and fulfillment of the basic requirements such as, e.g. retention time, isotope ratio according to the determination procedure as described in EPA method 1613 revision B. Additionally, the recoveries of the individual $^{13}C_{12}$-labelled internal standards were checked to be in the range of 60% to 120%. Lower or higher recoveries for individual congeners were accepted on the condition that their contribution to the TEQ value did not exceed 10% of the total TEQ value. The analytical system was calibrated using an eight-point calibration, followed by checking with single calibrations in regular intervals within each measuring sequence. Quantification of the individual PCDD/PCDF and PCB congeners was based on daily generated responses. Method blanks including extraction, cleanup and measuring were routinely monitored. Furthermore, precision and accuracy were checked by analysing in-house quality assurance pool samples within each batch of samples consisting of not more than 12 samples. The pool sample used in this project consisted of combined fish meal specimens having been determined beforehand with the quality of the analysis data being assured by means of control charts. In addition, precision and accuracy were weekly checked by analysing the certified reference material EDF-2525 (provided by Community Bureau of Reference - BCR, Belgium).

Quality control for the analysis of PBDEs was carried out as follows: next to the comparison of the retention time between native and $^{13}C_{12}$-labelled analytes (including $^{13}C_{12}$-BDE-209), relative isotope ratios were considered with a tolerance of 20% for peak identification. Limits of detection (LOD) were reached if peaks showed signals to be three times the baseline noise, and LOQ were set to be ten times the baseline noise. Additionally, the standard recovery rates of $^{13}C_{12}$-labelled quantification standards were checked to be in the range of 50% to 120%. Special attention with regard to recovery rates and peak shape was paid to BDE-209. The analytical system was calibrated using a ten-point calibration, followed by checking with single calibrations in regular intervals within each measuring sequence. Quantification of the individual PBDE congeners was based on daily generated responses. Method blanks including extraction, cleanup and measuring were monitored in parallel to each batch of samples consisting of not more than 12 single samples. Furthermore, precision and accuracy were checked by analysing in-house quality assurance pool samples within each batch of samples. The pool sample used in this project consisted of combined fish oil specimens having been determined beforehand with the quality of the analysis data being assured by means of control charts. In addition, precision and accuracy were weekly checked by analysing the certified reference material EDF-2525 (provided by Community Bureau of Reference - BCR, Belgium).

The quality control for the analysis of HCB and PFOS/PFOA was carried out similarly to the quality control measures applied to the analysis of PBDEs. For HBC, the analytical system was calibrated using a nine-point calibration. Precision and accuracy were checked by analysing in-house quality assurance pool samples consisting of combined feeding stuff specimens as well as certified standard solutions and also on a weekly basis by analysing the certified reference material EDF-2525 (provided by Community Bureau of Reference - BCR, Belgium). For PFOS/PFOA, in addition to retention time and relative isotope ratios, the ratio of the signal intensities of the two detected transitions was checked. The standard recovery rates of $^{13}C_{12}$-labelled quantification standards were checked to be in the range of 50% to 150%. Precision and accuracy were checked by analysing homogenised fish material of a previous inter-laboratory testing study.

For quality control of the analysis of mercury method blanks as well as internal reference material being monitored by means of a control chart were checked.

Analysis of suspended matter
Quality control for the analysis of PCBs and PCDD/PCDFs was carried out in accordance with DIN EN 1948 3-4. Here, next to the comparison of the retention time between native and $^{13}C_{12}$-labelled analytes, relative isotope ratios were considered with a tolerance of 20% for peak identification as additional criteria. LOD were reached if peaks showed to be three times the baseline noise, and LOQ were set to be ten times the baseline noise. Additionally, the standard recovery rates of $^{13}C_{12}$-labelled quantification standards were checked. The analytical system was calibrated using a five-point calibration, and quantification for the individual PCB and PCDD/PCDF congeners was based on daily generated response. The calibration and also the generated response factors were considered to be valid as long as a deviation under 20% for a single calibration was determined [DIN EN 19481]. Method blanks including extraction, cleanup and

analysis were routinely monitored. Furthermore, precision and accuracy were checked by analysing the NIST Standard Reference Material (SRM) 1649a.

Quality control for the analysis of PBDEs was carried out similarly to the quality control measures for PBDE analysis in biota with the following exceptions: limits of quantification were determined by comparing the signal/noise in 10 real suspended matter samples and a quantification limit $\geq 6^*$ baseline signal (600-ZUA-VA-007). The standard recovery rates of $^{13}C_{12}$-labelled quantification standards were checked to be in the range of 50% to 120%. The analytical system was calibrated using a five-point calibration, checked in each series of measurements by five standard solutions. Method blanks including extraction, cleanup and measuring were monitored in parallel to each batch of samples consisting of not more than 6 single samples with a tolerance of $\leq 1/3$ of quantification limit.

Statistical analysis

Statistical analysis was performed using the statistical software SAS 9.2 and SPSS 19. The Shapiro-Wilk test demonstrated that the data of the contaminants were not normally distributed. For differences between mean contaminant concentration (least squares means) of sampling sites, an analysis of covariance with age and lipid content as covariates and Tukey's test were performed (α =0.1). Significant differences between sampling sites were checked using the Kruskall-Wallis test. For correlations between biometric parameters and contaminants Pearson's correlation coefficients were used. As n >100, an approximately normal distribution was assumed and significance levels could be calculated. To test the influence of biometric parameters on contaminant concentration, an analysis of covariance was performed between each biometric parameter and the contaminants after eliminating the influence of the sampling sites.

To analyse the differences of mean contaminant concentrations in different parasite classes, an analysis of variance was performed followed by Tukey's test. To compare the parasite patterns of the sampling sites, a Monte Carlo simulation of Fisher's exact test was used. Logistic regressions between parasite infection and biometric parameters or pollutants were calculated. For all the tests, apart from Tukey's test on differences on mean contaminant concentration of sampling sites, a significance level of α =0.05 was set.

Competing interests

The authors declare that they have no competing interests.

Authors' contributions

BG compiled and analysed the data. FJS participated in the study coordination and carried out the biometric investigations. GS performed the statistical analyses. The final manuscript was read and approved by all authors.

Acknowledgements

The authors wish to acknowledge the help of Peter Perkons, LANUV, with preparation of the figures and tables. They would also like to thank Nina Lohmann, Eurofins GfA Lab Service, Germany, and Ernst Hiester, Marcel Klees, Klaus Sielex and Paul Bachhausen (all LANUV) for revision of the chapter on analytical methods.

References

1. ICES: *Report of the 2013 Session of the Joint EIFAC/ICES Working Group on Eels. 18-22 March 2013, Sukarietta, Spain, 4–10 September 2013, ICES CM 2013/ACOM:18.* Copenhagen: ICES; 2013.
2. ICES: *Report of the 2006 session of the Joint EIFAC/ICES Working Group on Eels, FAO European Inland Fisheries Advisory Commission; International Council for the Exploration of the Sea. Rome, 23-27 January 2006. EIFAC Occasional Papaer No. 38. ICES CM 2006/ACFM:16.* Rome: FAO/Copenhagen; 2006:352.
3. Belpaire C, Goemans G, Geeraerts C, Quataert P, Parmentiert K, Hagel P, De Boer J: **Decreasing eel stocks: survival of the fattest?** *Ecol Freshw Fish* 2009, **18**:297–314.
4. Palstra AP, Heppener D, van Ginneken V, Székely C, van den Thillart GEEJM: **Swimming performance of silver eels is severely impaired by the swim-bladder parasite Anguillicola crassus.** *J Exp Mar Biol Ecol* 2007, **352**:244–256.
5. Geeraerts C, Focant JF, Eppe G, de Pauw E, Belpaire C: **Reproduction of European eel jeopardised by high levels of dioxins and dioxin-like PCBs?** *Sci Total Environ* 2011, **409**:4039–4047.
6. Corsi I, Mariottini M, Badesso A, Caruso T, Borghesi N, Bonacci S, Iacocca A, Focardi S: **Contamination and sub-lethal toxicological effects of persistent organic pollutants in the European eel (Anguilla anguilla) in the Orbetello lagoon (Tuscany, Italy).** *Hydrobiologia* 2005, **550**:237–249.
7. Blanchet-Letrouvé I, Zalouk-Vergnoux A, Vernisseau A, Couderc M, Le Bizec B, Elie P, Herrenknecht C, Mouneyrac C, Poirier L: **Dioxin-like, non-dioxin like PCB and PCDD/F contamination in European eel (Anguilla anguilla) from the Loire estuarine continuum: spatial and biological variabilities.** *Sci Total Environ* 2014, **472**:562–571.
8. Geeraerts C, Belpaire C: **The effects of contaminants in European eel: a review.** *Ecotoxicology* 2010, **19**:239–266.
9. van Ginneken V, Palstra A, Leonards P, Nieveen M, van den Berg H, Flik G, Spanings T, Niemantsverdriet P, van den Thillart G, Murk A: **PCBs and the energy cost of migration in the European eel (Anguilla anguilla L.).** *Aquat Toxicol* 2009, **92**:213–220.
10. Hamers T, Kamstra JH, Sonneveld E, Murk AJ, Kester MH, Andersson PL, Legler J, Brouwer A: **In vitro profiling of the endocrine-disrupting potency of brominated flame retardants.** *Toxicol Sci* 2006, **92**:157–173.
11. Legler J: **New insights into the endocrine disrupting effects of brominated flame retardants.** *Chemosphere* 2008, **73**:216–222.
12. Ren X, Guo L: **Molecular toxicology of polybrominated diphenyl ethers: nuclear hormone receptor mediated pathways.** *Environ Sci Process Impacts* 2013, **15**:702.
13. Muirhead EK, Skillman AD, Hook SE, Schultz IR: **Oral exposure of BDE-47 in fish: toxicokinetics and reproductive effects in Japanese medaka (Oryzias latipes) and fathead minnows (Pimephales promelas).** *Environ Sci Technol* 2006, **40**:523–528.
14. Hoff P, van Dongen W, Esmans E, Blust R, de Coen W: **Evaluation of the toxicological effects of perfluorooctane sulfonic acid in the common carp (Cyprinus carpio).** *Aquat Toxicol* 2003, **62**:349–359.
15. Roland K, Kestemont P, Loos R, Tavazzi S, Paracchini B, Belpaire C, Dieu M, Raes M, Silvestre F: **Looking for protein expression signatures in European eel peripheral blood mononuclear cells after in vivo exposure to perfluorooctane sulfonate and a real world field study.** *Sci Total Environ* 2014, **468–469**:958–967.
16. Hoff P, Van Campenhout K, Van de Vijver K, Covaci A, Bervoets L, Moens L, Huyskens G, Goemans G, Belpaire C, Blust R: **Perfluorooctane sulfonic acid and organohalogen pollutants in liver of three freshwater fish species in Flanders (Belgium): relationships with biochemical and organismal effects.** *Environ Pollut* 2005, **137**:324–333.
17. Carlson E, Zelikoff JT: **The immune system of fish: a target organ of toxicity.** In *The Toxicology of Fishes.* Edited by Di Giulio RT, Hinton DE. Boca Raton: CRC Press; 2008:489–529.

18. Maes GE, Raeymaekers J, Hellemans B, Geeraerts C, Parmentier K, De Temmerman L, Volckaert F, Belpaire C: **Gene transcription reflects poor health status of resident European eel chronically exposed to environmental pollutants.** *Aquat Toxicol* 2013, **126:**242–255.

19. Stachel B, Christoph EH, Goetz R, Herrmann T, Krueger F, Kuehn T, Lay J, Loeffler J, Päpke O, Reincke H, Schröter-Kermani C, Schwartz R, Steeg E, Stehr D, Uhlig S, Umlauf G: **Dioxins and dioxin-like PCBs in different fish from the river Elbe and its tributaries, Germany.** *J Hazard Mater* 2007, **148:**199–209.

20. Tapie N, Le Ménach K, Pasquaud S, Elie P, Devier MH, Budzinski H: **PBDE and PCB contamination of eels from the Gironde estuary: from glass eels to silver eels.** *Chemosphere* 2011, **83:**175–185.

21. Pujolar JM, Milan M, Marino IAM, Capoccioni F, Ciccoti E, Belpaire C, Covaci A, Malarvannan G, Patarnello T, Bargelloni L, Zane L, Maes GE: **Detecting genome-wide gene transcription profiles associated with high pollution burden in the critically endangered European eel.** *Aquat Toxicol* 2013, **132–133:**157–164.

22. Sühring R, Möller A, Freese M, Pohlmann J-D, Wolschke H, Sturm R, Xie Z, Hanel R, Ebinghaus R: **Brominated flame retardants and dechloranes in eels from German rivers.** *Chemosphere* 2013, **90:**118–124.

23. Macgregor K, Oliver IW, Harris L, Ridgway IM: **Persistent organic pollutants (PCB, DDT, HCH, HCB & BDE) in eels (Anguilla anguilla) in Scotland: current levels and temporal trends.** *Environ Pollut* 2010, **158:**2402–2411.

24. Maes J, Belpaire C, Goemans G: **Spatial variations and temporal trends between 1994 and 2005 in polychlorinated biphenyls, organochlorine pesticides and heavy metals in European eel (Anguilla anguilla L.) in Flanders, Belgium.** *Environ Pollut* 2008, **153:**223–237.

25. Noël L, Chekri R, Millour S, Merlo M, Leblanc J, Guérin T: **Distribution and relationships of As, Cd, Pb and Hg in freshwater fish from five French fishing areas.** *Chemosphere* 2013, **90:**1900–1910.

26. Hloušková V, Lanková D, Kalachová K, Hrádková P, Poustka J, Hajšlová J, Pulkrabová J: **Occurrence of brominated flame retardants and perfluoroalkyl substances in fish from the Czech aquatic ecosystem.** *Sci Total Environ* 2013, **461–462:**88–98.

27. McHugh B, Poole R, Corcoran J, Anninou P, Boyle B, Joyce E, Barry Foley M, McGovern E: **The occurrence of persistent chlorinated and brominated organic contaminants in the European eel (Anguilla anguilla) in Irish waters.** *Chemosphere* 2010, **79:**305–313.

28. Miniero R, Guandalini E, Dellatte E, Iacovella N, Abate V, De Luca S, Iamiceli AL, di Domenico A, De Felip E: **Persistent organic pollutants (POPs) in fish collected from the urban tract of the river Tiber in Rome (Italy).** *Ann Ist Super Sanita* 2011, **47:**310–315.

29. Knutzen J, Bjerkeng B, Næs K, Schlabach M: **Polychlorinated dibenzofurans/dibenzo-p-dioxins (PCDF/PCDDs) and other dioxin-like substances in marine organisms from the Grenland fjords, S. Norway, 1975–2001: present contamination levels, trends and species specific accumulation of PCDF/PCDD congeners.** *Chemosphere* 2003, **52:**745–760.

30. Palstra AP, van Ginneken VJ, Murk AJ, van den Thillart GE: **Are dioxin-like contaminants responsible for the eel (Anguilla anguilla) drama?** *Naturwissenschaften* 2006, **93:**145–148.

31. Heinisch E, Kettrup A, Bergheim W, Wenzel S: **Persistent chlorinated hydrocarbons (PCHCs), source-orientated monitoring in aquatic media. 6. Strikingly high contaminated sites.** *Fresen Environ Bull* 2007, **16:**1248–1273.

32. Detzel A, Patyk A, Fehrenbach H, Franke B, Giegrich J, Lell M, Vogt R: *Ermittlungen von Emissionen und Minderungsmaßnahmen für persistente organische Schadstoffe in der Bundesrepublik Deutschland, Forschungsbericht 295 44 365, UBA-FB 98–115, UBA-Texte 74/98.* Berlin: Umweltbundesamt; 1998.

33. de Boer J, Dao QT, van Leeuwen SP, Kotterman MJ, Schobben JH: **Thirty year monitoring of PCBs, organochlorine pesticides and tetrabromodiphenylether in eel from The Netherlands.** *Environ Pollut* 2010, **158:**1228–1236.

34. Byer JD, Lebeuf M, Alaee M, Brown SR, Trottier S, Backus S, Keir M, Couillard CM, Casselman J, Hodson PV: **Spatial trends of organochlorinated pesticides, polychlorinated biphenyls, and polybrominated diphenyl ethers in Atlantic Anguillid eels.** *Chemosphere* 2013, **90:**1719–1728.

35. Malarvannan G, Belpaire C, Geeraerts C, Eulaers I, Neels H, Covaci A: **Assessment of persistent brominated and chlorinated organic contaminants in the European eel (Anguilla Anguilla) in Flanders, Belgium: levels, profiles and health risk.** *Sci Total Environ* 2014, **482–483:**222–233.

36. Kammann U, Brinkmann M, Freese M, Pohlmann J-D, Stoffels S, Hollert H, Hanel R: **PAH metabolites, GST and EROD in European eel (Anguilla**

37. anguilla) as possible indicators for eel habitat quality in German rivers. *Environ Sci Pollut Res* 2013, **21:**2519–2530.

37. Heinisch E, Kettrup A, Bergheim W, Martens D, Wenzel S: **Persistent chlorinated hydrocarbons (PCHC), source-orientated monitoring in aquatic media. 4. The chlorobenzenes.** *Fresen Environ Bull* 2006, **15:**148–169.

38. Roosens L, Dirtu AC, Goemans G, Belpaire C, Gheorghe A, Neels H, Blust R, Covaci A: **Brominated flame retardants and polychlorinated biphenyls in fish from the river Scheldt, Belgium.** *Environ Int* 2008, **34:**976–983.

39. Tomy GT, Palace VP, Halldorson T, Braekevelt E, Danell R, Wautier K, Evans B, Brinkworth L, Fisk AT: **Bioaccumulation, biotransformation and biochemical effects of brominated diphenyl ethers in juvenile Lake Trout (Salvelinus namaycush).** *Environ Sci Technol* 2004, **38:**1496–1504.

40. Kierkegaard A, Balk L, Tjärnlund U, De Wit CA, Jansson B: **Dietary uptake and biological effects of decabromodiphenyl ether in rainbow trout (Oncorhynchus mykiss).** *Environ Sci Technol* 1999, **33:**1612–1617.

41. LANUV: *Verbreitung von PFT in der Umwelt. Ursachen - Untersuchungsstrategie - Ergebnisse - Maßnahmen.* LANUV Fachbericht Nr. 34. Recklinghausen: LANUV; 2011:118.

42. Linde AR, Sanchez-Galan S, Garcia-Vazquez E: **Heavy metal contamination of European eel (Anguilla anguilla) and brown trout (Salmo trutta) caught in wild ecosystems in Spain.** *J Food Prot* 2004, **67:**2332–2336.

43. Eira C, Torres J, Miquel J, Vaqueiro J, Soares A, Vingada J: **Trace element concentrations in Proteocephalus macrocephalus (Cestoda) and Anguillicola crassus (Nematoda) in comparison to their fish host, Anguilla anguilla in Ria de Aveiro, Portugal.** *Sci Total Environ* 2009, **407:**991–998.

44. Sandheinrich MB, Wiener JG: **Methylmercury in freshwater fish: recent advances in assessing toxicity of environmentally relevant exposures.** In *Environmental Contaminants in Biota: Interpreting Tissue Concentrations.* Edited by Beyer BN, Meador JP. Boca Raton: CRC Press; 2011:169–190.

45. Müller L, Neugebauer F, Fromme H: **Levels of coplanar and non-coplanar polychlorinated biphenyls (PCB) in eel and sediment samples from Berlin/Germany.** *Organohalogen Compd* 1999, **43:**397–400.

46. Ashley JT, Horwitz R, Steinbacher JC, Ruppel B: **A comparison of congeneric PCB patterns in American eels and striped bass from the Hudson and Delaware River estuaries.** *Mar Pollut Bull* 2003, **46:**1294–1308.

47. Harrad SJ, Smith DJT: **Bioaccumulation factors (BAFs) and biota to sediment accumulation factors (BSAFs) for PCBs in pike and eels.** *Environ Sc Pollut Res Int* 1997, **4:**189–193.

48. Nunes M, Marchand P, Vernisseau A, Le Bizec B, Ramos F, Pardal MA: **PCDD/Fs and dioxin-like PCBs in sediment and biota from the Mondego estuary (Portugal).** *Chemosphere* 2011, **83:**1345–1352.

49. Lepom P, Karasyova T, Sawal G: **Occurrence of polybrominated diphenyl ethers in freshwater fish from Germany.** *Organohalogen Compd* 2002, **58:**209–212.

50. Sühring R, Byer J, Freese M, Pohlmann J-D, Wolschke H, Möller A, Hodson PV, Alaee M, Hanel R, Ebinghaus R: **Brominated flame retardants and Dechloranes in European and American eels from glass to silver life stages.** *Chemosphere* 2014, **116:**104–111.

51. Roosens L, Geeraerts C, Belpaire C, Van PI, Neels H, Covaci A: **Spatial variations in the levels and isomeric patterns of PBDEs and HBCDs in the European eel in Flanders.** *Environ Int* 2010, **36:**415–423.

52. Stapleton HM, Alaee M, Letcher RJ, Baker JE: **Debromination of the flame retardant decabromodiphenyl ether by juvenile carp (Cyprinus carpio) following dietary exposure.** *Environ Sci Technol* 2004, **38:**112–119.

53. Stapleton HM, Brazil B, Holbrook RD, Mitchelmore CL, Benedict R, Konstantinov A, Potter D: **In vivo and in vitro dibromination of decabromodiphenyl ether (BDE 209) by juvenile rainbow trout and common carp.** *Environ Sci Technol* 2005, **40:**4653–4658.

54. Dominguez AA, Law RJ, Herzke D, de Boer J: **Bioaccumulation of brominated flame retardants.** In *Brominted Flame Retardants.* Edited by Eljarrat E, Barceló D. Berlin: Springer; 2011:141–185.

55. Mariottini M, Corsi I, Della Torre C, Caruso T, Bianchini A, Nesi I, Focardi S: **Biomonitoring of polybrominated diphenyl ether (PBDE) pollution: a field study.** *Comp Biochem Physiol C Toxicol Pharmacol* 2008, **148:**80–86.

56. LANUV: **Verzehrsempfehlung Aale vom 16.07.2012.** [www.lanuv.nrw.de/verbraucher/warnungen/verzehr.htm]

57. ICES: **Annex 4. Manual for the Ageing of Atlantic Eel. Otolith preparation methodologies, age interpretation and image storage.** In *ICES Workshop on Age Reading of European and AmericanEel (WKAREA), 20–24 April 2009, Bordeaux, France. ICES CM 438 2009:ACOM 48.* Copenhagen: ICES; 2009:59.

58. Durif C, Dufour S, Elie P: The silvering process of *Anguilla anguilla*: a new classification from the yellow resident to the silver migrating stage. *J Fish Biol* 2005, **66**:1025–1043.

59. Neugebauer F, Schröter-Kermani C, Päpke O, Steeg W: Analytical experiences with the German environmental specimen bank: time trends of PCDD/F and DL_PCB in bream (*Abramis brama*) caught in German rivers. *Organohalogen Compd* 2011, **73**:1340–1343.

60. Van den Berg M, Birnbaum LS, Denison M, de Vito M, Fraland W, Feeley M, Fiedler H, Hakansson H, Hanberg A, Haws L, Roes M, Safe S, Schrenk D, Tohyama C, Tritscher A, Tuomisto J, Tysklind M, Walker N, Peterson RE: **RE:** World Health Organization reevaluation of human and mammalian toxic equivalency factors for dioxins and dioxin-like compounds. *Toxicol Sci* 2005, **2006**(93):223–241.

61. Päpke O, Schröter-Kermani C, Stegemann D, Neugebauer F, Ebsen F: Analytical experiences with the German Environmental Specimen Bank: polybrominated diphenyl ethers in deer liver samples and corresponding soils. *Organohalogen Compd* 2011, **73**:416–419.

62. Klees M, Hiester E, Bruckmann P, Schmidt TC: Determination of polychlorinated biphenyls and polychlorinated dibenzo-p-dioxins and dibenzofurans by pressurized liquid extraction and gas chromatography coupled to mass spectrometry in street dust samples. *J Chromatogr A* 2013, **1300**:17–23.

11

Effects of a textile azo dye on mortality, regeneration, and reproductive performance of the planarian, *Girardia tigrina*

Alyson Rogério Ribeiro and Gisela de Aragão Umbuzeiro[*]

Abstract

Background: Many dyes can be considered emerging contaminants. The most widely used dyes belong to the class of azo compounds, some of which are known to have toxic and genotoxic properties. They are used in great quantities in textile activities and are of environmental concern because of their potential discharge in water. Planarians have been successfully used as test organisms in hazard evaluation of different chemicals, and we demonstrate the suitability of *Girardia tigrina* for laboratory testing. The aim of this work was to evaluate the suitability of the planarian species *G. tigrina* to assess the ability of the azo textile dye disperse red 1 to cause acute toxicity and adverse effects in the regeneration and reproduction of newborn and adult specimens.

Results: Disperse red 1 presented a median LC_{50} of 75 and 152 mg/L, respectively, for newborns and adults of *G. tigrina*, showing that newborns are twice as susceptible to the dye. Uncoordinated movements, irregular twists, colored skin, increased mucous production, and regenerative delays were observed after dye exposure at sub-lethal concentrations.

Conclusions: A no-observed-adverse-effect concentration (NOAEC) of 0.1 mg/L could be determined for disperse red 1 based on the fecundity test. Zinc seems to be a suitable positive control for monitoring the sensitivity in *G. tigrina* tests within only 24 h of exposure. This study demonstrates the applicability of *G. tigrina* tests in the hazard evaluation of water contaminants, such as azo dyes.

Keywords: Regeneration; Planarians; Reproductive performance; Azo dye; Disperse red 1; *Girardia tigrina*

Background

Synthetic dyes can be considered emerging contaminants because they are potentially toxic and have been found in the aquatic environment and there are no regulations stating the maximum allowable concentrations in water to ensure the protection of aquatic biota and human health. The world production of key dyes is estimated to be more than 10 tons per year [1]. The most widely used dyes belong to the class of azo compounds derived from certain aromatic amines and are used in products such as textiles, foodstuffs, cosmetics, house products, paints, and inks. This is significant because some of these dyes are known to have potential toxic and genotoxic properties [2,3]. The use of azo dyes to the color of textiles is of considerable

interest, as this can raise environmental concerns because of the high volume of water involved in the dyeing process. When a dye is used in this process, a portion does not undergo bonding to the fibers thus remaining in the water bath [4]. As a consequence, high volumes of wastewaters containing dyes and related auxiliaries are produced and may be released into the environment. For instance, for 10 kg of polyester fabric, 100 g of disperse red 1 will be dissolved in 200 L of water, of which are round 1% of the dye will remain in the water bath at the end of the process. Hence, around 200 L of wastewater containing 1 g of the dye will be generated. It is also known that conventional treatments involving aerobic lagoons or activated sludge are not efficient in the removal or biological degradation of these dyes [5] and therefore, the dyes are still present in the final effluent or in the sludge of the treatment plant [6,7]. Disperse dyes used to dye synthetic fibers are generally sparingly water-soluble

* Correspondence: giselau@ft.unicamp.br

School of Technology, State University of Campinas - UNICAMP, Paschoal Marmo Street 1888, Limeira, SP 13484-332, Brazil

compounds, but they can be found in the water column, because of their commercial formulation that includes surfactants needed for the dyeing process. Some disperse dyes have been found in the aquatic environment [8], and their presence was related to the observed mutagenic activity of the water and sediments [6,7,9].

Recently, Ferraz et al. [10] showed that the textile dye disperse red 1 ((N-ethyl-N-(2-hydroxyethyl)-4-(4-nitrophenylazo) aniline) is highly toxic to aquatic invertebrates, in addition to being mutagenic. Vacchi et al. [11] studied a commercial disperse red 1 dye product and found that the ecotoxicity of the commercial preparation was similar to the dye itself. The median effective concentration (EC_{50}) for Daphnia was 0.1 mg/L; therefore, this dye can be classified as highly toxic to aquatic organisms according to the Globally Harmonized System of Classification and Labelling of Chemicals (GHS) [12]. More recently, studies have been performed to estimate the concentrations of dyes in surface waters [13], Zocollo et al.'s submitted article. The authors determined the concentrations of disperse azo dyes in river and effluent samples collected in Brazil and showed that they were present in the levels of nanograms to micrograms per liter. However, more studies are required to estimate the concentration levels in the aquatic environment, especially in countries where dyeing activities are more intense, like in India, China, and Brazil.

Planarians have been used as test organisms in the hazard evaluation of different chemicals using different endpoints such as mortality, regeneration, micronucleus (MN) frequency, and enzymatic activity [14-21]. Planarians generally reproduce by transverse fission, but some species, like Girardia tigrina [22], also have hermaphroditic sexual organs and can generate cocoons of fertilized eggs [23]. Therefore, reproduction impairment can also be used as an endpoint in ecotoxicity tests with planarians [20,22].

Because of its ability to regenerate [24,25], planarians have been used to verify if chemicals can interfere with this process [17,18,26-30]. A small body fragment can generate an intact planarian due the presence of totipotent stem cells called neoblasts, which migrate from the parenchymal tissue to the injured site and differentiate by mitosis to other planarian cell types [31,32].

There have been a number of publications that examined the response of planarians to chemical exposures using molecular and enzymatic approaches [21,33-35]. We focused our study towards three different low-cost and environmentally relevant endpoints: mortality, regeneration, and reproduction. The aim of this work was to evaluate the suitability of the planarian species G. tigrina as a test organism to assess the acute toxicity of the azo textile dye, disperse red 1, to newborns and adults and to verify its ability to cause adverse effects on regeneration and cocoon production in the exposed animals.

Disperse red 1 was selected because it has been found in river waters that receive textile effluents in the region of Americana, São Paulo, Brazil (Zocolo et al.'s submitted article).

Results and discussion

G. tigrina showed an acceptable sensitivity to $ZnSO_4$, with a 24-h median lethal concentration of LC_{50} of 1.6 ± 0.2 mg/L expressed in Zn^{2+} and a variation coefficient of 12% in seven replicate acute newborn toxicity tests (Table 1). Chromium salts have previously been used to monitor sensitivity of laboratorial culture [15,36,37]. We also used chromium for this purpose but zinc provided better repeatability and faster responses (24 h) (Table 1). Although the mean LC_{50} values after 48 and 96 h for chromium are similar to those reported by Preza and Smith [15], the coefficient of variation (66%) in our study was higher than obtained by those authors (22%). This higher variation could be explained by the higher instability of chromium in solutions compared to zinc or by response differences between the planarian populations.

The LC_{50} values for disperse red 1 dye obtained with newborns and adults are presented in Table 2, and as expected, toxicity increased with exposure time. Three independent tests were performed, and the coefficients of variation of the means were below 17%, which was considered acceptable for ecotoxicological assays according to the recommendation of Environment Canada [38]. No mortality was observed when adults were exposed during the first 24 h but several animals showed uncoordinated movements and increased mucus production at all tested concentrations. Mean 96-h LC_{50} values of 75 ± 7.2 mg/L ($n = 3$) and 152 ± 5.8 ($n = 3$) mg/L were obtained

Table 1 Mean lethal concentrations (50%) (LC_{50}) of chromium and zinc in newborn G. tigrina acute tests for sensitivity assessment

Test	LC_{50} (mg/L)							
	Cr^{6+}				Zn^{2+}			
	24 h	48 h	72 h	96 h	24 h	48 h	72 h	96 h
I	13	11	8	7	2	2	2	2
II	11	11	11	7	1.6	1.6	1.6	1.6
III	*	27	20	11	1.6	1.6	1.6	1.6
IV	*	29	18	15	1.6	1.6	1.6	1.6
V	18	8	5	3	1.3	1.3	1.3	1.3
VI	15	10	5	4	1.6	1.6	1.6	1.6
VII	14	5	2	5	1.7	1.7	1.7	1.7
Mean	14.2	14.4	9.8	7	1.6	1.6	1.6	1.6
SD^a	2.5	9.5	6.8	4.6	0.2	0.2	0.2	0.2
$CV (\%)^b$	18	66	69	66	12	12	12	12

Asterisk indicates LC_{50}, not calculable for the statistical method employed.
^{a}SD standard deviation of the mean. ^{b}Coefficient of variation. The tests indicated as I, II, III, IV, V, VI, and VII correspond to independent experiments.

Table 2 Mean lethal concentrations (50%) (LC$_{50}$) of commercial disperse red 1 in newborn and adult *G. tigrina* tests

| Test | LC$_{50}$ (mg/L) | | | | | | | |
| | Newborn | | | | Adult | | | |
	24 h	48 h	72 h	96 h	24 h	48 h	72 h	96 h
I	127	127	74	67	*	201	179	159
II	120	91	79	79	*	207	156	148
III	131	111	87	80	*	198	162	150
Mean	126	110	80	75		202	167	152
SD[a]	5.6	18	6.5	7.2		4.6	12	5.8
CV (%)[b]	4.4	16.4	8.1	9.6		2.2	7.2	3.8

Asterisk indicates no mortality was observed. [a]SD standard deviation of the mean. [b]Coefficient of variation. The tests indicate as I, II, and III correspond to independent assays.

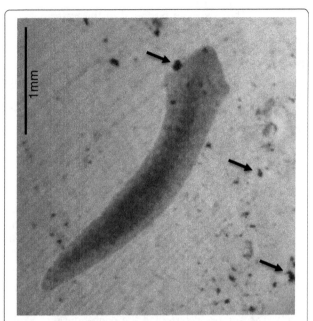

Figure 1 Picture of a newborn of *G. tigrina* after 96-h exposure to 10 mg/L of commercial disperse red 1 dye. Red-colored skin can be observed. Arrows indicate the precipitation and agglomeration of the dye. The image is 50 times increased.

for newborns and adults, respectively (Table 2). These results confirmed the findings of Preza and Smith [15] which showed that newborns were more sensitive than adults. Knakievicz and Ferreira [20] attributed this age response to different body surface/volume ratios of newborn organisms.

Ecotoxicological data of azo textile dyes are scarce in the literature. Ferraz et al. [10] tested disperse red 1 (95% purity) and obtained an EC$_{50}$ of 0.13 mg/L in a *Daphnia similis* acute toxicity test. Recently, the same commercial product that was analyzed in this study was used in acute toxicity tests with *D. similis* and *Hydra attenuata* [11]. The authors showed that the surfactant and other impurities present in the commercial dye did not influence the observed toxicity, at least for *Daphnia*. The observed effects were related to the main dye, which represents 60% of the commercial product. In that study, it was not possible to verify if the acute toxicity observed for Hydra was related also to the main dye, but Hydra was 15 times less sensitive than Daphnia to the commercial dye [11].

Aggregation and precipitation of the dye were observed at concentrations higher than 50 mg/L after the first 24 h of exposure. This is expected because the commercial dye contains surfactants. Therefore, when the concentration of the disperse dye increases, they can form aggregates and settle on the bottom. This behavior is also expected to occur in the aquatic systems [39]. Because planarians slide on the surface of the testing containers [40], the animals were exposed to the precipitated/agglomerated dye in addition to the dye remaining in the solution. This natural surface-contact behavior of the planarians is an important consideration when assessing the toxicity of compounds that are sparingly soluble in water.

After 24 h of exposure in the acute toxicity tests, newborns (Figure 1) and adults (not shown) exposed to concentrations higher than 10 mg/L showed red-colored skin, especially at the encephalic region. This could be explained by dermal or cilia dye adsorption during locomotion and/or skin respiration. The exposed organisms showed uncoordinated movements, irregular twists and increased mucous production after the first 24-h exposure. The mucous resulted in dye precipitation and aggregation around the animal bodies, forming what appeared to be a body capsule, which was abandoned when the planarians started to move. Planarian behavioral effects, such as increased mucus production, irregular twisting, and body contractions, have previously been associated with animal responses to toxicants [26,41,42].

In the regeneration assay, the first changes were observed after 24 h at the highest concentration tested, 200 mg/L. After 120 h, 95% of control planarians fully regenerated (see Figure 2e). The no-observed-adverse-effect concentration (NOAEC) for the dye was 10 mg/L and the lowest-observed-adverse-effect concentration (LOAEC) was 50 mg/L (Table 2). The regenerated planarians showed also red-colored skin, behavioral changes, and increased mucous production at concentrations higher than 10 mg/L. Regenerative delay started to occur at 50 mg/L, while at the highest concentration (200 mg/L), eight of the 20 organisms died within 96 h of exposure, and the ones that survived presented severe regenerative delay after 120 h of exposure (Table 3) with parenchymal loss due to lack of wound cicatrization (see Figure 2c,d). When cicatrization occurred, some delays in posterior regenerative stages were noticed. In seven animals the auricle

Figure 2 Morphological effects of different concentrations of commercial disperse red 1 dye in *G. tigrina*. (a) Individual without full body cicatrization, showing eyespots and no auricles (phase z) - after 120-h exposure at 200 mg/L. **(b)** Late development of the auricle (phase d) after 120 h at 50 mg/L. **(c, d)** Individuals with delayed and abnormal cicatrization (phase t) - after 120 h at 200 mg/L. **(e)** A fully developed planarian with no effect (negative control). Arrows are guides to the eye and the bar on the left of each figure represent 1 mm.

development was inhibited or delayed (see Figure 2a and 2b). For successful regeneration of the lost body parts, planarians need an intact nervous system to control the migration of the normal neoblasts [43,44] that may have been affected by the dye. Also, the mutagenic properties of the dye [10,11] could affect the normal mitotic process required for the cell replication, which is needed for regeneration. The planarian regeneration assays are useful for analyzing the effects of chemicals on cell differentiation and molecular organization [45]. Thus the migration and proliferation of neoblasts as well as other steps required to *G. tigrina* regeneration may have been affected by the azo dye.

Freshwater planarian reproduction is a complex process and some details are still unknown [22]. These animals have a many-sided reproductive organ with ovaries and hormones. Animals can be fertilized after mating [23,46]. Eggs, formed after fertilization, as well as vitelline cells, are enveloped by an oval wall called cocoons [47]. Hatching occurs after different incubation periods depending on the species and temperature [40,48]. The time between crosses and cocoon lay is still unknown. In our experiments, the fecundity index recorded for the animals exposed to 1 mg/L of disperse red 1 (stage 2, exposure) was statistically significantly lower than the control (Figure 3). Also, after the first week of exposure, animals from the treated group G4 (1 mg/L) presented colored skin and alterations in behavior during feeding,

such as continuous exposure of the pharynx and limited mobility. The pharynx is only exposed during a planarian's feeding process, and after feeding this organ returns to planarian's gastric cavity [40]. However, we observed that planarians, when exposed to the dye, showed retractile pharynx movements even when they were far from the food. Planarian reproductive success is related to food availability, demographic density, and temperature [22,48,49]. During the entire exposure time, the animals from G4 group were not able to eat and the cocoons produced during that time may have been generated using their remaining energy [36,50]

During stages 1 (before exposure) and 3 (recovery following exposure) no statistical differences were observed in the fecundity indexes among the groups, although fecundity capacity did not seem to be fully recovered when animals were exposed to 1 mg/L during stage 3 (Figure 3). The addition of the recovery stage was also useful to analyze if the planarians were able to remove the impregnated or absorbed dye from their bodies and if the effect on the pharynx's retractability was permanent. After 2 weeks of recovering in maintenance water more than 80% of the exposed animals showed normal skin color as well as normal pharynx function and feeding behavior. Therefore, disperse red 1 interferes with *G. tigrina*'s mobility and capacity for food identification at 1 mg/L exposure concentration. In sum, disperse red 1 dye adversely affects survival, fecundity, regeneration, and feeding behavior in

Table 3 Regenerative results for *G. tigrina* exposed to different concentrations of the commercial disperse red 1

Exposure time (h)	Planarians/regeneration steps (mg/L)					
	Control	10	50	100	150	200
12	20[a] (a)[b]	20(a)	20(a)	20(a)	20(a)	20(a)
24	20(b)	20(b)	20(b)	20(b)	20(b)	3(t) 17(b)
48	20(c)	20(c)	20(c)	20(c)	20(c)	3(t) 17(c)
72	20(de)	20(de)	20(de)	2(c) 18(de)	2(c) 18(de)	3(t) 17(de)
96	20(f)	20(f)	3(d) 17(f)	5(d) 15(f)	2(d) 18(f)	3(t) 9(d) 8(f)
120	1(f) 19(g)	20(g)	2(f) 18(g)	5(f) 15(g)	5(f) 15 (g)	8 dead 3(t) 3(z) 4(d) 2(g)
Pd[c]$_{120h}$		−1	1	4	4	9

[a]Number of animals. [b]Regeneration steps: (a) decapitated bodies, (b) beginning of blastema formation, (c) regular blastema, (d) beginning of auricle formation, (e) beginning of eyespots formation, (f) auricle and eyespots clearly defined, (g) full head formation, (t) body section without cicatrization, (z) without body cicatrization, with eyespots, and without auricle. [c]Pd is the number of planarians with delayed regeneration subtracted from the number of delayed planarians in the control, Pd greater or equal to 1 is considered as adverse, for more information see 'Methods' section.

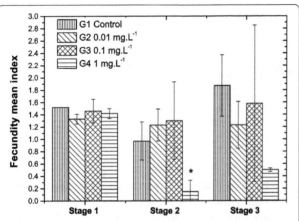

Figure 3 Mean fecundity index of *G. tigrina* planarian exposed to commercial textile dye disperse red 1 (control and exposed groups). At stage 1, exposure system adaptation, organisms were monitored for 2 weeks in maintenance medium. At stage 2, organisms were exposed for 5 weeks. At stage 3, recovery, exposed organisms were monitored for 2 weeks more in maintenance water. Bars represent the standard deviation of the mean. *Significant at $p \leq 0.05$.

planarians. Based on the most critical endpoint (fecundity index) it is possible to suggest a no-observed-adverse-effect-concentration (NOAEC) of 0.1 mg/L for disperse red 1. As observed in other studies [15,17,21,22,49], the viability of *G. tigrina* culture under low-cost laboratorial conditions was confirmed. Our results also show the repeatability of the applied protocol at least for zinc and disperse red 1.

Planarians can be considered an interesting model for ecotoxicological assessment of environmental contaminants because it allows the assessment of different sublethal effects, such as increase in mucus production and changes in behavior besides the evaluation of adverse effects on the regeneration process and reproduction as showed in this work.

Conclusions

Disperse red 1 presented a median LC_{50} of 75 and 152 mg/L, respectively, for newborns and adults of *G. tigrina*, showing that newborns are more susceptible to the dye than adults. Uncoordinated movements, irregular twists, colored skin, and increased mucous production were observed after dye exposure at sub-lethal concentrations. A no-observed-adverse-effect concentration (NOAEC) of 0.1 mg/L of disperse red 1 could be determined based on the fecundity test. Zinc seems to be a suitable reference substance for monitoring the sensitivity in *G. tigrina* tests within only 24 h of exposure. This study successfully demonstrates the applicability of *G. tigrina* tests in the hazard evaluation of azo dyes, a relevant example of an emerging contaminant. The results obtained in this work may also

be used for the determination of safe environmental concentrations of disperse red 1 for the aquatic system.

Methods
Chemical
The powdered commercial textile dye product disperse red 1 dye (*N*-Ethyl-*N*-(2-hydroxyethyl)-4-(4-nitrophenylazo) aniline; CAS No 2872-52-8), was purchased from PCIL© Dyes, São Paulo, Brazil. The product was chemically characterized by Vacchi et al. [11] and contains 60% of the main dye disperse red 1, 20% of other azo dye products, and 20% of a surfactant. Zinc sulfate ($ZnSO_4.7H_2O$; CAS No 7440-20-0; CAQ Casa de Química©, Diadema, São Paulo, Brazil) and potassium dichromate ($K_2Cr_2O_7$; CAS No 7778-50-9; Cromoline Química Fina©, Diadema, São Paulo, Brazil) in analytical grade were used as reference substances to monitor the sensitivity of the planarian culture. Dye and metal testing solutions were prepared in planarian maintenance water (see description below) in concentrations determined by preliminary tests.

Animals
Specimens of *G. tigrina* collected in a pristine pond in Rio Claro, São Paulo, Brazil [17], have been cultivated in the Ecotoxicology and Environmental Microbiology Laboratory (LEAL) in the University of Campinas, Limeira, Brazil, since 2005. Descendant animals were cultured by sexual reproduction at room temperature (22 to 25°C) under a light/dark photoperiod of 16:8 h, and fed once a week with bovine liver. The water for culturing the planarians and conducting experiments was designated as maintenance water and was prepared with dechlorinated tap water and 45 mg $CaCO_3$/L, pH 7.6 ± 0.3, and with gentle aeration [37] using an aquarium pump. Water was completely renewed each week after feeding.

Toxicity tests
Three environmentally relevant endpoints were investigated in this study through different toxicity tests: acute toxicity (mortality), regeneration assay and reproductive performance. Acute toxicity tests were carried out with adult and newborn planarians. Animals of more than 3 months of age and visible cocoon production capacity were considered adults, and animals with a maximum age of 10 days were considered newborns [15]. Each independent newborn toxicity test was carried out in duplicate using seven animals placed in 50 mL of test solution for 96 h [15]. In the adult toxicity tests healthy animals were randomly selected and five planarians were exposed to 100 mL of test solution. Each independent experiment was performed in triplicate [37]. In both the newborn and adult tests, the test solution was not renewed and the animals were not fed. Negative controls with maintenance water and positive controls with $ZnSO_4$

and $K_2Cr_2O_7$ were also performed. The nominal concentration used in the positive controls using chrome were 0.9, 1.8, 3.5, 7, 10, 18 and 35 mg/L expressed as Cr^{6+} and using zinc were 0.1, 0.2, 0.5, 1.1, 2.2, 4.5, and 6.8 mg/L, expressed as Zn^{2+}. Mortality, body degeneration, change in behavior and morphological alterations were monitored every 24 hours under a stereo-microscope (Stemi 2000-C, Zeiss©, Oberkochen, Germany). The nominal dye concentrations tested were 10, 50, 100, 150, and 200 mg/L for newborns and 50, 100, 150, 200, and 250 mg/L for adults.

For the regeneration assay, the heads of adult planarians were removed in the region behind the auricles using blades, and the decapitated animals were used immediately [14]. The animals were fed 1 day before the test started to ensure that they would have sufficient nutrients to regenerate. This is important because G. tigrina has the ability to store vital nutrients [36] and the digestion and assimilation of food, in particular liver, takes about 197 h [50]. The concentrations of disperse red 1 used (10, 50, 100, 150 and 200 mg/L) were selected based on the toxicity test results. In each independent experiment, each concentration was tested using four replicates; five decapitated animals per replicate were put in 20 mL of solution. Each head regeneration step [17,31] (designated, a decapitated bodies; b beginning of blastema formation; c regular blastema; d beginning of auricle formation; e beginning of eyespot formation; f auricle and eyespots clearly defined; g full head formation; and t body section without cicatrization, z without body cicatrization, with eyespots, and without auricle) were observed under the stereoscope daily, followed by renewal of the dye test solution [20]. Regeneration of decapitated planarians in maintenance water was used as a negative control. Animals were not fed, and they were kept in cultivation conditions for 120 h for full regeneration of the encephalic region [17].

The reproductive performance of adult specimens was evaluated based on a method described elsewhere [20]. The single experiment included a preexposure stage for adaptation purposes. We also followed the recovery capacity in a postexposure period, without the presence of the toxicant. Twenty-five healthy adult animals were randomly placed into plastic vessels containing 1 L of each test solution, without replicates. The test was carried out in three stages: at stage 1 (adaptation), all organisms were exposed only to maintenance medium and were monitored for 2 weeks. At stage 2 (toxicant exposure), three groups of organisms were exposed to sub-lethal concentrations (0.01, 0.1, or 1 mg/L) of the commercial product. We also included a negative control, without the dye. Concentrations were selected based on the acute toxicity test results. This stage lasted five weeks. At stage 3 (recovery), chemical exposure was interrupted at the end of the 5 weeks; animals were transferred into maintenance water and monitored for an additional 2 weeks.

Observations on cocoon production, animal behavior, and mortality were made during all stages. Identical cultivation maintenance procedures were used for all groups, with weekly feeding followed by total test solution renewal, and the counting and segregation of cocoons.

During the toxicity, reproductive, and regeneration experiments, any changes in animal behavior, skin color, and mucous production were registered on a daily basis. The retractability of the pharynx during feeding was also monitored.

Statistical analysis and expression of the results

For acute toxicity tests, the trimmed Spearman-Karber method [51] was used to calculate the LC_{50}, with 95% confidence interval, for every 24 h of exposure, until 96 h. In the regeneration tests, a 120-h delayed regeneration value (Pd120h) was calculated for each concentration. To obtain this value, the number of delayed regeneration planarians in the control was subtracted from the treated number. The NOAEC was the higher concentration that presented a Pd120h <1, and the LOAEC was the lowest concentration that presented a Pd120h ≥1.

In the reproductive performance assay, the mean fecundity index for each group in each stage was calculated using a method adapted from Knakievicz et al. [22]. Mean fecundity indexes were calculated by the total number of cocoons divided by the number of live planarians in every analyzed week, divided by the total experimental time, in weeks, at the end of each step. To compare the different treatments, analysis of variance (ANOVA) followed by the Tukey test was performed using the Origin® Pro8 program. The significance level was $p \le 0.05$.

Abbreviations

LC_{50}: median lethal concentration; EC_{50}: median effective concentration; GHS: Global Harmonization System; NOAEC: no-observed-adverse-effect concentration; LOAEC: lowest-observed-adverse-effect concentration; LEAL: Ecotoxicology and Environmental Microbiology Laboratory; CAS: Chemical Abstracts Service; Pd: delayed regeneration value; ANOVA: analysis of variance.

Competing interests

The authors declare that they have no competing interests.

Authors' contributions

ARR performed all experiments and statistical analysis and drafted the manuscript. GAU participated in the experimental design, supervised the research work, and was responsible for the final elaboration of the final manuscript. GAU also provided technical and financial support for this study. Both authors read and approved the final manuscript.

Acknowledgements

This work was part of the Thematic Project no. 2008/10449-7 financed by FAPESP - Fundação de Amparo a Pesquisa do Estado de São Paulo. ARR thanks FAPESP also for the master fellowship, process no. 2009/12737-2. ARR and GAU thank Alain Devaux, Sylvie Bony, Maria Beatriz Bohrer-Morel, Errol Zeiger, and Harold Freeman for the valuable suggestions, Francine Inforçato Vacchi for helping with the manuscript organization, and Erika Rabello Moretti for helping with the reproductive assay.

References

1. ETAD: *The Ecological and Toxicological Association of Dyes and Organic Pigments Manufacturers, Annual Review.* 2012. Issued April 2013 [www.etad.com/documents/Downloads/publications/etad_+ar+2012_final_web.pdf]

2. Novotný C, Dias N, Kapanen A, Malachová K, Vándrovcová M, Itävaara M, Lima N: **Comparative use of bacterial, algal and protozoan test to study toxicity of azo- and anthraquinone dyes.** *Chemosphere* 2006, **63**:1436–1442.

3. Chung KT, Cerniglia CE: **Mutagenicity of azo dyes: structure-activity relationships.** *Mutat Res-Rev Genet* 1992, **277**(Suppl 3):201–220.

4. Hanger K: *Industrial dyes: Chemistry, Properties and Applications, Health and Safety Aspects.* Germany: Wiley-VCH; 2003.

5. USEPA: *Aerobic and Aerobic Treatment of C.I. Disperse Blue 79.* US Department of Commerce. National Technical Information Service (NTIS) vols I and II. EPA/600/2-89/051 (PB 90–111642); 1989. http://nepis.epa.gov/Exe/ZyPURL.cgi?Dockey=2000TM97.txt.

6. Umbuzeiro GA, Roubicek DA, Rech CM, Sato MIZ, Claxton LD: **Investigating the sources of the mutagenic activity found in a river using the Salmonella assay and different water extraction procedures.** *Chemosphere* 2004, **54**(Suppl 11):1589–1597.

7. Umbuzeiro GA, Freeman HS, Warren SH, Oliveira DP, Terao Y, Watanabe T, Claxton LD: **The contribution of azo dyes to the mutagenic activity of Cristais River.** *Chemosphere* 2005, **60**(Suppl1):55–64.

8. Maguire RJ: **Occurrence and persistence of dyes in a Canadian River.** *Water Sci Technol* 1992, **25**:264–270.

9. Oliveira DP, Carneiro PA, Rech CM, Zanoni MV, Clauton LD, Umbuzeiro GA: **Mutagenic compounds generated from the chlorination of disperse azo dyes and their presence in drinking water.** *Environ Sci Technol* 2006, **40**(Suppl 21):6682–6689.

10. Ferraz ERA, Umbuzeiro GA, De Almeida G, Caloto Oliveira A, Chequer FMD, Zanoni MVB, Dorta DJ, Oliveira DP: **Differential toxicity of disperse red 1 and disperse red 13 in the Ames test, HepG2 cytotoxicity assay, and Daphnia acute toxicity test.** *Environ Toxicol* 2011, **26**:489–497.

11. Vacchi FI, Albuquerque AF, Vendemiatti JAS, Morales DA, Ormond AB, Freeman HS, Zocolo GJ, Zanoni MVB, Umbuzeiro GA: **Chlorine disinfection of dye wastewater: Implications for a commercial azo dye mixture.** *Sci Total Environ* 2013, **442**:302–309.

12. OECD (Organization for Economic Co-operation and Development): *Harmonized Integrated Hazard Classification System for Chemical Substances and Mixtures.* Paris: OECD; 2001.

13. Carneiro PA, Umbuzeiro GA, Oliveira DP, Zanoni MVB: **Assessment of water contamination caused by a mutagenic textile effluent/dyehouse effluent bearing disperse dyes.** *J Hazard Mater* 2010, **174**:694–699.

14. Calevro F, Filipi Deri P, Albertosi C, Batistoni R: **Toxic effects of aluminum, chromium and cadmium in intact and regenerating freshwater planarians.** *Chemosphere* 1998, **37**:651–659.

15. Preza DLC, Smith D: **Use of newborn *Girardia tigrina* (Girard, 1850) in acute toxicity tests.** *Ecotoxicol Environ Safety* 2001, **50**:1–3.

16. Horvat T, Kalafatić M, Kopjar N, Kavačević G: **Toxicity testing of herbicide norflurazon on an aquatic bioindicator species - the planarians *Polycelis feline* (Daly).** *Aquat Toxicol* 2005, **73**:342–352.

17. Barros GS, Angelis DF, Furlan LT, Corrêa-Junior B: **Use of freshwater planarians Dugesia (Girardia) tigrina testing toxicity of a petroleum refinery wastewater.** *J Braz Soc Ecotoxicol* 2006, **1**:67–70 (in Portuguese).

18. Kalafatić M, Kopjar N, Besendorfer V: **The impairments of neoblast division in regenerating planarian *Polycelis felina* (Daly.) caused by in vitro treatment with cadmium sulfate.** *Toxicol in Vitro* 2004, **19**:99–107.

19. Kalafatić M, Kovačević G, Franjević D: **Resistance of two planarians species to UV-irradiation.** *Folia Biol-Prague* 2006, **54**:103–108.

20. Knakievicz T, Ferreira HB: **Evaluation of copper effects upon *Girardia tigrina* freshwater planarians based on a set of biomarkers.** *Chemosphere* 2008, **71**:419–428.

21. Mei-Hui L: **Effects of nonionic and ionic surfactants on survival, oxidative stress, and cholinesterase activity of planarian.** *Chemosphere* 2008, **70**:1796–1803.

22. Knakievicz T, Vieira SM, Erdtmann B, Ferreira HB: **Reproductive modes and life cycles of freshwater planarians (Platyhelminthes, Tricladida, Paludicula) from southern Brazil.** *Invertebr Biol* 2006, **125**:212–221.

23. Fukushima M, Funabiki I, Hashizume T, Osada K, Yoshida W, Ishida S: **Detection and changes in levels of testosterone during spermatogenesis in the freshwater planarian *Bdellocephala brunnea*.** *Zool Sci* 2008, **25**:760–765.

24. Newmark PA, Sánchez-Alvarado A: **Not your father's planarian: a classic model enters the era of functional genomics.** *Nat Rev Genet* 2002, **3**:210–219.

25. Buttarelli FR, Pellicano C, Pontieri FE: **Neuropharmacology and behavior in planarians: translations to mammals.** *Comp Biochem Physiol C Toxicol Pharmacol* 2008, **147**(Suppl 4):399–408.

26. Franjević D, Krajna A, Kalafatić M, Ljubešić N: **The effects of zinc upon the survival and regeneration of planarian *Polycelis feline*.** *Biologia* 2000, **55**:689–694.

27. Kato C, Mihashi K, Ishida S: **Motility recovery during the process of regeneration in freshwater planarians.** *Behav Brain Res* 2004, **150**:9–14.

28. Iannaconne J, Tejada M: **Employment of regeneration of freshwater planarian *Girardia festae* (Borelli, 1898) (Tricladida: Dugesiidae) to evaluate toxicity of Carbofuran.** *Neotrop Helminthol* 2007, **1**:7–14.

29. Medvedev IV: **Regeneration in two freshwater planarian species exposed to methylmercury compounds.** *Russian J Devel Biol* 2008, **39**:232–235.

30. Novikov VV, Shieman IM, Fesenko EE: **Effect of weak static and low-frequency alternation magnetic fields on the fission and regeneration of the planarian *Dugesia (Girardia) tigrina*.** *Bioelectromagnetics* 2008, **29**:387–393.

31. Reddien PW, Sánchez-Alvarado A: **Fundamentals of planarian regeneration.** *Annu Rev Cell Dev Biol* 2004, **20**:725–757.

32. Saló E: **The power of regeneration and the stem-cell kingdom: freshwater planarians (Platyhelminthes).** *Biogeosciences* 2006, **28**:546–559.

33. Guecheva T, Henriques JAP, Erdtmann B: **Genotoxic effects of copper sulphate in freshwater planarian in vivo, studied with the single-cell gel test (comet assay).** *Mutat Res* 2001, **497**:19–27.

34. Gonzáles-Estévez C, Saló E: **Autophagy and apoptosis in planarians.** *Apoptosis* 2010, **15**:279–292.

35. Aboobaker AA: **Planarian stem cell: a simple paradigm for regeneration.** *Trends Cell Biol* 2011, **21**(Suppl 5):304–311.

36. Piontek M: **Application of *Dugesia tigrina* Girard in toxicological studies of aquatic environments.** *Pol Arch Hydrobiol* 1998, **45**(Suppl 4):565–572.

37. Sáfadi RS: **The use of freshwater planarians in acute toxicity tests with heavy metals.** *Verh Internat Limnol* 1998, **26**:2391–2392.

38. Orr PL, Hart DR, Craig GR: *Guidance document on control of toxicity test precision using reference toxicants.* Ottawa: Environmental Canada; 1990.

39. He L, Freeman HS, Lu L, Zhang S: **Spectroscopic study of anthraquinone dye/amphiphile systems in binary aqueous/organic solvent mixtures.** *Dyes Pigments* 2011, **91**:389–395.

40. Kolasa J, Tyler S: **Flatworms: Tuberllarians and Nemertea.** In *Ecology and classification of North American freshwater invertebrates.* 2nd edition. Edited by Thorp JH, Covich AP. New York: AP; 2011.

41. Žnidaric D, Kalafatić M, Kopjar N: **The survival of *Hydra oligactis* Pallas in unpleasant conditions.** *Zeitsch Angewan Zool* 1995, **81**:157–163.

42. Raffa RB, Desai P: **Description and quantification of cocaine withdrawal signs in Planaria.** *Brain Res* 2005, **1032**:200–201.

43. Villar D, Schaeffer DJ: **Morphogenetic action of neurotransmitters on regenerating planarians - a review.** *Biomed Environ Sci* 1993, **6**(Suppl 4):327–347.

44. Baguñà J: **The planarian neoblast: the rambling history of its origin and some current black boxes.** *Int J Dev Biol* 2012, **56**(Suppl1-3):19–37.

45. Cebrià F, Guo T, Jopek J, Newmark PA: **Regeneration and maintenance of the planarian midline is regulated by a slit orthologue.** *Dev Biol* 2007, **307**:394–406.

46. Miyashita H, Nakagawa H, Kobayashi K, Hoshi M, Matsumoto M: **Effects of 17β-estradiol and bisphenol A on the formation of reproductive organs in planarians.** *Biol Bull* 2011, **220**:47–56.

47. Vara DC, Leal-Zanchet AM, Lizardo-Daudt HM: **Enbryonic development of *Girardia tigrina* (Girard, 1850) (Platyhelminthes, Tricladida, Paludicola).** *Bras J Biol* 2008, **68**(Suppl 4):889–895.

48. Benazzi M, Gremigni V: **Developmental biology of triclad turberllarians (Planaria).** In *Developmental Biology of Freshwater Invertebrates.* Edited by Harrison FW, Cowden RR. New York: Alan R Liss Inc; 1982:121–151.

49. Dolci-Palma IA: **Reproduction, eating and reaction to food in planarians under different maintenance conditions.** *Psicol USP* 1995, **6**(Suppl 1):173–196 (in Portuguese).

50. Lima OS: **Total preparation of planarians and observations on digestion.** *Sci Cult* 1954, **6**:162–164 (in Portuguese).

51. Hamilton MA, Russo RC, Thurston RV: **Trimmed Spearman-Karber method for estimating median lethal concentrations in toxicity bioassays.** *Environ Sci Technol* 1977, **11**:714–719.

12

Will climate change increase irrigation requirements in agriculture of Central Europe? A simulation study for Northern Germany

Jan Riediger[1*], Broder Breckling[1], Robert S Nuske[2] and Winfried Schröder[1]

Abstract

Background: By example of a region in Northern Germany (County of Uelzen), this study investigates whether climate change is likely to require adaption of agricultural practices such as irrigation in Central Europe. Due to sandy soils with low water retention capacity and occasional insufficient rainfall, irrigation is a basic condition for agricultural production in the county of Uelzen. Thus, in the framework of the comprehensive research cluster *Nachhaltiges Landmanagement im Norddeutschen Tiefland (NaLaMa-nT)*, we investigated whether irrigation might need to be adapted to changing climatic conditions. To this end, results from regionalised climate change modelling were coupled with soil- and crop-specific evapotranspiration models to calculate potential amounts of irrigation to prevent crop failures. Three different runs of the climate change scenario RCP 8.5 were used for the time period until 2070.

Results: The results show that the extent of probable necessary irrigation will likely increase in the future. For the scenario run with the highest temperature rise, the results suggest that the amount of ground water presently allowed to be extracted for irrigation might not be sufficient in the future to retain common agricultural pattern.

Conclusions: The investigation at hand exemplifies data requirements and methods to estimate irrigation needs under climate change conditions. Restriction of ground water withdrawal by German environmental regulation may require an adaptation of crop selection and alterations in agricultural practice also in regions with comparable conditions.

Keywords: European Water Framework Directive; Evapotranspiration; Soil moisture; Water availability; Water and Substance Simulation Model; Uelzen

Background

Current projections of global climatic change strongly support the expectation that during the next few decades, global temperatures will continue to increase as well as spatial and temporal patterns of temperature and precipitation will be shifting [1,2]. This may have implications for many environmental processes which are influenced by thermal and soil moisture conditions. The increase in temperature for instance could lead to soil moisture deficits and a growing risk of vegetation desiccation due to increasing evapotranspiration and decreasing soil moisture [3,4]. Ecological and economic consequences for European agricultural ecosystems are expected to vary widely according to the spatial patterns of land cover, land use practise and regional climate change [5–7]. Anyway, referring for instance to Europe, according to the EU Water Framework Directive (WFD 2000/60/EC) [8], a good ecological status of surface and ground water has to be achieved. To safeguard sustainable water management, vulnerability assessments [9] and adaptation strategies based on estimates of irrigation demands are needed [10–14]. Such appraisals should be, as far as possible, spatially explicit across scales - form the local to the global [15–19] - and should regard spatial variability in terms of agricultural regions [20] and natural landscapes. As holds true for the estimation of the meteorological aspects of climate change, modelling techniques of its ecological (and economical) impacts should be used [21].

* Correspondence: jan.riediger@uni-vechta.de
[1]Chair of Landscape Ecology, University of Vechta, Driverstraße 22, PO Box 15 53, Vechta 49377, Germany
Full list of author information is available at the end of the article

Global agriculture used about 2,600 km^3 of water each year since the year 2000, i.e. 2% of annual precipitation over land and 17 mm of water spread evenly over the global land surface. This is a +75% increase from 1960 levels and a +400% increase from 1900 levels of irrigation. Out of the world's croplands, 18%, i.e. about 2% of the total land surface, are irrigated and produced 40% of the world's food. On average, the irrigated areas receive an addition of 800 mm of water each year [22]. About 70% of all water withdrawn worldwide from rivers and aquifers are used for agriculture [19]. To estimate the pressure of irrigation on the available water resources, irrigation water requirement and irrigation water withdrawal have to be assessed [23], including strategies for enhancing the water use efficiency [24]. Irrigation water requirement depends on the crop water requirement and the water naturally available to the crops (effective precipitation, soil moisture, etc.). About 2% of the global land area and 17% of the cultivated area, respectively, are irrigated. In Europe, 44% of the total water withdrawal is used for agriculture [25]. The total area equipped for irrigation, i.e. the total irrigable area in EU-27 accounts for roughly 16 million ha in 2003 and 15 million ha in 2007 on a total of 172 million ha of agricultural land; however, about 10 million ha was actually irrigated in 2007 [26]. In Germany, the area equipped for irrigation totals about 516,000 ha, and ca. 235,000 ha of them were irrigated [23]. In the federal state of Lower Saxony, comprising the county of Uelzen, about 3,000,000 ha, i.e. about 50% of the irrigated area of Germany is located. This irrigated area covers 11.5% of the agricultural land of Lower Saxony. In the county of Uelzen, roughly 58,000 ha are reported to be equipped for irrigation and 90% of the crop land is currently irrigated [27–31].

It is a current research task to assess whether for specific processes, regions and crops adaptation requirements will emerge under climate change conditions [32–34] to consolidate decisions on adaptation strategies to be developed and set in action. For the Northern German Lowlands, the research project 'Sustainable Land Management in the North German Lowland' (acronym, NaLaMa-nT, see http://www.nalama-nt.de/, funded by the Federal Ministry of Education and Research) currently evaluates agricultural implications.

The county of Uelzen belongs to the landscape unit of the Lüneburg Heath which is located in the federal state of Lower Saxony in Northern Germany (Figure 1) [35]. Due to sub-continental climate, the county of Uelzen is characterised by comparatively low precipitation amounts (in autumn and winter 7.5 mm per month lower than the average in Lower Saxony) and relatively high daily average temperatures and slightly higher evapotranspiration amounts (in summer 3 mm per month higher compared to the whole area of Lower Saxony) [36].

Because of a low annual rainfall of approximately 600 mm per year, Uelzen is one of the counties in Germany where irrigation agriculture plays a significant role [33].

On average, 73 mm/m^2 and year of ground water are used for irrigation in the county of Uelzen [27]. The guarantee of adequate water availability became a fundamental condition for the local economy [33]. The amount of ground water allowed to be taken for irrigation differs within the county. The local authorities have currently set ground water use for irrigation to a maximum of 79 mm/m^2 and year with a moving average over 7 years in order to avoid depletion [27]. On average, 59.2% of the permitted amount of extracted ground water was used per year in the time period 1997 to 2004. In the low-rainfall year 2003, 125% was used. Increasing abstraction of ground water could cause negative ecological consequences, e.g. the endangerment of wetlands in the region [33]. The county of Uelzen belongs to the Elbe river basin which is the driest amongst the five largest river basins in Germany. There, the vulnerability against water stress in dry periods is currently a problem for agriculture. This scarcity is expected to increase with subsequent adverse impacts [37].

Climate change not only affects water availability but also the demand for water. If the climate in a given region gets drier and warmer, water availability will decrease and be exacerbated by increasing water demand [16]. Soil water content and therewith irrigation in agriculture is likely to be affected by higher temperature. Evapotranspiration sensitively depends on temperature regimes [38]. For temperate regions with a highly variable rainfall pattern, it is difficult to predict to which extent temperature changes will require a modification of agricultural practices. We use regional climate model data generated by STARS II [39] based on the RCP 8.5 scenario together with information on regional soil conditions [40], crop types and pattern of the current agricultural practice (by courtesy of J. Hufnagel and N. Svoboda, Leibnitz Centre for Agricultural Landscape Research (ZALF), Müncheberg, Germany) to assess whether it would be possible to continue the current cultivation pattern in Uelzen or whether adaptations and changes of the current irrigation pattern and crop management would be necessary to avoid critical conditions and crop losses - based on the assumption of a validity of the climate change scenario. Therefore, we developed the evaporation calculation model BewUe (Bewässerung (irrigation) Uelzen) to estimate the irrigation requirement using soil water content as an indicator. The presented work is part of a larger crop rotation simulation endeavour covering further Northern German regions. The computed irrigation data are essential for preparing the study of current and future water, carbon and nitrogen balances within the NaLaMa-nT project. For

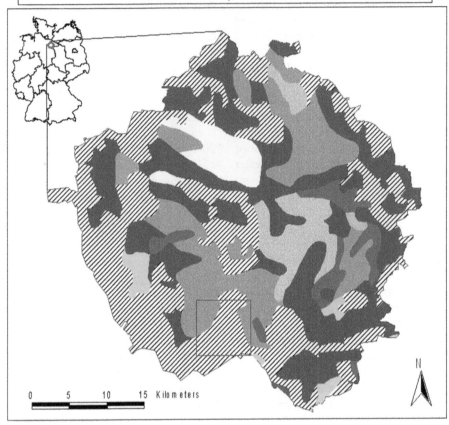

Main soil types according to the use-specific soil map of Germany (BÜK1000)

Figure 1 Uelzen soil map [40]. 6 = low moor; 7 = raised moor; 10 = gleys and alluvial soils out of sand and/or loamy sand; 11 = gleys and alluvial soils out of sand, often mixed with loam and/or clay; 19 = luvisols out of loamy sand and/or loamy silt, partly stony; 25 = luvisols and pale leached soils out of slightly loamy sand; 28 = pseudogley-luvisols and pseudogley-pale leached soils out of slightly loamy sand; 31 = brown podzolic soils out of sand; 33 = podzols out of sand; 46 = lessivés, pale leached and brown soils out of fine sandy silt.

this purpose, the Water and Substance Simulation Model (WASMOD, [41]) was used.

Results

Irrigation requirements were calculated for all years in the reference period 1991 to 2010 and in the scenario time span 2011 to 2070. Irrigation will be applied as soon as the soil water content is lower than 20% of the available water capacity. The amount of water applied in a single irrigation event was set to 20 mm/m^2. Typical examples illustrating the variability of soil water content and required irrigation are shown in Figure 2.

As climate conditions in terms of daily average temperature, precipitation and evapotranspiration show a high

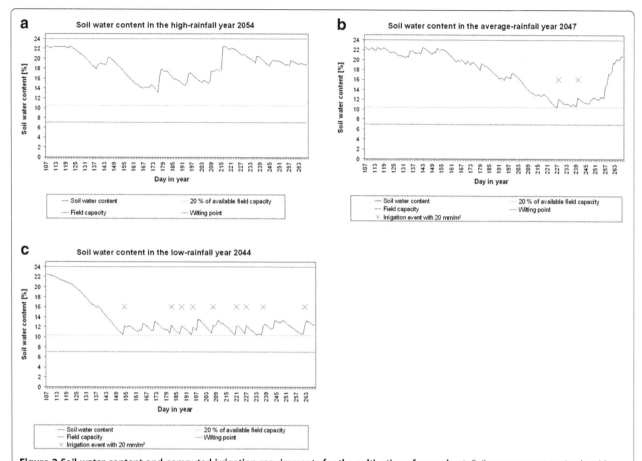

Figure 2 Soil water content and computed irrigation requirements for the cultivation of sugar beet. Soil water content as simulated by model BewUe and computed irrigation requirements for the cultivation of sugar beet. Results are shown for **(a)** the high-rainfall year 2054, **(b)** the average-rainfall year 2047 and **(c)** the low-rainfall year 2044 of the Tmax scenario run. The years were selected by the 10, 50 and 90% quantile for the annual rainfall in the time period 1951 to 2070 for scenario run Tmax of the RCP 8.5 scenario.

variance between different years (Figure 3) in all scenario runs; irrigation requirements also differ between the considered years (Figure 2). There was no irrigation necessary in the high-rainfall year 2054 of scenario run Tmax. The precipitation water was sufficient to grow sugar beet in this year. Two irrigation arrangements were necessary in the average-rainfall year 2047 of scenario run Tmax. With every irrigation arrangement, the soil water content increases by +2%. In the dry year 2044 of scenario run Tmax, nine irrigation arrangements were necessary to keep the soil water content sufficiently above 20% of the available field capacity. Irrigation would be still required late in the year on day 262.

The irrigation requirement is also related to the crop-specific evapotranspiration during the cultivation phase: crop species with a high irrigation requirement like sugar beet and potato show the highest total evapotranspiration amount (330 and 287 mm) during the irrigation phase compared to winter barley and winter rye (182 and 225 mm) with a lower irrigation requirement (Figure 4).

Figure 5 summarises the irrigation requirements during the considered time period. Accordingly, the irrigation requirement will increase for sugar beet, potato, winter wheat, winter rye and summer barley in the scenario runs Tmed (Kendall's tau b = +0.306, +0.266 and +0.262) and Tmax (b = +0.374, +0.385 and 0. + 311) as average temperatures increase. The longer the crop remains on the field, the higher the irrigation requirement is and will be, respectively. For field crops, which are harvested early (e.g. winter barley), the irrigation requirement remains on a comparatively low level (b = −0.13 in scenario run Tmin, +0.87 in scenario run Tmed and +0.103 in scenario run Tmax). The irrigation requirement is decreasing in the scenario run Tmin for all field crops because of a decreasing evapotranspiration and a slight increase in precipitation (b = +0.006 for sugar beet, −0.022 for potato and −0.139 for winter wheat, winter rye and summer barley). Because of a high variance, the Kendall's tau b as correlation coefficient and the coefficients of determination (R^2) show low or, at most, medium values for all different field crops and scenario runs. The irrigation

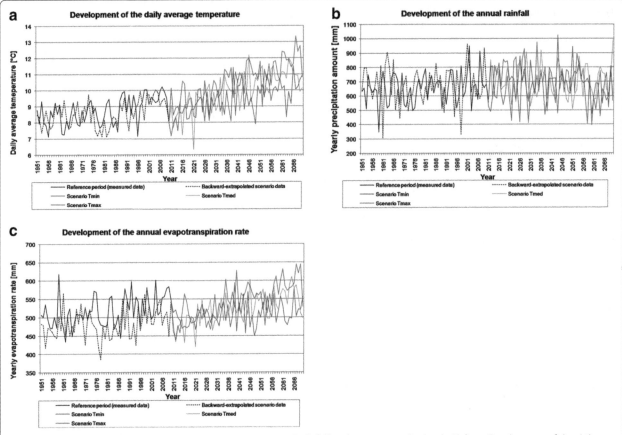

Figure 3 Development of the daily average temperature, annual rainfall and evapotranspiration in Uelzen. Development of the daily average temperature **(a)**, annual rainfall **(b)** and evapotranspiration **(c)** in the selected grid cell in the County of Uelzen (Northern Germany) during the reference period 1991 to 2010 and the different scenario runs Tmin, Tmed and Tmax (time period 2011 to 2070).

requirement is directly linked to the climate conditions (temperature, precipitation and evapotranspiration), which also show a high variance between the different years in all scenario runs (Figure 3).

To compute the moving average for the year 1991, the values for the six previous years 1985 to 1990 have been considered. For the year 2011, the values for the years 2005 to 2010 of the reference period 1991 to 2010 were considered. The simulated crop rotation starts with the cultivation of sugar beet in the year 1991.

The crop rotation sugar beet-potato-winter rye-winter barley is most common in the county of Uelzen and in particular on areas labelled as soil type 31 (brown podzolic soil out of slightly loamy, slightly silty sand and sand) (Figure 1; J. Hufnagel and N. Svoboda, ZALF, Müncheberg, Germany, personal communication). In particular, the cultivation of sugar beet and potato required large amounts of irrigation water. Figure 6 demonstrates that the irrigation requirement is likely to increase in the years after 2030 in the scenario run Tmax for this crop rotation compared to the reference period 1991 to 2010. The allowed amount of irrigation water of 79 mm/m^2 and year in a moving average over 7 years is likely to be

exhausted in several years after the year 2064. In the reference period 1991 to 2010 and in the scenario runs Tmin and Tmed, the required amount of irrigation water was always lower than 60 mm/m^2 and year in a moving average over 7 years. In scenario run Tmed, the irrigation requirement was similar to the reference period, but in some years (2052, 2063, 2065), the moving average was slightly higher than during the reference period 1991 to 2010. In scenario run Tmin, the irrigation requirement is decreasing compared to the reference period 1991 to 2010 because of a slight increase in precipitation and a decreasing evapotranspiration.

Discussion

The amount of water extracted for irrigation depends on climate and soil characteristics, on the political and economic boundary conditions, and on farmers' management decisions such as crops cultivated and techniques applied, as for instance, irrigation [42]. The results of this simulation study conducted in Northern Germany (County of Uelzen) show that together with increasing temperature and evapotranspiration, the irrigation requirements are likely to increase also (Figures 3,

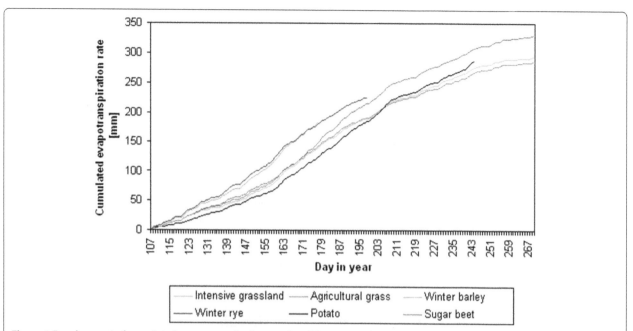

Figure 4 Development of cumulated evapotranspiration rate for different grassland types and crop species during irrigation phase. Evapotranspiration was simulated for a soil depth of 0 to 60 cm of soil type 31 (Figure 1) using WASMOD [41] in the average-rainfall year 2047 of the scenario run Tmax with two irrigation arrangements on days 228 and 240. Crop management specifications were set according to regional standards (N. Svoboda, personal communication, based on crop surveys executed at the Leibnitz Centre for Agricultural Landscape Research (ZALF) and provided to the NaLaMa-nT project data base). Curves are drawn for the irrigation relevant time span (until 3 weeks before harvest, see Methods section and Table 1).

5, and 6). On the basis of the scenario calculations, higher irrigation levels will most likely be required in the future (except scenario run Tmin). This could exceed local availability of ground water. Implications could be, on the one hand, the endangerment of regional wetlands [33] when the ground water level should substantially decrease due to irrigation water extraction. On the other hand, crop damage or crop failure could become more common - if the land use management would not be adapted.

Both, changes in average climate conditions during the main growing season and climate variability, particularly heat waves and droughts, affect crop growth. The fitting capacity to cope with climate change depends on the crop type and ecological boundary conditions [43]. Thus, additional irrigation requirements can be avoided or reduced by a reasonably adapted choice of the cultivated field crops [33]. With an accompanying climate change, dates for sowing could be shifted - to a limited extent - to earlier times in the year to use the natural water supply during the wet months in spring more efficiently [34]. At present, irrigation is mainly applied using long-throw sprinkler [33]. If irrigation efficiency would be improved by the application of techniques with lower evaporation losses (e.g. drip irrigation), the potential problems might be reduced. In some places the use of surface water for irrigation could also be adequately adjusted and contribute to the irrigation requirements [33].

In general, scenario data - like the results shown in this paper - go along with uncertainties and are just adequate to point out possible future trends. The uncertainties in the presented calculation approach originate amongst others from climate projection data, which are linked to uncertainties themselves. For example, the used scenario data seems to predict too low evapotranspiration amounts (Figure 3). This implies that the future irrigation requirement could be even higher than the calculated values indicate (Figures 5 and 6). The presented results in particular show how irrigation requirement in the county of Uelzen could develop in the future based on the employed scenario specifications.

The presented calculation approach can be understood as being based on conservative assumptions. In reality, irrigation arrangements are likely to be done at even higher soil water contents (e.g. at 30% to 50% of the available field capacity) and with higher amounts of irrigation water than 20 mm/m^2 and irrigation event. This leads to the assumption that the computed irrigation levels are even lower than they actually could be in the future.

In a few exceptional years, the soil water content might be lower at starting day 107 than the default start value due to a noticeably dry winter. In these cases the irrigation requirement could be higher than

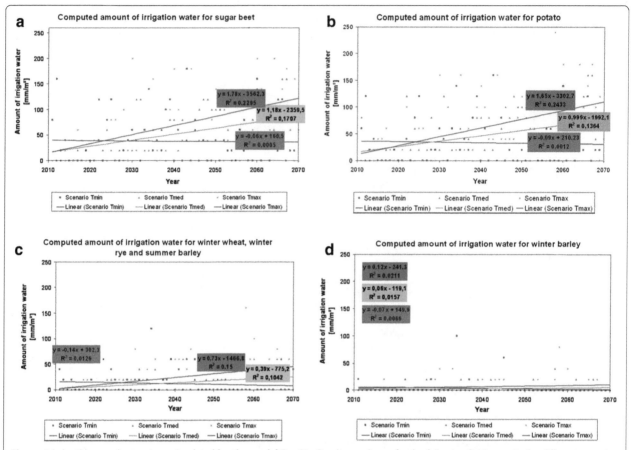

Figure 5 Irrigation requirements as simulated by the model BewUe. Results are shown for the following field crops in the different scenario runs: **(a)** sugar beet, **(b)** potato, **(c)** winter wheat, winter rye, summer barley and **(d)** winter barley. The linear regression functions and the coefficients of determination (R^2) obtained for the different crop species in the different scenario runs are also shown.

the calculated values indicate. Nevertheless, it could be shown that as by 2010, assuming the worst case scenario from down-scaled global climate change models, 62% to 80% of agricultural land within a Central European region could shift towards a new agroclimatic class and 98% by 2050, respectively [44].

It is important to emphasise that irrigation requirements cannot be directly derived from climate modelling. In fact, outcomes of climate modelling need to be linked with data on soil conditions, cultivation patterns and eventually also economic considerations in order to assess future sustainability of particular crops to be grown and the resulting requirements of external water input. Calculations were done with evapotranspiration data for grassland [45] based on simulated climate data [39]. This leads to uncertainties because the evapotranspiration for grassland can differ from agricultural sites depending on plant development stage and crop species. We checked this by computing evapotranspiration with a standalone model approach (Figure 4). The WASMOD results show that the evapotranspiration for

sugar beet and potato differs only to a relatively low extent compared to grassland (approximately +9% and −10% on average during the irrigation phase). The variance for winter barley and winter rye is higher compared to grassland (approximately +34% and +39% on average during the irrigation phase). As winter barley and winter rye are harvested early in the year (days 177 and 198), this affects a time period when irrigation requirement is relatively low as the soil retains water from the winter period. The calculation of crop-specific transpiration in combination with plant phenology is appropriate [46]; however, it would only gradually reduce remaining uncertainties. The current approach was based on Penman-Monteith calculations [45] for site-specific conditions.

In future research activities, it will be intended to expand the calculations to additional regions, soil types and crop rotations. This would be of high relevance since the European agriculture is characterised by high productivity [47] and accounts for 50% of the global trade food products [7].

Will climate change increase irrigation requirements in agriculture of Central Europe? A simulation study...

135

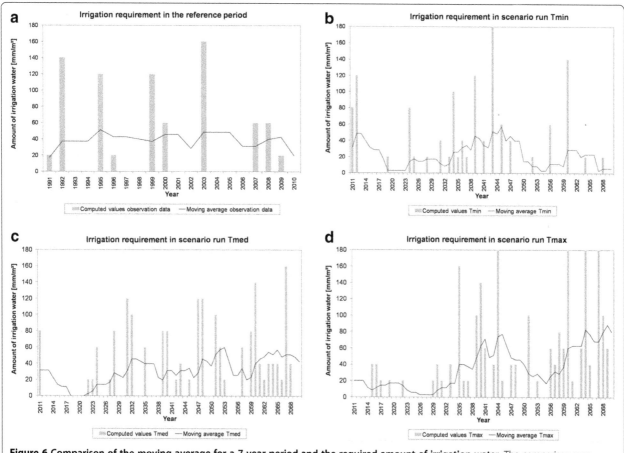

Figure 6 Comparison of the moving average for a 7-year period and the required amount of irrigation water. The comparison was simulated by the model BewUe for the crop rotation sugar beet-potato-winter rye-winter barley. Results are shown for **(a)** the reference period (1991 to 2010) and scenario runs Tmin **(b)**, Tmed **(c)** and Tmax **(d)** (time period 2011 to 2070).

Furthermore, the presented results for irrigation amounts obtained using the BewUe programme should be cross-checked with observation data for single field crops and different years. But such observation data are not available yet. Additionally, further climate change scenarios are of interest and should be used to improve understanding of the possible future irrigation requirement.

Conclusions

The results show that it is reasonable to expect that regional implications of global climate change will affect evapotranspiration as an important aspect in crop cultivation. Ground water availability for irrigation allows, under the given conditions, a short-term buffering towards extremes. Based on the scenario calculations, it can be expected that the current agricultural practice in the county of Uelzen will not be directly limited by regional climatic alterations. However, in the more distant future, where climate change is on the one side to become more pronounced and more uncertain to be

predicted on the other side, additional measures might be necessary to prevent higher frequencies of crop failures in some years. Changes in irrigation techniques or adaptation of crop rotation types are amongst these measures. The presented work shows how irrigation requirement can be calculated combining regional projections of climate conditions specified as long-term 'scenario-weather', combined with crop requirements and management, and soil conditions. The results show that it is likely that agriculture in the county of Uelzen has to be adapted to increasing irrigation requirement in the future, if conditions develop as assumed in the employed scenarios.

However, further research will give evidence whether changes either in institutional and market conditions or in climatic conditions will dominantly influence the development of agriculture illustrated by a comparable example of adaptation for maize production in Switzerland [42]. Since soil climate is expected to change significantly even in Central Europe, 'more attention should be paid to studying the impacts of climate change on soil climate' [48], i.e. soil temperature and hydric soil regimes.

Methods
Data basis

Homogenised daily climate observation and scenario data for the climate change scenario Representative Concentration Pathways (RCP) 8.5 were provided by the Potsdam Institut für Klimafolgenforschung (PIK) to the NaLaMa-nT project consortium for a 10 × 10 km grid and the time periods 1951 to 2010 [36] and 2011 to 2070 [39], respectively (see Figure 3). Observation data of the German weather service [36] were checked, homogenised and interpolated to the 10-km grid [49]. The scenario data are generated by STARS II [39] and interpolated to the same grid. Our calculations were done for one grid cell located in the County of Uelzen (Figure 1), which was selected because it has the highest representativity for the county of Uelzen. Data for this grid cell show the lowest sum of deviations for the parameters of daily mean temperature, precipitation, and global radiation compared to the county average. To assess the RCP 8.5 scenario of the STARS II model scenario, backward-extrapolated scenario data for the time period 1951 to 2010 were provided and were compared to the observation data. The RCP 8.5 scenario is the most fierce emission scenario in the recent IPCC assessment [1,50,51]. Nevertheless, the RCP 8.5 scenario is surpassed by actually observed emissions [52]. It is up to the year 2060 the most similar emission scenario to the so far widely used SRES A1B scenario [1]. The used STARS II projections (Tmin, Tmed, Tmax) were driven by different temperature gradients. For each projection, the median run - based on the change of climate water balance - of 100 model runs was selected. The three projections Tmin, Tmed and Tmax are based on results from the GCM models INM-CM4 (Institute of Numerical Mathematics, Russian Academy of Sciences, Moscow, Russia, 2009), ECHAM6 (Max Planck Institute for Meteorology, Hamburg, Germany, 2012) and ACCESS1.0 (Commonwealth Scientific and Industrial Research Organisation, Melbourne, Australia, 2013). They represent a low, median and high temperature increase for the RCP 8.5 scenario.

To calculate changes in soil humidity, we used additional data for evapotranspiration, which were available for grassland correspondent to the guidelines of the Food and Agriculture Organization of the United Nations (FAO) computed by the Penman-Monteith equation [45] but not for single field crops (homogenised data source for the time period 1951 to 2010 [36] and scenario data for the time period 2011 to 2070 [39]). The scenario input data are shown in Figure 3. We used WASMOD [41] to compare the evapotranspiration

amounts of different grassland types and crop species. They are in the expected order of magnitude, considering the condition that actual plant transpiration is a highly variable process leading to large standard deviations when measured in repeated experiments [53] (Figure 4).

The daily average temperature, precipitation and evapotranspiration show a high variance within the different years for the reference period 1991 to 2010 and in all scenario runs (Figure 3). This means that e.g. the general increase in temperature in scenario run Tmax is also interrupted by several years with low daily mean temperatures. The daily mean temperature is increasing in the time period 2011 to 2070 in the scenario runs Tmed and Tmax with +1.3 and +1.7°C in the time period 2011 to 2070, respectively. In the scenario run Tmin, there is a minor increase of +0.9°C on average. The annual evapotranspiration is +5.6 and +28.4 mm per year on average higher in the scenario runs Tmed and Tmax compared to the reference period 1991 to 2010. In the scenario run Tmin, the annual evapotranspiration is decreasing with −8.7 mm per year on average compared to the reference period 1991 to 2010. Concerning the annual rainfall, there is no clear trend indicating changes compared to the current situation. The annual rainfall is 654.4 mm per year during the reference period 1991 to 2010. On average, the annual rainfall is increasing in all scenario runs with +44.7 in Tmin, +45.0 in Tmed and +14.1 mm per year in Tmax compared to the reference period 1991 to 2010. Several climate scenarios also show that the main amount of the yearly precipitation will shift from the growing season (spring and summer) towards the time outside the vegetation period (i.e. winter) [33] or that the annual precipitation will even decrease [1]. This could also influence irrigation requirement.

Figure 3 also show that the modelled scenario data seem to be lower than in reality by comparing the measured climate data and the backward-extrapolated scenario data for the reference period 1991 to 2010. Temperature (Figure 3a) and especially evapotranspiration (Figure 3c) are decreasing after 2010 compared to the measured data in the reference period. Thus, irrigation requirement - calculated with observation data for the reference period and scenario data for the time period 2011 to 2070 - could be even higher in years after 2010 than the calculated values indicate.

Soil data of the land use-specific soil map of Germany [40] were used. The county of Uelzen is dominated by sandy soil types with low amounts of silt and/or clay (Figure 1). Because of the coarse-grained texture, these soils have a low available field capacity (approximately 17% to 22% in a soil depth of 0 to 30 cm) [40]. Soils in

Table 1 Crop plants grown in Uelzen with dates for harvesting and starting time of maturity

Crop	Share of agricultural area (%)	Harvest time (day in year)	Start time of maturity (day in year)
Spring barley	9	219	198
Triticale	4	219	198
Winter barley	12	198	177
Winter rye	5	219	198
Winter wheat	14	219	198
Winter rapeseed	4	183	162
Winter rapeseed with organic fertilisation		208	187
Silage maize	5	261	240
Potato	22	265	244
Sugar beet	15	290	269

Information courtesy of N. Svoboda, Leibnitz Centre for Agricultural Landscape Research (ZALF), Müncheberg, Germany.

eastern Lower Saxony show a soil water deficit in some places during the growing season [54].

The information on the current agricultural practice was provided by the Leibnitz Centre for Agricultural Landscape Research (N. Svoboda, personal communication, based on crop surveys executed at the ZALF and provided to the NaLaMa-nT project data base). The dominant crops are listed in Table 1. The most common crop rotation is sugar beet-potato-winter rye-winter barley, covering 11.9% of the agricultural area.

Methodological approach

For the calculations we used data for soil type 31 (brown podzolic soil out of slightly loamy, slightly silty sand and sand), which is the soil type most common in the county's agricultural area (47%) (Figure 1). Soil type 31 has a field capacity of 24% (which is equivalent to a soil water content of 240 l/m³), an available water capacity of 17% and a wilting point of 7% (which is equivalent to 70 l/m³ of stagnant water) in a soil depth of 0 to 60 cm [40]. The soil depth of 0 to 60 cm was assumed to be the rooting zone of the soil and therefore the relevant part of the soil considered in our calculations.

Soil and climate data were used in the model BewUe written by B. Breckling using the programming language SIMULA [55]. The model functionality encompasses reading input data for a number of successive years and writing results into according annual output files conditionally adding irrigation when soil water content falls beneath the given threshold, i.e. it computes irrigation requirements including the day(s) of the year on which irrigation is required for single field crops and different scenario runs (Tmin, Tmed and Tmax) of the RCP 8.5 climate scenario. Irrigation requirements depend on the soil water content. We assumed that irrigation will be applied as soon as the soil water content becomes lower than 20% of the available water capacity, equivalent to a soil water content of 10.4% (which is equivalent to a soil water content of 104 l/m³ including 70 l/m³ of stagnant water and 34 l/m³ = 20% of available field capacity). The amount of water applied in a single irrigation event was set

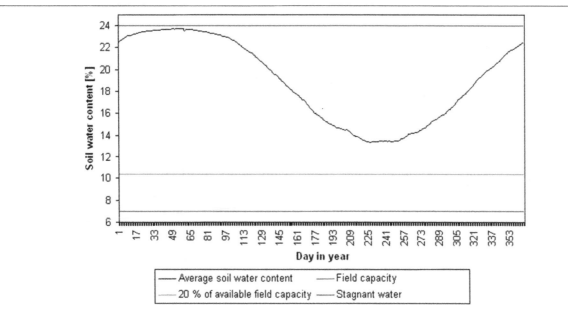

Figure 7 Average soil water content for the soil type 31 of the BÜK 1000. Average soil water content for the soil type 31 (brown podzolic soil out of slightly loamy, slightly silty sand and sand) of the BÜK 1000 [40] in a soil depth of 0 to 60 cm in the time period 1951 to 2070 (for observation data and scenario run Tmed).

to 20 mm/m^2. The main model functional is outlined in the pseudocode given in Equation 1.

Begin
Soil water content (day_{n+1}) = soil water content (day_n)
+ precipitation amount (day_n)
(read from climate data)
− evapotranspiration amount (day_n)
(read from climate data)
If soil water content (day_n) < (threshold) then soil water content (day_n) =
soil water content (day_n) +
irrigation amount of 20 mm / m^2
End

$$(1)$$

The maximum soil water content is defined as ≤ 240 l/m^3 (soil water content at field capacity) and the minimum soil water content as ≥ 70 l/m^3 (amount of stagnant water).

In the first step, we determined the time period, in which irrigation is necessary (irrigation phase) and in which the soil water content should be computed by the use of the SIMULA model. Therefore, the soil water content was calculated for average conditions with the climate observation data for the time period 1951 to 2010 and the Tmed scenario run for the time period 2011 to 2070. For this calculation, the soil water content was set to 240 l/m^3 on the first day (1 January 1951) assuming that the soil is fully water saturated at this time. To compute the average development of the soil water content, the daily values were averaged over the time period 1951 to 2070 (Figure 7).

The soil water content decreases noticeably after day 107 in an average year (Figure 7). In average years, irrigation should not be required before day 107. The climatic water balance gets negative from day 107 onwards because of an increasing evapotranspiration that is not compensated by precipitation. Thus, the calculation of the soil water content within the BewUe model was set to start at day 107 with an actual value of 224.9 l/m^3. Starting time and start value are considered representative for average climate conditions in the county of Uelzen and were used to compute the irrigation requirement for all different scenario runs, years and crop species.

Irrigation was considered unnecessary during the maturation phase of the field crops. In general, maturation takes between 2 and 4 weeks depending on the yearly course of the weather (J. Hufnagel and N. Svoboda, ZALF, Müncheberg, Germany, personal communication). In our calculations, we assumed that maturation starts 3 weeks before harvesting in all years and different scenario runs. Harvest dates differ between the relevant crops (Table 1). Thus, the soil water content and irrigation requirement for e.g. sugar beet were computed from day 107 with a start value of 224.9 l/m^3 until day 269 (i.e. start of maturation). The crop plants listed in Table 1 grow on 90% of the currently cultivated agricultural area in the

county of Uelzen. In particular, the cultivation of sugar beet and potato, which are grown on 46% of the whole agricultural crop land, is linked to a high irrigation requirement [27].

Temporal trends for irrigation requirements are secured statistically by computing the Kendall's tau b as correlation coefficient (Figure 5). The Mann-Kendall test is the adequate method for non-parametric tests and it is not influenced by seasonal effects.

Competing interests
The authors declare that they have no competing interests.

Authors' contributions
JR and BB participated in generating the methodological approach and in writing of the article. BB evolved the simulation model and JR did the calculation. WS is the leader of the research project in which the simulation study was embedded. He gave conceptual advices, conducted comprehensive literature research and revised the article. RN helped with the preparation of the climate data and contributed to the description of the employed climate and scenario data. All authors read and approved the final manuscript.

Acknowledgements
The NaLaMa-nT project and the presented work are funded by the Federal Ministry of Education and Research under research grant 033L029H. We thank N. Svoboda and J. Hufnagel (Leibnitz Centre for Agricultural Landscape Research (ZALF), Müncheberg, Germany) for the provision of crop rotation and crop management information.

Author details
^{1}Chair of Landscape Ecology, University of Vechta, Driverstraße 22, PO Box 15 53, Vechta 49377, Germany. ^{2}Northwest German Forest Research Station, Grätzelstraße 2, Göttingen 37079, Germany.

References
1. Intergovernmental Panel on Climate Change (IPCC): *Climate Change 2013: The Physical Science Basis. Contribution of Working Group I to the Fifth Assessment Report of the Intergovernmental Panel on Climate Change*. In Edited by Stocker TF, Qin D, Plattner G-K, Tignor M, Allen SK, Boschung J, Nauels A, Xia Y, Bex V, Midgley PM. Cambridge, United Kingdom and New York, NY, USA: Cambridge University Press; 2013:1535.
2. Intergovernmental Panel on Climate Change (IPCC), Solomon S, Qin D, Manning M, Chen Z, Marquis M, Averyt KB, Tignor M, Miller HL (Eds): *Climate Change 2007: The Physical Science Basis. Contribution of Working Group I to the Fourth Assessment Report of the Intergovernmental Panel on Climate Change*. Cambridge: Cambridge University Press; 2007:996.
3. Goyal RK: Sensitivity of evapotranspiration to global warming. A case study of arid zone of Rajasthan (India). *Agric Water Manag* 2004, 69:1–11.
4. Rind D, Goldberg R, Hansen J, Rosenzweig C, Ruedy R: Potential evapotranspiration and the likelihood of future drought. *J Geophys Res Atmos* 1990, 95(7):9983–10004.
5. Bindi M, Olesen JE: The responses of agriculture in Europe to climate change. *Reg Environ Chang* 2011, 11(Suppl 1):151–158.
6. Ciscar J-C: The impacts of climate change in Europe (the PESETA research project). *Clim Chang* 2012, 112:1–6.
7. Iglesias A, Garrote L, Quiroga S, Moneo M: A regional comparison of the effects of climate change on agricultural crops in Europe. *Clim Chang* 2012, 112:29–46.
8. Directive 2000/60/EC of the European Parliament and of the Council of 23 October 2000 establishing a framework for community action in the field of water policy. *OJ L* 2000, 327:1–73.
9. Tzilivakis J, Warner DJ, Green A, Lewis KA: Adapting to climate change: assessing the vulnerability of ecosystem services in Europe in the context of rural development. *Mitig Adapt Strateg Glob Chang* 2013, 1–26. doi:10.1007/s11027-013-9507-6.

10. Wriedt G, Van der Velde M, Aloe A, Bouraoui F: *Water requirements for irrigation in the European Union. A model based assessment of irrigation water requirements and regional water demands in Europe, Scientific and Technical Research series.* Ispra: European Commission, Joint Research Centre Institute for Environment and Sustainability; 2008.

11. Calzadilla A, Rehdanz K, Betts R, Falloon P, Wiltshire A, Tol RSJ: **Climate change impacts on global agriculture.** *Clim Chang* 2013, **120**:357–374.

12. Chaturvedi V, Hejazi M, Edmonds J, Clarke L, Kyle P, Davies E, Wise M: **Climate mitigation policy implications for global irrigation water demand.** *Mitig Adapt Strateg Glob Chan* 2013, 1–19. doi:10.1007/s11027-013-9497-4.

13. Christensen OB, Goodess CM, Ciscar J-C: **Methodological framework of the PESETA project on the impacts of climate change in Europe.** *Clim Chang* 2012, **112**:7–28.

14. Heumesser C, Fuss S, Szolgayová J, Strauss F, Schmid E: **Investment in irrigation systems under precipitation uncertainty.** *Water Resour Manag* 2012, **26**:3113–3137.

15. González-Zeas D, Quiroga S, Iglesias A, Garrote L: **Looking beyond the average agricultural impacts defining adaption needs in Europe.** *Reg Environ Chang* 2013, **2013**:1–11. doi:10.1007/s10113-012-0388-0.

16. Döll P: **Impact of climate change and variability on irrigation requirements: a global perspective.** *Clim Chang* 2002, **54**:269–293.

17. Lehner B, Döll P, Alcamo J, Henrichs T, Kaspar F: **Estimating the impact of global change on flood and drought risks in Europe: a continental, integrated analysis.** *Clim Chang* 2006, **75**:273–299.

18. Olesen JE, Carter TR, Díaz-Ambrona CH, Fronzek S, Heidmann T, Hickler T, Holt T, Minguez MI, Morales P, Palutikof JP, Quemada M, Ruiz-Ramos M, Rubaek GH, Sau F, Smith B, Sykes MT: **Uncertainties in projected impacts of climate change on European agriculture and terrestrial ecosystems based on scenarios from regional climate models.** *Clim Chang* 2007, **81**:123–143.

19. Siebert S, Henrich V, Frenken K, Burke J: *Update of the global map of irrigation areas to version 5. Project report.* Bonn, Germany: Institute of Crop Science and Resource Conservation, Rheinische Friedrich-Wilhelms-Universität; 2013:1–178.

20. Maracchi G, Sirotenko O, Bindi M: **Impacts of present and future climate variability on agriculture and forestry in the temperate regions: Europe.** *Clim Chang* 2005, **70**:117–135.

21. Jopp F, Reuter H, Breckling B: *Modelling Complex Ecological Dynamics.* Heidelberg: Springer; 2011.

22. Sacks WJ, Cook BI, Buenning N, Levis S, Helkowski JH: **Effects of global irrigation on the near-surface climate.** *Clim Dyn* 2009, **33**:159–175.

23. Frenken K, Gillet V: **Aquastat. Irrigation water requirement and water withdrawal by country.** [http://www.fao.org/nr/water/aquastat/water_use_agr/index.stm]

24. Iglesias A, Quiroga S, Moneo M, Garrote: **From climate change impacts tot he development of adaption strategies: challenges for agriculture in Europe.** *Clim Chang* 2012, **112**:143–168.

25. European Commission: **Agriculture and rural development. Agriculture and environment. Agriculture and water.** [http://ec.europa.eu/agriculture/envir/water/]

26. European Commission: **Agri-environmental indicator. irrigation.** [http://epp.eurostat.ec.europa.eu/statistics_explained/index.php/Agri-environmental_indicator_-_irrigation#Further_Eurostat_information]

27. Battermann HW, Theuvsen L: *Feldberegnung in Nordost-Niedersachsen. Regionale Bedeutung und Auswirkungen differenzierter Wasserentnahmeerlaubnisse, Endbericht. Studie im Auftrag des Fachverbandes Feldberegnung 2007–2009.* Georg-August-Unsiversität Göttingen, Department für Agrarökonomie und Rurale Entwicklung; [http://www.lwk-niedersachsen.de/index.cfm/portal/6/nav/203/article/14250.html]

28. Food and Agriculture Organization of the United Nations (FAO): **Global map of irrigation areas. Germany.** [http://www.fao.org/nr/water/aquastat/irrigationmap/deu/index.stm]

29. Fricke E: **Optimierte Nährstoffausnutzung durch Bewässerung.** [http://www.fachverband-feldberegnung.de/basisinfo.htm]

30. Heidt L: **Auswirkungen des Klimawandels auf die potentielle Beregnungsbedürfigkeit Nordost-Niedersachsens.** *Geoberichte* 2009, **13**:60–67.

31. Schaller M, Weigel H-J: **Analyse des Sachstands zu Auswirkungen von Klimaveränderungen auf die deutsche Landwirtschaft und Maßnahmen zur Anpassung.** *FAL Landbauforschung Völkenrode* 2007, **316**:247p.

32. Weichselgartner W: *Risiko – Wissen – Wandel.* Strukturen und Diskurse problemorientierter Umweltforschung. München: Oeko; 2013.

33. Landwirtschaftskammer Niedersachsen (NLWKN): **Projektbericht. Aquarius. Dem Wasser kluge Wege ebnen.** 2012, [http://www.lwk-niedersachsen.de/index.cfm/portal/6/nav/203/article/12396.html]

34. Landwirtschaftskammer Niedersachsen (NLWKN): *Projektbericht. No Regret. Genug Wasser für die Landwirtschaft?!*; 2008. [http://www.lwk-niedersachsen.de/index.cfm/portal/6/nav/203/article/14250.html]

35. Schröder W, Schmidt G, Pesch R, Matejka H, Eckstein T: *Konkretisierung des Umweltbeobachtungsprogrammes im Rahmen eines Stufenkonzeptes der Umweltbeobachtung des Bundes und der Länder Teilvorhaben 3.* Berlin: Umweltforschungsplan des Bundesministers für Umwelt, Naturschutz und Reaktorsicherheit; 2001.

36. *2010: German Weather Service (DWD);* 2010.

37. Krysanova V, Dickens C, Timmerman J, Varela-Ortega C, Schlüter M, Roest K, Huntjens P, Jaspers F, Buiteveld H, Moreno E, de Pedraza Carrera J, Slámová R, Martínková M, Blanco I, Esteve P, Pringle K, Pahl-Wostl C, Kabat P: **Cross-comparison of climate change adaption strategies across large river basins in Europe, Africa and Asia.** *Water Resour Manag* 2010, **24**:4121–4160.

38. Bloemer S: **Ingenieurbiologie und Klimawandel. Worauf sich Planer und Unternehmen einstellen müssen.** *Neue Landschaft* 2008, **8**:46–53.

39. Orlowsky B, Gerstengarbe FW, Werner PC: **A resampling scheme for regional climate simulations and its performance compared to a dynamical RCM.** *Theor Appl Climatol* 2008, **92**(3–4):209–223.

40. Federal Institute for Geosciences and Natural Resources (BGR): *Use-specific soil map 1:1.000.000 (BÜK 1000 N2.3).* Hannover and Berlin: Digit. Archiv FISBo BGR; 2007.

41. Reiche EW: **Entwicklung, Validierung und Anwendung eines Modellsystems zur Beschreibung und flächenhaften Bilanzierung der Wasser- und Stickstoffdynamik in Böden.** *Kieler Geographische Schriften* 1991, **79**:1–150.

42. Finger R, Hediger W, Schmid S: **Irrigation as adaption strategy to climate change—a biophysical and economic appraisal for Swiss maize production.** *Clim Chang* 2011, **105**:509–528.

43. Moriondo M, Bindi M, Kundzewicz ZW, Szwed M, Chorynski A, Matczak P, Radziejewski M, McEvoy D, Wreford A: **Impact and adaptation opportunities for European agriculture in response to climatic change and variability.** *Mitigation Adaptation Strategies Global Change* 2010, **15**:657–679.

44. Trnka M, Eitzinger J, Semerádivá D, Hlavinka P, Balek J, Dubrovský M, Kubu G, Štěpánek P, Thaler S, Možný M, Žalud Z: **Expected changes in agroclimatic conditions in Central Europe.** *Clim Chang* 2011, **108**:261–289.

45. Allen RG, Pereira LS, Raes D, Smith M: *Crop Evapotranspiration. Guidelines for Computing Crop Water Requirements. FAO irrigation and drainage paper. vol 56.* 1998:1–15.

46. Ma S, Churkina G, Trusilova K: **Investigating the impact of climate change on crop phonological events in Europe with a plant phenology model.** *Int J Biometeorol* 2012, **56**:749–763.

47. Mariani L, Parisi SG, Cola G, Failla O: **Climate change in Europe and effects on thermal resources for crops.** *Int J Biometeorol* 2012, 1–12. doi:10.1007/s00484-012-0528-8.

48. Trnka M, Kersebaum KC, Eitzinger J, Hayes M, Hlavinka P, Svoboda M, Dubrovský M, Semerádová D, Wardlow B, Pokorný E, Možný M, Wilhite D, Zdeněk Ž: **Consequences of climate change for the soil climate in Central Europe and the central plains of the United States.** *Clim Chang* 2013, **120**:405–418.

49. Österle H, Gerstengarbe F-W, Werner PC: *Ein neuer meteorologischer Datensatz für Deutschland, 1951 – 2003,* Proceedings der 7. Deutschen Klimatagung 2006. Klimatrends: Vergangenheit und Zukunft, Meteorologisches Institut der Ludwig-Maximilians-Universität, München: 3p.

50. Moss RH, Edmonds JA, Hibbard KA, Manning MR, Rose S: **The next generation of scenarios for climate change research and assessment.** *Nature* 2010, **463**:747–756.

51. van Vuuren DP, Edmonds J, Kainuma M, Riahi K, Thomson A, Hibbard K, Hurtt GC, Kram T, Krey V, Lamarque J-F, Masui T, Meinshausen M, Nakicenovic N, Smith SJ, Rose SK: **The representative concentration pathways: an overview.** *Clim Chang* 2011, **109**:5–31.

52. Peters GP, Andrew RM, Boden T, Josep Canadell G, Ciais P, Le Quéré C, Marland G, Raupach MR, Wilson C: **The challenge to keep global warming below 2°C.** *Nat Clim Chang* 2013, **3**:4–6.

53. Korres W: *Analysis of surface soil moisture patterns in an agricultural landscape utilizing measurements and ecohydrological modelling. PhD Thesis.* Germany: University of Cologne; 2013.

54. Sutmöller J, Spellmann H, Fiebiger C, Albert M: **Der Klimawandel und seine Auswirkungen auf die Buchenwälder in Deutschland.** *Beiträge aus der NW-FVA* 2008, **3:**135–158.

55. Dahl O, Myrhaug G, Nygaard K: *SIMULA 67. Common base language. Technical report publication,* Volume 1; 1968:172 pages.

13

Environmental impacts of nanomaterials: providing comprehensive information on exposure, transport and ecotoxicity - the project DaNa2.0

Dana Kühnel[1*], Clarissa Marquardt[2], Katja Nau[2], Harald F Krug[3], Björn Mathes[4] and Christoph Steinbach[4]

Abstract

Background: Assessing the impact of new technologies or newly developed substances on our environment is a challenge, even more so if the applied test methods - both toxicological and analytical - are often found to be inadequate and need amendments or even new developments as it is in the case of nanotechnology. This is illustrated by numerous publications in the field of nano-ecotoxicology which although they have been investigating the impact of a number of nanomaterials on several organisms almost never allow for explicit statements on potential hazards of these nanomaterials. This fact not only hampers the knowledge communication to all non-scientists (e.g. consumers) but it also complicates the transfer of the obtained results for other scientists.

Results: Risk communication is an essential and thus integral part of risk management. For this purpose, the project DaNa2.0 (Data and knowledge on nanomaterials - processing of socially relevant scientific facts) provides processed and hence easy accessible information on the potential safety issues of nanomaterials, mainly via the project website www.nanoobjects.info. This will allow various stakeholder groups to get impartial information on potential effects of nanomaterials and help consumers to make informed decisions. DaNa2.0 is funded by the German Federal Ministry of Education and Research (BMBF) and supported by the Swiss government as well as by the European InterReg IVb programme. The DaNa2.0 team is an interdisciplinary group of scientists from different areas such as materials science, chemistry, biology and human and environmental toxicology. Extending the project team in DaNa2.0 with European experts allows for broadening of the existing knowledge portfolio by adding further cross-cutting topics and increasing our expertise, e.g. in the field of environmental exposure and fate.

Conclusions: On the project website www.nanoobjects.info, a unique link between nanomaterials in practical applications (e.g. environmental remediation) and their potential impacts is provided. The focus of this publication will be on all issues with environmental relevance, which are addressed in the 'Knowledge Base Nanomaterials' on the project website. These issues include environmental exposure and behaviour of nanomaterials and nano-ecotoxicology.

Keywords: Nanomaterials; Nano-objects; Nano-ecotoxicity; Knowledge base; Knowledge dissemination; Nanotechnology; Environmental impact

Background

Since the start of the DaNa project with its corresponding website www.nanoobjects.info in 2009 and its continuation in DaNa2.0 since 2013, the annual number of publications dealing with toxicological aspects of nanomaterials has been increasing exponentially. Numerous research papers on various nanomaterials and organisms have been published (as demonstrated by extensive reviews, e.g. [1,2]), resulting in a growing number of nanomaterials (NM) and their related information on the website. Besides the scientific output of these literature sources, there is an increasing demand for more information on this new technology, reflected in the growing number of visitors to the project website.

However, it is still difficult to provide simple and explicit statements on potential environmental hazards of nanomaterials. This is either due to a lack of information, e.g. regarding the physico-chemical properties of nanomaterials,

* Correspondence: dana.kuehnel@ufz.de
[1]Department of Bioanalytical Ecotoxicology, Helmholtz Centre for Environmental Research - UFZ, 04318 Leipzig, Germany
Full list of author information is available at the end of the article

or due to a high variation in test procedures and consequently to a high variability in test results. Beyond that, the transfer of nano-toxicity-related information to scientist from other fields and also to non-scientists is difficult. To bridge this gap, the project DaNa2.0 is working on improving these communication pathways from nano-safety-related science to various stakeholder groups. The major goal of the DaNa Knowledge Base Nanomaterials is to provide condensed, up-to-date and easy comprehensible information on the safety of nanomaterials for different recipient groups simultaneously [3-5]. The website provides a unique combination of material science-based information and nano-applications with toxicological information, helping, e.g. consumers in making informed decisions.

In order to assess the quality of the published studies and to consequently achieve an aggregation to the ones meeting the requirements of good scientific practice, the DaNa expert team developed the so-called Literature Criteria Checklist. The list of chosen criteria allows the side-by-side evaluation of background information on the applied nanomaterial together with the selected toxicological assays [4]. Ultimately, only those scientific findings complying with the checklist will be integrated in the knowledge base. This evaluation process along with the writing of the respective articles is achieved by a multi-disciplinary core team that is supported since the start of the follow-up project DaNa2.0 (2013) by several external experts bringing in their expertise in specific fields.

In order to cover the full scope of potential interactions of nano-related applications with the various organisms, the DaNa2.0 expert team evaluates all scientific research papers with regard to human and ecotoxicological aspects thereby continuously updating the articles on the 25 market-relevant nanomaterials listed on the website [3].

This paper focuses on the DaNa2.0 activities with regard to environmental aspects, namely exposure and behaviour in the environment, as well as potential hazards towards environmental organisms. Furthermore, the website provides an overview of the German Nanosafety Research Landscape. On the DaNa2.0 internet platform, research projects funded by the German Federal Ministry of Education and Research (BMBF) are provided with the opportunity to present their project results related to environmental issues of nanotechnology in a recipient-friendly way.

Results and discussion
DaNa2.0 - general structure of the platform (www.nanoobjects.info)
The content of the website is structured into different areas providing different types of information on the opportunities and potential risks of nanotechnology:

1. Knowledge base - provides comprehensive articles on application, material properties, exposure, behaviour and toxicity of currently 25 engineered nanomaterials (ENMs)
2. Basics - provides short information on basic issues from release to uptake behaviour, risk and fate of ENM
3. Cross-cutting section - provides comprehensive articles on over-arching issues in nanotechnology, significant to more than one ENM
4. Methodology - provides a collection of nano-relevant standard operating procedures (SOPs) (template and finalised documents) together with documents used for quality management of cited literature (Literature Criteria Checklist)
5. Projects section - provides an overview on relevant funded national projects
6. Frequently asked questions (FAQ) section - provides a collection of questions and answers to relevant nano-related issues and opportunity to directly ask questions to the DaNa2.0 experts
7. Glossary - provides short definitions of related terms used in the articles
8. News section - provides announcements of new developments in nanotechnology, important publications, conferences and workshops, and more
9. Links section - provides links to websites, networks, information platforms as well as national and international projects on nanotechnology and safety issues

The content of the articles within the knowledge base and also in the cross-cutting section is presented in three levels of complexity according to needs and interest of different recipient groups:

1. Short and basic - interested citizens/consumers (public oriented)
2. Detailed - journalists, stakeholders, scientists from other fields
3. Complex - scientists from related fields, regulators

DaNa2.0 Knowledge Base Nanomaterials
The core section of the DaNa2.0 web platform is the knowledge base which provides a wealth of facts and data for currently 25 market-relevant nanomaterials. The actual state of knowledge is derived from evaluation of the scientific literature, from project reports and official reports from governmental agencies or organisations. An application-based database was established, as most visitors rather seek information related to a specific product than to a specific type of nanomaterial (Figure 1). The table-based approach allows the selection either of an application, e.g. 'environmental remediation' and

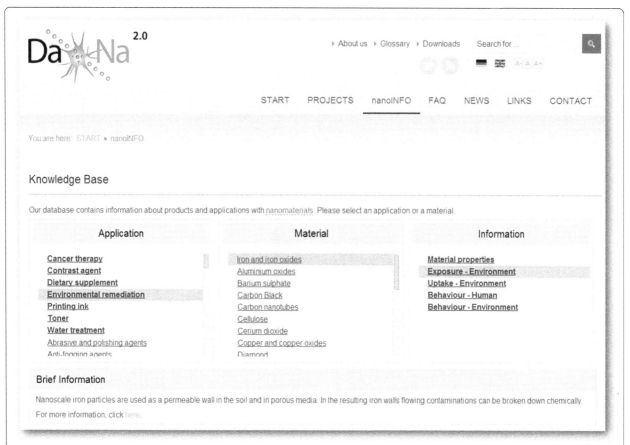

Figure 1 Knowledge base. The knowledge base is the core of the DaNa2.0 website and is organised in a table-based approach to provide links between nanomaterials, nano-products, and additional information on the material and potential toxicological effects [6].

'textiles' (Figure 1, column 1), or of an ENM of interest, e.g. 'iron oxide' (Figure 1, column 2). In turn, the nanomaterial used in the respective type of application or the type of application belonging to an ENM is highlighted. Several applications with an environmental background are already included and the total list of applications which is updated on a regular basis. Having selected the desired application together with a material, the availability of further information on exposure, uptake, behaviour and the material is indicated (Figure 1, column 3). By selecting an issue of interest, the brief information box below the table not only provides a comprehensive statement on the state of knowledge but also allows accessing more complex information. The interested visitor is then navigated to the more in-depth content of the main knowledge base. Likewise, all 25 ENMs included in the knowledge base can also be accessed directly via the material menu.

The in-depth content of the knowledge base provides besides general material information (production, usage, properties) sections on exposure, uptake and behaviour with regard to human and environmental toxicology and potential risks for environmental organisms (e.g. fish,

plants, mussels; see Table 1). Figure 2 shows an exemplary excerpt from the website for the nanomaterials 'iron and iron oxides' that deals with 'uptake and potential risks for environmental organisms'. The introduction of organism icons is currently introduced into the articles to facilitate the navigation and information access.

An overview on the huge variety of organisms, primary cells and cell lines used to study potential hazardous effects of nanomaterials is given in Table 1, summarising the 25 market-relevant ENMs currently listed in the knowledge base together with the corresponding environmental test systems. For 23 out of the 25 listed ENMs in the knowledge base, there is ecotoxicological information available. As for the human toxicology, the body of literature on environmental effects of nanomaterials is growing rapidly resulting in the aforementioned problems with comparability. In most cases, different types of the same nanomaterials were used or experimental conditions varied from study to study making it increasingly difficult to interpret and compare the results [9]. In order to deal with these issues, the DaNa expert team developed the 'Literature Criteria Checklist' to manage and monitor the quality

Table 1 Overview of all 25 market-relevant nanomaterials listed currently in the DaNa knowledge base

Nanomaterial	Environmentally relevant application or product with likely release to the environment	Anticipated most relevant release path/way	Studied organisms/cells/cell lines considered in the knowledge base at www.nanoobjects.info
Aluminium oxide	Abrasive and polishing agents	Aerosols	Mud tube worm, shrimp, earthworm, basket shell, nematode, bacteria, daphnids, zebrafish, thale cress, rye, lettuce, corn, carrots, soy, cabbage, cucumber, radish, rapeseed, ryegrass
Barium sulphate	Contrast agent	Waste water	No ecotoxicity studies available
Carbon black	Printing ink, toner	Aerosol	Common mussel, brown algae (toothed wrack), fruit fly, amphipods
Carbon nanotubes (CNT)	All applications anticipated so far involve CNT embedded in a matrix		Rainbow trout, earthworm, lugworm, tomato, cabbage, carrot, onion, radish, rape, lettuce, ryegrass, corn and cucumber
Cellulose	Wound dressing	Solid waste	Bacteria, fungi, daphnids, fish, algae, rainbow trout liver cells
Cerium dioxide	Diesel additive	Car exhaust	Bacteria, algae, zebrafish (embryo, adult), daphnids, rainbow trout liver cells, soybean
Copper and copper oxide	Wood preservatives	Waste	Bacteria, protozoa, worms, snail, frog embryo, zebrafish, green algae, radish, ryegrass, duckweed, corn
Diamond	Abrasive and polishing agents	Aerosol	Nematodes, frog embryos, daphnids
Fullerenes	Cosmetics, sports equipment, lubricant	Direct release to surface or waste water	Bacteria, algae, daphnids, common mussel blood cells, oyster (larvae, adult, liver cells), zebrafish embryo
Gold	Diagnostics and therapy	Waste water	Bacteria, daphnids, blue mussel, basket shell, zebrafish embryo, rainbow trout liver cell, cucumber, lettuce
Graphene	No applications on the market, yet		Bacteria, nematodes, zebrafish embryo, tomato, spinach, cabbage
Indium tin oxide (ITO)	All applications anticipated so far involve ITO embedded in a matrix		Various aquatic organisms (vertebrates, algae)
Iron and iron oxides	Ground and waste water remediation	Intended release to ground water	Bacteria, medaka embryos, rainbow trout gill cells, pumpkin, lima bean, ryegrass
Platinum	Catalytic exhaust converter	Aerosol formation	Zebrafish embryo, rainbow trout gill cells
Quantum dots (QD)	All applications anticipated so far involve quantum dots embedded in a matrix		Bacteria, algae, daphnids, zebrafish embryo, rainbow trout
Silicon dioxide	Skin care, textiles, therapeutics	Surface and waste water	Bacteria, green algae, mussel blood cells
Silver	Textiles, antimicrobial applications	Wash-off, release of ions from silver-coated surfaces	Bacteria, fungi, green algae, daphnids, zebrafish embryo, medaka cell line
Strontium carbonate	Pyrotechniques	Aerosol	No ecotoxicity studies available
Titanium dioxide	Sunscreen, surface coating	Direct release to surface waters, run-off from coated surfaces	Bacteria, daphnids, nematodes, lugworm, mussel, snail, woodlice, rainbow trout, zebrafish (embryo and adult), green algae, onion, willow, tobacco
Titanium nitride	All applications anticipated so far involve TiN embedded in a matrix		Rainbow trout gill and intestinal cells

Table 1 Overview of all 25 market-relevant nanomaterials listed currently in the DaNa knowledge base *(Continued)*

Tungsten carbide (WC)	WC nanoparticles are used for tool production and are tightly bound in the tools		Rainbow trout gill cells
Tungsten carbide cobalt (WC-Co)	WC-Co nanoparticles are used for tool production and are tightly bound in the tools		Rainbow trout gill cells
Zeolite/clays	Fertiliser	Soil	No ecotoxicity studies available
Zinc oxide	Sunscreen, textiles	Direct release to surface waters, wash-off	Bacteria, protozoa, woodlice, zebrafish, rainbow trout, daphnids, green algae, ryegrass, corn, soybean, zucchini, thale cress
Zirconium dioxide	All applications anticipated so far involve ZrO_2 embedded in a matrix		Bacteria, algae, zebrafish embryo

Relevant environmental applications, the anticipated most relevant release path into the environment and the test organisms used in the studies were considered in this summary. As the whole ENM life cycle was not considered, disposal as relevant release at end of life was left out.

Figure 2 DaNa2.0 knowledge base. The DaNa2.0 knowledge base provides a constantly growing source of information regarding the environmental impact of nanomaterials [7]. For 23 out of the 25 nanomaterials included in the knowledge base, there is ecotoxicological information available. Currently, the accessibility of information is improved by using organism icons to facilitate the navigation throughout the articles [8].

of all used literature prior to inclusion and publication in the knowledge base (see respective chapter).

In general, consumer products with a high likelihood of the applied ENM to get in close contact to the user such as sunscreen or textiles are of biggest concern for humans whilst at the same time these ENMs are also the most prone to enter our environment. Therefore, these ENMs have also a great potential to be hazardous for the environment and its inhabitants, and ultimately, a polluted environment will always act back on humans and their conditions of living. Keeping this in mind, the DaNa2.0 team supported by the external experts is working continuously on constantly updating the state of knowledge in these fields. Specifically for environmental issues, the external experts provide background in (environmental) risk assessment, ecotoxicity, NM release and transport, and exposure assessment.

Other sections with environmental relevance
Basics section
The basics section provides fundamental introductions into nanotechnology and safety issues, such as general information on release of nanomaterials, uptake and behaviour in organisms and environmental media, as well as a section on how to assess the potential risk NM pose to humans and the environment (Figure 3).

Cross-cutting section
This section deals with overarching issues with significance to ENMs in general. The two articles currently published on the website deal with coatings and with different crystal structures of ENM, respectively. Both issues are not only of relevance to materials scientists but also need consideration in (eco)toxicological testing. This section will be further expanded during the term of

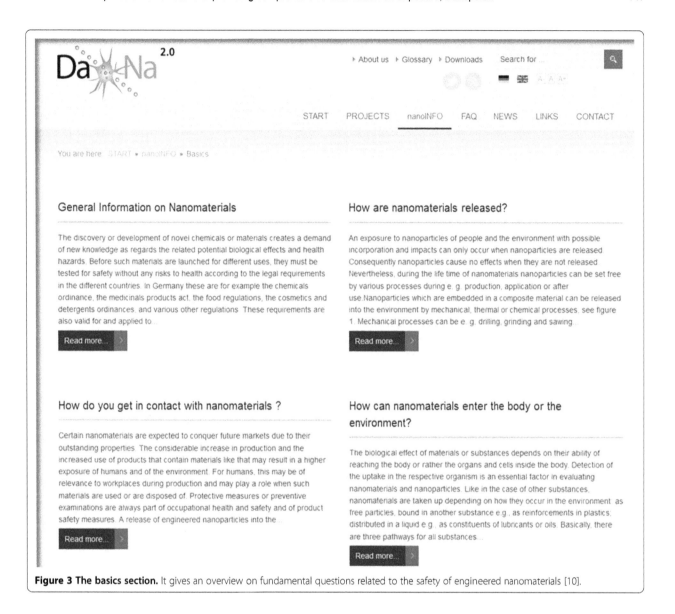

Figure 3 The basics section. It gives an overview on fundamental questions related to the safety of engineered nanomaterials [10].

the project by, e.g. including articles on ENMs in paints or on ENM detection in the environment.

Projects and standard operating procedures

Besides the knowledge base, another main focus of DaNa2.0 is to give an overview of the German Nanosafety Research Landscape including previous, current and future nano-(eco)toxicology-related projects funded by the BMBF. Twelve projects were part of the funding action NanoNature (running from 2009 to 2014), referring mostly to environmental applications. The projects conducted research on, e.g. the improvement of catalytic processes or filtration techniques by applying nanotechnology. Some projects also covered the assessment of potential ecotoxicological hazards of the nanomaterials investigated (e.g. Fe-Nanosit). Three projects dealt with iron-based nanoparticles or nanocomposites applied

in environmental remediation, specifically the decontamination of ground water (e.g. Fe-Nanosit, see Figure 4). Most of the funded projects have been completed until early 2014. Hence, future tasks for the DaNa2.0 team are to first display these results (reports, publications, etc.) on the respective project pages and next to implement these new findings, if having met the necessary quality criteria to enter into the knowledge base (Figure 2). Furthermore, SOPs developed within the projects will be published via the website. New projects will start soon and will be likewise presented.

Additionally, the DaNa2.0 team supported the exchange and interaction between projects from the funding initiatives NanoCare and NanoNature with annual cluster meetings which are planned to continue once the new projects have started. One successful outcome of these cluster meetings was the creation of an interest group

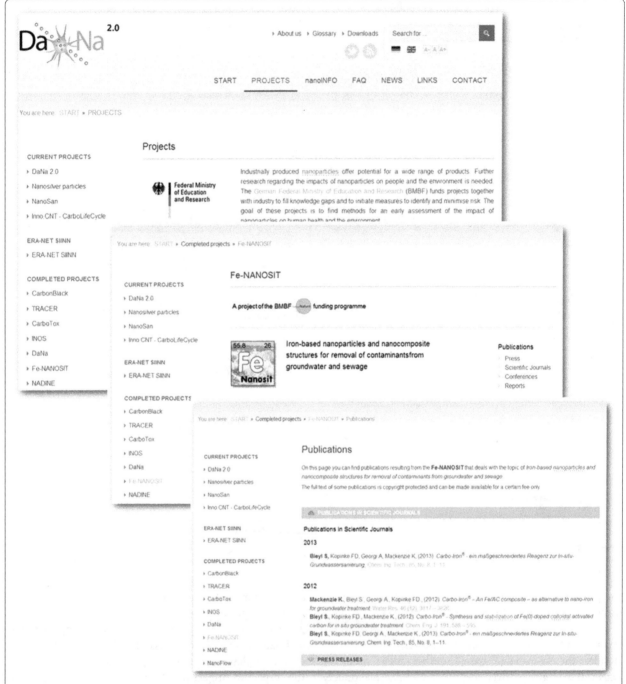

Figure 4 The projects section. It gives an overview on the German Nanosafety Research Landscape and presents all relevant projects within the funding actions NanoCare and NanoNature [11]. Most projects have been completed by early 2014, and the inclusion of results is currently under way. On the individual project sites - here shown for the project Fe-Nanosit - the main scope of each project is presented, together with all project partners, and major outcomes (reports, publication) are listed and linked to the respective sources [12,13].

of ecotoxicologists focussing on the specific peculiarities of iron-based nanomaterials that need to be considered in ecotoxicity testing. This interest group for example organised a workshop focusing on physicochemical characterisation of these nanomaterials before conducting ecotoxicological tests.

Accounting for the need to harmonise experimental practices, the DaNa2.0 team in close cooperation with project members from the nano-funding initiatives developed a template for SOPs. The aim was to provide a common format to share, compare and describe methodologies specific to nanomaterials. But the template may also be used by other projects or serve as a basis for future projects and publications. The pdf form is available for download via the website [14].

Frequently asked questions

The *Frequently asked questions* (FAQ) section gives answers to the most common and important questions on nanomaterials and nanotechnology. These include questions specifically relating to environmental issues, e.g. *How are nanoparticles displayed in the context of recycling management? How can I recognize whether a product contains nanoparticles? How dangerous are nanoscale particles that are already present in the environment?* There is also the possibility to directly interact with the DaNa2.0 experts by submitting questions related to nanomaterials via email. The DaNa2.0 team will respond and publish the answer on the website if considered to be of public relevance.

Glossary

In some instances, the usage of terms very specific to nanotechnology and related fields is hard to avoid. In order to provide further explanation, the glossary not only gives quick and easy definitions for general terms like 'zeta

potential', 'agglomerate' or 'surface charge' but also covers terms from environmental sciences, such as 'bioaccumulation' or 'predicted environmental concentration'. It is assessable either directly in alphabetic order or via tooltip by choosing the marked terms in the text (see Figure 5).

Literature criteria checklist

The need to develop the DaNa Literature Criteria Checklist arose from the necessity to manage the evaluation of all nano-toxicity data regarding its scientific value prior to including it in the knowledge base. In this way, all data with insufficient background information on the investigated ENMs or on the applied methodology will be excluded from the knowledge base, as these data will lead to false conclusions on potential toxic effects of nanomaterials. The assessment criteria cover (1) the extent of the physical-chemical characterisation of the ENMs, (2) the toxicity testing procedures and (3) general issues (e.g. data evaluation). In addition, mandatory assessment criteria are specified, in distinction to desirable criteria. The complete list of quality criteria is accessible to all interested scientists via the DaNa2.0 website [17]. Evaluating every scientific publication according to the criteria finally allows for selection of solely those papers that provide sound

Figure 5 The glossary. It provides definitions of commonly used terms in nanotechnology and related research. It is assessable either directly [15] or via tooltip by choosing the marked terms in the text (mouse-over) [16].

background information on the nanomaterials and the toxicological tests applied [4]. Hence, only scientific facts complying with the checklist will be included in the knowledge base.

Dissemination of nanotechnology-related information

The main aim of the teams' dissemination activities is to provide the general public with sound and up-to-date information related to nanotechnology. The most important instrument for this task is the website www.nanoobjects. info. Since the launch of the website in 2009, a constant increase in the visitor numbers was observed, demonstrating a high need for information (Figure 6). The statistical data shown in Figure 6 clearly demonstrate that not only

the overall visitor numbers are increasing but also there is a growing international awareness level for a national project. Amongst the top 10 visitor countries, the USA and China come in second and third place, respectively, after Germany but ahead of other European countries such as the UK or Switzerland. Besides the DaNa2.0 website, the expert team uses additional channels to communicate the collected knowledge on nanotechnology-related issues to interested laymen. Members of the DaNa2.0 team participate regularly in group discussions, public dialogue events, conferences (e.g. SETAC) and fairs, and interact with journalists (interviews, commentaries, etc.). Beyond that, the annual cluster meetings, the place where partners of the

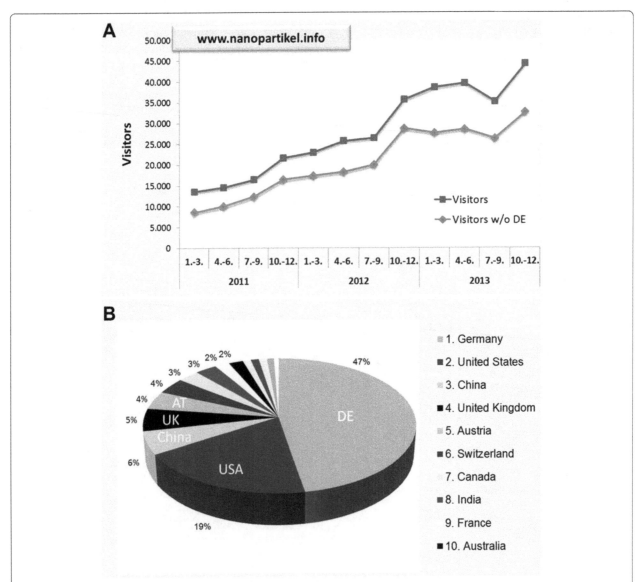

Figure 6 Access statistics from the website. Increasing national and international demand for sound and up-to-date information on nanotechnology. **(A)** Visitor numbers from 2011 to 2013 summarised quarterly in total (red) and without Germany (blue). **(B)** Top 10 visitor countries from 2013.

BMBF-funded projects are given the opportunity to exchange and interact, are also partly open to the public.

Conclusions

The DaNa2.0 website offers easy access to complex scientific issues related to the environmental impact of nanomaterials. The information provided is constantly extended, updated and adjusted to the latest developments in the field of nanotechnology. Taking into account the rapid development in the field of nanotechnology, keeping all the content of www.nanoobjects.info constantly updated is a challenge. Hence, the involvement of European environmental experts in the DaNa2.0 consortium will not only provide further input by, e.g. including further issues with environmental relevance, such as risk assessment and realistic environmental exposure, but will also facilitate the inclusion of latest research results. Likewise, the 'entry port' for scientific facts to be integrated into the knowledge base, the Literature Criteria Checklist, is currently being critically revised. With regard to ecotoxicity testing, specific criteria to, e.g. evaluate the application of nanomaterials to the test media (e.g. soil) and also concerning analytic methods have to be developed and included.

The DaNa2.0 consortium will further strengthen the cooperation and exchange with other dissemination organisations (databases, websites, projects) on the national and international level. Examples are the associations to EU projects such as eNanoMapper (www.enanomapper.net), NanoValid (www.nanovalid.eu) and the InterRegIVb-funded project NANORA (www.nanora.eu), the latter is initiating a French version of the knowledge base. International contributions include also the dialogue with the European NanoSafetyCluster and contributions to an international activity concerning curation of nanotechnology data coordinated by the US NCIP Nanotechnology Working Group [18]. Additionally, members of the DaNa2.0 team participate in several public dialogues or discussions on nanotechnology and nano-toxicology in order to inform the public and raise awareness for their activities. The team is also contributing to other international activities such as the OECD WPMN activities [19]. Citations of the DaNa2.0 website as single national database in the European Commission's 'Commission Staff Working Paper' [20], in 'The JRC Web Platform on Nanomaterials' [21] and in the commentary 'Focusing the research efforts' [22] clearly demonstrate the importance and quality of the provided data, respectively. Finally, the comprehensive collection of environmental relevant facts on nanomaterials will assist in defining knowledge gaps and further research needs.

Methods

This article describes how information of nanotechnology and its potential adverse effects are collected, evaluated and edited in order to provide generally understandable facts and information on a web-based platform. A more detailed description of the methodology applied can be retrieved from the paragraphs introducing the single sections of the website.

Abbreviations

BMBF: German Federal Ministry of Education and Research; DaNa2.0: Data and knowledge on nanomaterials - evaluation of socially relevant scientific facts; ENM: engineered nanomaterial; JRC: Joint Research Centre; NM: nanomaterials; OECD: Organisation for Economic Co-operation and Development; WPMN: Working Party on Manufactured Nanomaterials; SETAC: Society of Environmental Toxicology and Chemistry; SOP: standard operating procedures.

Competing interests

The authors of this article form the core team of DaNa2.0. The core team and the external experts are funded by the German Federal Ministry for Education and Research (BMBF) under grant no. 03X0131. They are also the authors of the content published on the project website www.nanoobjects.info.

Authors' contributions

DK and CM drafted and revised the manuscript. KN included information regarding the website and access statistics. BM and CS provided input regarding materials science and cooperation and synergisms to other projects. HK and CS included information on cooperation and synergisms to other projects. All authors read and approved the final manuscript.

Acknowledgements

We thank the German Federal Ministry for Education and Research (BMBF) for funding the DaNa2.0 project (Data and knowledge on nanomaterials - evaluation of socially relevant scientific facts), grant no. 03X0131 (August 2013 to July 2017).

Author details

[1]Department of Bioanalytical Ecotoxicology, Helmholtz Centre for Environmental Research - UFZ, 04318 Leipzig, Germany. [2]Institute for Applied Computer Science, Karlsruhe Institute of Technology (KIT), 76344 Eggenstein-Leopoldshafen, Germany. [3]Research Focus Area Health and Performance, Empa - Swiss Federal Laboratories for Materials Science and Technology, 9014 St. Gallen, Switzerland. [4]Society for Chemical Engineering and Biotechnology (DECHEMA), 60486 Frankfurt am Main, Germany.

References

1. Bondarenko O, Juganson K, Ivask A, Kasemets K, Mortimer M, Kahru A: **Toxicity of Ag, CuO and ZnO nanoparticles to selected environmentally relevant test organisms and mammalian cells in vitro: a critical review.** *Arch Toxicol* 2013, **87**(7):1181–1200.

2. Jackson P, Jacobsen NR, Baun A, Birkedal R, Kuhnel D, Jensen KA, Vogel U, Wallin H: **Bioaccumulation and ecotoxicity of carbon nanotubes.** *Chem Cent J* 2013, **7**.

3. Steinbach C, Mathes B, Krug HF, Wick P, Kühnel D, Nau K: **Die Sicherheit von Nanomaterialien in der Diskussion: DaNa, eine Internet-Wissensplattform für Interessierte.** *Deutsche Apotheker Zeitschrift* 2012, **152**(16):90–92.

4. Marquardt C, Kühnel D, Richter V, Krug HF, Mathes B, Steinbach C, Nau K: **Latest research results on the effects of nanomaterials on humans and the environment: DaNa - Knowledge Base Nanomaterials.** *J Phys Conf Ser* 2013, **429**:012060.

5. Krug HF, Wick P, Hirsch C, Kühnel D, Marquardt C, Nau K, Mathes B, Steinbach C: **Im Gleichgewicht? Risikoforschung zu Nanomaterialien.** *Arbeitsmed Sozialmed Umweltmed* 2014, **49**:6–18.

6. **The DaNa2.0 Knowledge Base.** [http://www.nanopartikel.info/en/nanoinfo/knowledge-base]

7. **Uptake and risk for environmental organisms (iron and iron oxides).** [http://www.nanopartikel.info/en/nanoinfo/materials/iron-and-iron-oxides/uptake/1351-uptake-and-risk-for-environmental-organisms]

8.   Uptake and risk for environmental organisms (copper and copper oxides). [http://www.nanopartikel.info/en/nanoinfo/materials/copper-and-copper-oxides/uptake/1413-uptake-and-risk-for-environmental-organisms]

9.   Hansen S, Jensen K, Baun A: NanoRiskCat: a conceptual tool for categorization and communication of exposure potentials and hazards of nanomaterials in consumer products. *J Nanopart Res* 2013, **16**(1):1–25.

10. Basics section. [http://www.nanopartikel.info/en/nanoinfo/basics]

11. Projects section. [http://www.nanopartikel.info/en/projects]

12. Projects (Fe-Nanosit). [http://www.nanopartikel.info/en/projects/completed-projects/fenanosit]

13. Project (Fe-Nanosit) publications. [http://www.nanopartikel.info/en/projects/completed-projects/fenanosit/ver-fe-nanosit]

14. SOPs section. [http://www.nanopartikel.info/en/nanoinfo/methods/992-standard-operating-procedures]

15. Glossary section. [http://www.nanopartikel.info/en/glossary/E]

16. Exposure environment. [http://nanopartikel.info/en/nanoinfo/materials/carbon-nanotubes/exposure/1400-exposure-environment]

17. Criteria checklist. [http://www.nanopartikel.info/en/nanoinfo/methods/991-literature-criteria-checklist]

18. The Nanotechnology Working Group. [https://wiki.nci.nih.gov/display/ICR/Nanotechnology+Working+Group]

19. Kühnel D, Nickel C: **The OECD expert meeting on ecotoxicology and environmental fate — towards the development of improved OECD guidelines for the testing of nanomaterials.** *Sci Total Environ* 2014, **472**:347–353.

20. EC: **Commission Staff Working Paper - types and uses of nanomaterials, including safety aspects.** *SWD* 2012, (288). European Commission.

21. The JRC Web Platform on Nanomaterials. [http://ihcp.jrc.ec.europa.eu/our_activities/nanotechnology/jrcreleases-online-web-platform-on-nanomaterials/]

22. Schrurs F, Lison D: **Focusing the research efforts.** *Nat Nanotechnol* 2012, **7**(9):546–548.

14

Microplastics in freshwater ecosystems: what we know and what we need to know

Martin Wagner[1*], Christian Scherer[1], Diana Alvarez-Muñoz[2], Nicole Brennholt[3], Xavier Bourrain[4], Sebastian Buchinger[3], Elke Fries[5], Cécile Grosbois[6], Jörg Klasmeier[7], Teresa Marti[8], Sara Rodriguez-Mozaz[2], Ralph Urbatzka[9], A Dick Vethaak[10], Margrethe Winther-Nielsen[11] and Georg Reifferscheid[3]

Abstract

Background: While the use of plastic materials has generated huge societal benefits, the 'plastic age' comes with downsides: One issue of emerging concern is the accumulation of plastics in the aquatic environment. Here, so-called microplastics (MP), fragments smaller than 5 mm, are of special concern because they can be ingested throughout the food web more readily than larger particles. Focusing on freshwater MP, we briefly review the state of the science to identify gaps of knowledge and deduce research needs.

State of the science: Environmental scientists started investigating marine (micro)plastics in the early 2000s. Today, a wealth of studies demonstrates that MP have ubiquitously permeated the marine ecosystem, including the polar regions and the deep sea. MP ingestion has been documented for an increasing number of marine species. However, to date, only few studies investigate their biological effects.

The majority of marine plastics are considered to originate from land-based sources, including surface waters. Although they may be important transport pathways of MP, data from freshwater ecosystems is scarce. So far, only few studies provide evidence for the presence of MP in rivers and lakes. Data on MP uptake by freshwater invertebrates and fish is very limited.

Knowledge gaps: While the research on marine MP is more advanced, there are immense gaps of knowledge regarding freshwater MP. Data on their abundance is fragmentary for large and absent for small surface waters. Likewise, relevant sources and the environmental fate remain to be investigated. Data on the biological effects of MP in freshwater species is completely lacking. The accumulation of other freshwater contaminants on MP is of special interest because ingestion might increase the chemical exposure. Again, data is unavailable on this important issue.

Conclusions: MP represent freshwater contaminants of emerging concern. However, to assess the environmental risk associated with MP, comprehensive data on their abundance, fate, sources, and biological effects in freshwater ecosystems are needed. Establishing such data critically depends on a collaborative effort by environmental scientists from diverse disciplines (chemistry, hydrology, ecotoxicology, etc.) and, unsurprisingly, on the allocation of sufficient public funding.

Keywords: Chemistry; Ecotoxicology; Environmental quality; Litter; Microplastics; Monitoring; Plastics; Polymers; Review; Water framework directive

* Correspondence: wagner@bio.uni-frankfurt.de
[1]Department of Aquatic Ecotoxicology, Goethe University Frankfurt am Main, Max-von-Laue-Str. 13, Frankfurt 60438, Germany
Full list of author information is available at the end of the article

Background

Microplastics are freshwater contaminants of emerging concern

Among the multiple human pressures on aquatic ecosystems, the accumulation of plastic debris is one of the most obvious but least studied. While plastics generate remarkable societal benefits [1], there are downsides to our 'plastic age'. Durability, unsustainable use, and inappropriate waste management cause an extensive accumulation of plastics in natural habitats [2]. In the marine environment, plastics of various size classes and origins are ubiquitous and affect numerous species that become entangled in or ingest plastics [3].

Under environmental conditions, larger plastic items degrade to so-called microplastics (MP), fragments typically smaller than 5 mm in diameter (see Table 1 for further information). Besides these degradation products (secondary MP), MP can also be produced as such (primary MP). For instance, MP are intentionally used as resin pellets (raw material for the production of plastic products) or as ingredient of personal care products (e.g., peelings and shower gels).

MP are of special concern since their bioaccumulation potential increases with decreasing size. MP may be ingested by various organisms ranging from plankton and fish to birds and even mammals, and accumulate throughout the aquatic food web [4]. In addition, plastics contain a multitude of chemical additives [5] and adsorb organic contaminants from the surrounding media [6]. Since these compounds can transfer to organisms upon ingestion, MP act as vectors for other organic pollutants [7] and are, therefore, a source of wildlife exposure to these chemicals [8,9].

Accordingly, MP are considered an emerging global issue by various experts [10,11] and international institutions [12,13]. These concerns and the public interest, however, focus almost exclusively on marine plastic debris. However, we argue that *microplastics are also freshwater contaminants of emerging concern*. This is supported by three arguments. First, although data is scarce, MP are present in freshwater ecosystems. Second, MP contain and adsorb micropollutants and pathogens. Third, laboratory studies demonstrate that marine organisms ingest MP and suffer adverse effect. While data on freshwater species is scarce, there is no reason to suppose that they remain unaffected. Thus, concerns about the impact of MP on freshwater ecosystems are legitimate and should receive more scientific attention.

State of the science: focus on marine microplastics

So far, scientific efforts focus on marine MP, and studies on their abundance and effects become increasingly available. Because of its high mobility, plastic debris has practically permeated the global marine environment [14,15], including the polar regions [2], mid-ocean islands [16], and the deep sea [17]. Because of their specific hydrology, the large oceanic gyres are hot spots of plastic pollution (colloquially termed 'garbage patches'), accumulating buoyant plastic debris. Here, the plastic abundance often exceeds that of zooplankton [18-21]. With respect to Europe's regional seas, MP have been reported for the Baltic, North, and Mediterranean Sea [22-25].

Most of the studies investigate neustonic and pelagic MP. However, MP are also present in sediments and have been detected on the shorelines and seafloors of six continents [15,26,27] with typical concentrations ranging from 1 to 100 items kg^{-1} [28]. A Belgian study reports a maximum of 400 items kg^{-1} in coastal harbor sediments [29]. Higher concentrations were reported in a Dutch study with 770 and 3,300 items kg^{-1} dry weight sediment in the Wadden Sea and the Rhine estuary, respectively [30]. Although abundant ubiquitously, the spatial distribution of

Table 1 Classification of environmental (micro)plastics

Category	Description
Classification	Environmental plastics are a *very heterogeneous* group of litter that can be characterized by various descriptors. In the literature, they are frequently stratified according to size, origin, shape, polymer type, and color. So far, there is no common classification system. Recently, the European MSFD Working Group on Good Environmental Status (WG-GES) provided a 'Monitoring Guidance for Marine Litter in European Seas' [76], which represents an important step towards a standardized sampling and monitoring of marine microplastics.
Size	The WG-GES defines size classes for plastic litter as follows: macroplastics (>25 mm), mesoplastics (5 to 25 mm), large microplastics (1 to 5 mm), and small microplastics (20 μm to 1 mm). Accordingly, items smaller than 20 μm will classify as nanoplastics.
Origin	Microplastics can also be categorized according to its origin: Primary microplastics are produced as such, for instance as resin pellets (raw materials for plastic products) or as additives for personal care products (e.g., shower gels and peelings). Secondary microplastics are degradation products of larger plastic items, which are broken down by UV radiation and physical abrasion to smaller fragments.
Polymers	The polymer type of environmental (micro)plastics can be determined by Fourier transformed infrared spectroscopy (FT-IR) or Raman spectroscopy. In concordance to global production rates, high- and low-density polyethylene (HD/LD-PE), polyethylene terephthalate (PET), polypropylene (PP), polystyrene (PS), and polyvinyl chloride (PVC) are the most common polymers found in the environment. In addition, polyamide fibers (nylon) from fishing gears are frequent.
Shape	The shape can be described according to the main categories: fragments (rounded, angular), pellets (cylinders, disks, spherules), filaments (fibers), and granules [76].

MP in the marine environment is very heterogeneous [14]. This might be partly due to differences in methodology [28].

Field reports on detrimental interactions of plastics with biota (e.g., entanglement) are manifold [4]. However, only about a dozen studies have investigated MP uptake and effects under laboratory conditions, including two studies on freshwater species (literature search on ISI Web of Science, search term 'microplastic*', manual filtering). With nine of these papers published since 2012, this is a very recent area of research. The ingestion of MP by marine invertebrates has been demonstrated in the laboratory for a broad spectrum of marine species: zooplankton [31-33], the lugworm *Arenicola marina* [34], the Blue mussel *Mytilus edulis* [35-37], and the sandhopper *Talitrus saltator* [38]. *M. edulis* is the only invertebrate in which the transfer of MP from the digestive tract to tissue has been studied and documented [35,36].

Data on the effects of MP exposure is limited. For zooplankton, a reduced algal feeding has been observed [31]. MP increased the mortality and decreased the fertility in copepods [32]. In the lugworm, MP reduced the weight and feeding and increased the bioaccumulation of plastic-associated polychlorinated biphenyls (PCBs) [34]. Reduced filtering activity and histological changes as response to inflammation have been reported for *M. edulis* [36,37], although another study did not find significant effects [35]. In the only study with marine vertebrates, the common goby *Pomatoschistus microps* was exposed to MP and pyrene [39]. MP delayed the pyrene-induced mortality but induced several toxicity biomarkers. In addition, two recent studies demonstrate the trophic transfer of MP along the marine food web from meso- to macrozooplankton [33] and from mussels to crabs [40].

Discussion
Presence of microplastics in freshwater ecosystems
Despite of the wealth of data on marine MP, to date, only a handful of studies investigate MP in a freshwater context. MP have been detected in the surface waters of the Laurentian Great Lakes [41]. The average abundance in the neuston was 43,000 items km^{-2}, with a hotspot near metropolitan areas, which may represent important sources.

Three studies report the occurrence of MP in the sediments of lakes. Zbyszewski and Corcoran [42] found 0 to 34 plastic fragments m^{-2} on the shorelines of Lake Huron (Canada). Here, MP accumulation may be attributed to the lake's currents and nearby plastic manufacturers. Extending their shoreline monitoring to the Lakes Erie and St. Clair, Zbyszewski et al. [43] report 0.2 to 8 items m^{-2}. Sampling two beaches of Lake Garda (Italy), Imhof et al. [44] found 100 and 1,100 MP items m^{-2} at the southern and northern shores, respectively. Similar to the Great Lakes, MP here consisted mainly of low-density polymers (polystyrene (PS), polyethylene (PE), and polypropylene (PP)).

Moore et al. [45] provide the first, non-peer-reviewed report on MP in rivers. In three Californian rivers, they found, on average, 30 to 109 items m^{-3}. The midstream of the Los Angeles River carried 12,000 items m^{-3} and will discharge > 1 billion MP items day^{-1} into the Pacific Ocean. Although very limited, this data indicates that rivers transport relevant amounts of MP.

According to a recent study, the same is true for the second largest European rivers: Lechner et al. [46] used stationary driftnets and visual inspection to monitor plastic debris in the Austrian Danube. The authors report approximately 900 (2010) and 50 (2012) plastic items 1,000 m^{-3} in the size class of 0.5 to 50 mm. In a worst-case scenario, the Danube would discharge 4.2 t plastics day^{-1} and 1,500 t plastics $year^{-1}$ to the Black Sea. The latter is more than the total plastic load of the whole North Atlantic Gyre [47]. Lechner et al. provide first evidence that large rivers transport significant amounts of (micro)plastics and thus contribute substantially to the marine plastics pollution.

Because data on the presence of MP in river sediments is lacking, the Federal Institute of Hydrology and the Goethe University carried out a small, exploratory study with sediments from the rivers Elbe, Mosel, Neckar, and Rhine (Germany). Using density separation and visual inspection, we found 34 to 64 MP items kg^{-1} dry weight, with the River Rhine containing the highest load. Plastic fragments accounted for 60% of the total MP; the remaining particles were synthetic fibers (Figure 1). Thus, as is the case for marine and estuarine sediments, river and lake sediments may be sinks for MP, deserving further investigation.

Sources of microplastics
To date, the sources of marine MP are still not very well characterized. A rough estimation predicts that 70% to 80% of marine litter, most of it plastics, originate from inland sources and are emitted by rivers to the oceans [12]. Potential sources include wastewater treatment plants (WWTPs), beach litter, fishery, cargo shipping, and harbors [12,23,25,29,48]. Although data is so far unavailable, runoff from industrial plastic production sites may be an additional source. Taken together, most marine studies tentatively refer to inland waters as relevant sources (indeed they are rather transport pathways), while actual data is still scarce.

Inland sources of MP have not been investigated thoroughly. In analogy to the marine systems, major contributors will likely include WWTPs and runoff from urban, agricultural, touristic, and industrial areas, as well as shipping activities. Another potential source is sewage sludge that typically contains more MP than effluents

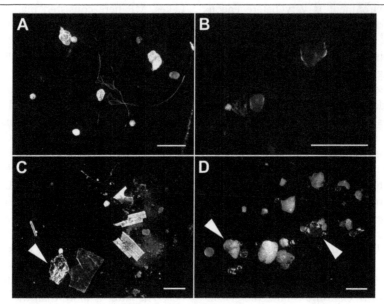

Figure 1 Microplastics in sediments from the rivers Elbe (A), Mosel (B), Neckar (C), and Rhine (D). Note the diverse shapes (filaments, fragments, and spheres) and that not all items are microplastics (e.g., aluminum foil **(C)** and glass spheres and sand **(D)**, white arrowheads). The white bars represent 1 mm.

[49]. Sewage sludge is still frequently used for landfilling and as fertilizer in agriculture, and surface runoff may transfer MP to rivers and lakes and ultimately river basins and the sea. Washing clothes [26] and personal care products [50] are sources of MP in WWTPs. Since the retention capacity of conventional wastewater treatment processes appears to be limited [14], a characterization of MP emission by WWTPs and other sources is urgently needed to understand where freshwater MP is coming from.

Impact of microplastics on freshwater species

In a field report, Sanchez et al. [51] provide the only data on MP in freshwater fish so far. They investigated gudgeon (*Gobio gobio*) caught in 11 French streams and found MP in the digestive tract of 12% of the fish. Although again very preliminary, this field report shows that freshwater species ingest MP. However, the rate of MP ingestion in different fish species will certainly depend on their feeding strategy. Rosenkranz et al. [52] demonstrate that the water flea *Daphnia magna* rapidly ingests MP under laboratory conditions. MP (0.02 and 1 mm) appear to cross the gut epithelium and accumulate in lipid storage droplets. This is of specific concern because MP infiltrating tissues might induce more severe effects. Imhof et al. [44] report the uptake of MP by annelids (*Lumbriculus variegatus*), crustaceans (*D. magna* and *Gammarus pulex*), ostracods (*Notodromas monacha*), and gastropods (*Potamopyrgus antipodarum*). While the available studies demonstrate that a broad spectrum of aquatic taxa is prone to MP ingestion, the toxicological effects remain uninvestigated for freshwater species.

Microplastics as vector for other contaminants

Due to their large surface-to-volume ratio and chemical composition, MP accumulate waterborne contaminants including metals [53] and persistent, bioaccumulative, and toxic compounds (PBTs) [54]. A review on the relationship between plastic debris and PBTs (e.g., PCBs and DDT) has been published recently [55], and a number of studies exist for polycyclic aromatic hydrocarbons (PAHs) [56-61]. However, there is a lack of information on other important contaminants like pharmaceuticals and endocrine-disrupting compounds (EDCs). Nonylphenol and bisphenol A have been detected in MP [60,62,63]. Fries et al. [24] detected various plastic additives in MP, including some well-known EDCs (e.g., phthalates). In addition, Wagner and Oehlmann [64,65] demonstrated that plastics leach EDCs. Since the spectrum of contaminants is different in freshwater and marine systems, the chemical burden of freshwater MP remains to be studied.

The interaction of MP and chemicals has been studied in adsorption-desorption experiments [6,57]. While there is significant complexity in this interaction, MP may act as vector transferring environmental contaminants from water to biota. While different modeling studies arrive at contrasting conclusions [54,66,67], a recent experimental study demonstrates that fish exposed to contaminants sorbed to MP bioaccumulate these compounds and suffer adverse effects (glycogen depletion and histopathological alterations [68]). However, to date, there are too few studies investigating whether MP are indeed vectors that facilitate the transfer of organic contaminants to biota. Because a verification of the 'vector hypothesis' would have major ecological implications, it deserves further investigation, especially in a freshwater context.

Microplastics as vector for exotic species and pathogens

Not only the complex mix of chemicals contained in and sorbed to MP and/or ingestion of MP by biota is a cause for concern but also microorganisms developing biofilms on MP particles. Only very few studies have been conducted on this issue with marine ecosystems being the focal point of interest [69-72]. Zettler et al. [72] described a highly diverse microbial community ('plastisphere') attaching plastic marine debris in the North Atlantic. Several plastisphere members are hydrocarbon-degrading bacteria which may potentially influence plastic debris fragmentation and degradation. But they also found opportunistic (human) pathogens like specific members of the genus *Vibrio* dominating plastic particles. Therefore, MP can act as a vector for waterborne (human) pathogens influencing the hygienic water quality. The fact that the microbial communities on MP are distinct from surrounding water (only some marine bacteria develop biofilms on microplastic particles (e.g., [71,72])) suggests that MP serve as a kind of new habitat. Until now, the complex interaction between microorganisms/microbial communities as a key player in aquatic ecosystems/food webs and MP, especially in freshwater, is poorly understood and needs to be further investigated.

Microplastics in connection to European water policies

The issue of (micro)plastics connects to several European water policies. The European Marine Strategy Framework Directive (MSFD, 2008/56/EC) addresses the issue of marine litter, including plastics. Here, MP are covered by Descriptor 10 of Commission Decision 2010/477/EU, which defines the good environmental status of marine waters [73].

In contrast, the Water Framework Directive (WFD, 20/60/EC) applying to European inland waters does not specifically refer to plastic litter. However, the Member States have the obligation to monitor anthropogenic pressures. Here, MP are promising candidates, especially because they might act as vectors for a wide range of freshwater contaminants. For instance, MP have been shown to contain the WFD priority substances di(ethylhexyl) phthalate (DEHP), nonylphenol, octylphenol, and PAHs (2008/105/EC, Annex II).

Several other European Directives relate to the potential sources of freshwater MP, including the Directives on packaging waste (2004/12/EC), waste (2008/98/EC), landfills (1999/31/EC), urban wastewater (91/271/EEC), sewage sludge (86/278/EEC), and ship-source pollution (2005/35/EC). In addition, the Union's chemicals legislation (REACH, 1907/2006/EC) will apply to plastic monomers and additives of relevant production volumes.

In a recent 'Green paper on a European strategy on plastic waste in the environment', the European Commission addresses the issue as part of a wider review of its waste legislation [74]. While the Green Paper focuses on potential mitigation strategies for plastic litter at the source, it also expresses 'particular concern' about MP.

Conclusions
Knowledge gaps and research needs

The investigation of (micro)plastics in aquatic environments is a highly dynamic and interdisciplinary area of research covering and bringing together the disciplines of oceanography and hydrology as well as environmental monitoring, modeling, chemistry, and toxicology. In recent years, this collaborative effort advanced our understanding of the environmental impact of MP, especially by providing extensive monitoring data. Ongoing research activities focus, however, almost exclusively on marine MP.

Data on freshwater ecosystems is at best fragmentary if not absent. This lack of knowledge hampers a science-based environmental risk assessment of freshwater MP. Such assessment is needed to facilitate a societal and political discussion at national and European levels on the issue, which, depending on the outcome, will result in mitigation measures eventually. For instance, MP could be integrated as descriptor of environmental status in the WFD. However, environmental scientists first need to close the gaps of knowledge with regard to exposure and hazard of freshwater MP and the associated chemicals. Based on the current state of the science, the following research needs emerge (Figure 2):

1. Monitoring the presence of microplastics in freshwater systems. While few studies on large lakes and rivers are available, we have no clear picture on the magnitude of the plastics pollution in surface waters. Generating comprehensive monitoring data on the abundance of freshwater MP is needed to understand their environmental impact.
2. Investigating the sources and fate of freshwater microplastics. Currently, we still do not understand the behavior of MP in aquatic ecosystems. Based on data on their abundance, modeling approaches are needed to identify hotspots and sinks and quantify loads. One important aspect of understanding the environmental fate is also to identify relevant inland sources of MP and determine the fragmentation rates of large plastic debris.
3. Assessing the exposure to microplastics. With evidence coming from marine species, it appears plausible that freshwater organisms will ingest MP, too. However, actual data is scarce. Environmental toxicologists need to determine the intake of MP by freshwater key species. It will be crucial to

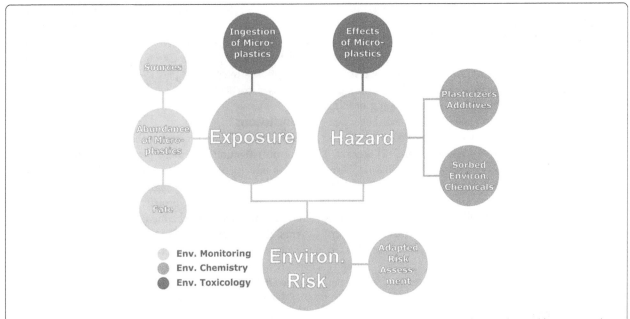

Figure 2 Research aspects with regard to freshwater microplastics. All areas need to be investigated more thoroughly to assess the environmental risk associated with microplastics in freshwater ecosystems.

understand which plastic characteristics (size, material, and shape) promote an uptake and what is the fate of MP in the biota (e.g., excretion, accumulation, and infiltration of tissues). These aspects need to be studied under laboratory conditions and in the field to determine the actual exposure.

4. Evaluating the biological effects of microplastics exposure. Besides abundance and exposure, the question whether MP induce adverse effects in organisms is crucial to determine their environmental hazard. In the absence of effect studies on freshwater species, one can only speculate on potential sensitive endpoints: Ingested plastic fragments may most likely affect the metabolism (starvation due to decreased energy intake) and induce inflammation (when transferring to tissues). Because this is an area of research where the least progress has been made so far, the investigation of MP effects on marine and freshwater species need to be intensified considerably.

5. Understanding the interaction between microplastics and other freshwater contaminants. Plastics itself can contain and release toxic chemicals (e.g., monomers or plastic additives [75]). In addition, they can accumulate environmental chemicals from the surrounding. This may increase the chemical exposure of the ingesting organism and, thus, toxicity. The findings on chemicals associated with marine MP (mostly POPs) cannot be transferred to freshwaters because here the spectrum and concentrations of

pollutants is very different. Therefore, it is important to investigate the chemical burden of freshwater MP, including the absorption/desorption kinetics and the transfer of chemicals from plastics to biota.

6. Develop a novel framework for the risk assessment of microplastics. MP can be direct and indirect stressors for the aquatic environment: They are contaminants of emerging concern *per se* and, in addition, may serve as vectors for invasive species and for other pollutants. To account for that, the classical risk assessment framework needs to be adapted. For instance, the mixture toxicity of MP-associated compounds and the modulation of the compounds' bioavailability need to be integrated.

There are some challenges in investigating these aspects: To generate commensurable data on the abundance of freshwater MP, harmonized monitoring procedures, including sampling, identification, and characterization, are needed. For that, the 'Monitoring Guidance for Marine Litter in European Seas' developed by the European MSFD Working Group on Good Environmental Status [76] provides an excellent starting point. The separation of MP from the sample materials (sediments or suspended particulate matter) and the confirmation of the plastics' identity to avoid misclassification is still a very resource-consuming and biased process (e.g., when visually identifying MP in complex samples). Here, sample throughput and accuracy need to be increased. Likewise, we need to improve the capability to detect very small MP in the low micrometer range. Boosting technological innovation in the area of MP research (e.g., coupling of microscopy and spectroscopy to identify very small MP) will help meet those challenges.

In conclusion, based on our knowledge on the environmental impact of marine MP, their freshwater counterparts should be considered contaminants of emerging concern. However, there is a considerable lack of knowledge on MP in surface waters worldwide. Data on their presence, sources, and fate is scarce if not absent. The same is true for their chemical burden and biological effects. To enable science-based environmental risk assessment of freshwater MP, it is imperative to initiate coordinated and collaborative research programs that close these gaps of knowledge.

Abbreviations

EDCs: endocrine-disrupting compounds; FT-IR: Fourier transformed Infrared spectroscopy; HD/LD-PE: high-/low-density polyethylene; MP: microplastics; MSFD: European Marine Strategy Framework Directive; PAHs: polycyclic aromatic hydrocarbons; PBTs: persistent, bioaccumulative, and toxic compounds; PCBs: perchlorinated biphenyls; PET: polyethylene terephthalate; POPs: persistent organic pollutants; PP: polypropylene; PS: polystyrene; PVC: polyvinyl chloride; WFD: European Water Framework Directive; WWTPs: wastewater treatment plants.

Competing interests

The authors declare that they have no competing interests.

Authors' contributions

MW conceived the manuscript. CS, MW, SB, and GR performed the pilot study on microplastics in river sediments. All authors (MW, CS, DAM, NB, XB, SB, EF, CG, JK, TM, SRM, RU, ADV, MWN, and GR) provided substantial input to, read, and approved the final manuscript.

Acknowledgements

This work was, in part, funded by the German Federal Institute of Hydrology.

Author details

[1]Department of Aquatic Ecotoxicology, Goethe University Frankfurt am Main, Max-von-Laue-Str. 13, Frankfurt 60438, Germany. [2]Catalan Institute for Water Research (ICRA), Girona 17003, Spain. [3]Department Biochemistry and Ecotoxicology, Federal Institute of Hydrology, Koblenz 56002, Germany. [4]Service Etat des Eaux Evaluation Ecologique, Agence de l'Eau Loire-Bretagne, Ploufragan 22440, France. [5]Water, Environment and Eco-technologies Division, Bureau de Recherches Géologiques et Minières (BRGM), Orléans 45100, France. [6]GéoHydrosystèmes Continentaux (GéHCO), Université Francois Rabelais de Tours, Tours 37000, France. [7]Institute of Environmental Systems Research, Universität Osnabrück, Osnabrück 49074, Germany. [8]Investigación y Proyectos Medio Ambiente S.L. (IPROMA), Castellón de la Plana 12005, Spain. [9]Interdisciplinary Centre of Marine and Environmental Research (CIIMAR), Porto 4050-123, Portugal. [10]Unit Marine and Coastal Systems, Deltares and Institute for Environmental Studies, VU University Amsterdam, Amsterdam 1081, The Netherlands. [11]Environment and Toxicology, DHI, Hørsholm 2970, Denmark.

References

1. Andrady AL, Neal MA: Applications and societal benefits of plastics. *Philos Trans R Soc Lond B Biol Sci* 2009, 364:1977–1984.
2. Barnes DK, Galgani F, Thompson RC, Barlaz M: Accumulation and fragmentation of plastic debris in global environments. *Philos Trans R Soc Lond B Biol Sci* 2009, 364:1985–1998.
3. Gregory MR: Environmental implications of plastic debris in marine settings- entanglement, ingestion, smothering, hangers-on, hitch-hiking and alien invasions. *Philos Trans R Soc Lond B Biol Sci* 2009, 364:2013–2025.
4. Wright SL, Thompson RC, Galloway TS: The physical impacts of microplastics on marine organisms: a review. *Environ Pollut* 2013, 178:483–492.
5. Dekiff JH, Remy D, Klasmeier J, Fries E: Occurrence and spatial distribution of microplastics in sediments from Norderney. *Environ Pollut* 2014, 186:248–256.
6. Bakir A, Rowland SJ, Thompson RC: Competitive sorption of persistent organic pollutants onto microplastics in the marine environment. *Mar Pollut Bull* 2012, 64:2782–2789.
7. Zarfl C, Matthies M: Are marine plastic particles transport vectors for organic pollutants to the Arctic? *Mar Pollut Bull* 2010, 60:1810–1814.
8. Oehlmann J, Schulte-Oehlmann U, Kloas W, Jagnytsch O, Lutz I, Kusk KO, Wollenberger L, Santos EM, Paull GC, Van Look KJ, Tyler CR: A critical analysis of the biological impacts of plasticizers on wildlife. *Philos Trans R Soc Lond B Biol Sci* 2009, 364:2047–2062.
9. Teuten EL, Saquing JM, Knappe DR, Barlaz MA, Jonsson S, Bjorn A, Rowland SJ, Thompson RC, Galloway TS, Yamashita R, Ochi D, Watanuki Y, Moore C, Viet PH, Tana TS, Prudente M, Boonyatumanond R, Zakaria MP, Akkhavong K, Ogata Y, Hirai H, Iwasa S, Mizukawa K, Hagino Y, Imamura A, Saha M, Takada H: Transport and release of chemicals from plastics to the environment and to wildlife. *Philos Trans R Soc Lond B Biol Sci* 2009, 364:2027–2045.
10. Depledge MH, Galgani F, Panti C, Caliani I, Casini S, Fossi MC: Plastic litter in the sea. *Mar Environ Res* 2013, 92:279–281.
11. Sutherland WJ, Clout M, Cote IM, Daszak P, Depledge MH, Fellman L, Fleishman E, Garthwaite R, Gibbons DW, De Lurio J, Impey AJ, Lickorish F, Lindenmayer D, Madgwick J, Margerison C, Maynard T, Peck LS, Pretty J, Prior S, Redford KH, Scharlemann JPW, Spalding M, Watkinson AR: A horizon scan of global conservation issues for 2010. *Trends Ecol Evol* 2010, 25:1–7.
12. GESAMP (IMO/FAO/UNESCO-IOC/UNIDO/WMO/IAEA/UN/UNEP Joint Group of Experts on the Scientific Aspects of Marine Environmental Protection): Proceedings of the GESAMP international workshop on micro-plastic particles as a vector in transporting persistent, bio-accumulating and toxic substances in the oceans. In *GESAMP Reports & Studies*. Edited by Bowmer T, Kershaw P. Paris: UNESCO-IOC; 2010. 68 pp.
13. United Nations Environment Programme: *UNEP Yearbook: Emerging Issues in Our Global Environment*. Nairobi: UNEP Division of Early Warning and Assessment; 2011.
14. Cole M, Lindeque P, Halsband C, Galloway TS: Microplastics as contaminants in the marine environment: a review. *Mar Pollut Bull* 2011, 62:2588–2597.
15. Ivar do Sul JA, Costa MF: The present and future of microplastic pollution in the marine environment. *Environ Pollut* 2014, 185:352–364.
16. do Ivar do Sul JA, Costa MF, Barletta M, Cysneiros FJ: Pelagic microplastics around an archipelago of the Equatorial Atlantic. *Mar Pollut Bull* 2013, 75:305–309.
17. Van Cauwenberghe L, Vanreusel A, Mees J, Janssen CR: Microplastic pollution in deep-sea sediments. *Environ Pollut* 2013, 182:495–499.
18. Eriksen M, Maximenko N, Thiel M, Cummins A, Lattin G, Wilson S, Hafner J, Zellers A, Rifman S: Plastic pollution in the South Pacific subtropical gyre. *Mar Pollut Bull* 2013, 68:71–76.
19. Goldstein MC, Titmus AJ, Ford M: Scales of spatial heterogeneity of plastic marine debris in the northeast Pacific Ocean. *PLoS One* 2013, 8:e80020.
20. Ryan PG: Litter survey detects the South Atlantic 'garbage patch'. *Mar Pollut Bull* 2014, 79:220–224.
21. Moore CJ, Moore SL, Leecaster MK, Weisberg SB: A comparison of plastic and plankton in the North Pacific Central Gyre. *Mar Pollut Bull* 2001, 42:1297–1300.
22. Collignon A, Hecq JH, Glagani F, Voisin P, Collard F, Goffart A: Neustonic microplastic and zooplankton in the North Western Mediterranean Sea. *Mar Pollut Bull* 2012, 64:861–864.
23. Dubaish F, Liebezeit G: Suspended microplastics and black carbon particles in the Jade system, southern North Sea. *Water Air Soil Pollut* 2013, 224:1352.
24. Fries E, Dekiff JH, Willmeyer J, Nuelle MT, Ebert M, Remy D: Identification of polymer types and additives in marine microplastic particles using pyrolysis-GC/MS and scanning electron microscopy. *ESPI* 2013, 15:1949–1956.
25. Norén F: *Small Plastic Particles in Coastal Swedish Waters*. Sweden: KIMO; 2007.

26. Browne MA, Crump P, Niven SJ, Teuten E, Tonkin A, Galloway T, Thompson R: Accumulation of microplastic on shorelines worldwide: sources and sinks. *Environ Sci Technol* 2011, **45**:9175–9179.

27. Thompson RC, Olsen Y, Mitchell RP, Davis A, Rowland SJ, John AWG, McGonigle D, Russell AE: Lost at sea: where is all the plastic? *Science* 2004, **304**:838.

28. Hidalgo-Ruz V, Gutow L, Thompson RC, Thiel M: Microplastics in the marine environment: a review of the methods used for identification and quantification. *Environ Sci Technol* 2012, **46**:3060–3075.

29. Claessens M, De Meester S, Van Landuyt L, De Clerck K, Janssen CR: Occurrence and distribution of microplastics in marine sediments along the Belgian coast. *Mar Pollut Bull* 2011, **62**:2199–2204.

30. Leslie HA, van Velzen MJM, Vethaak AD: *Microplastic Survey of the Dutch Environment. Novel Data Set of Microplastics in North Sea Sediments, Treated Wastewater Effluents and Marine Miota*. Amsterdam: Institute for Environmental Studies, VU University Amsterdam; 2013.

31. Cole M, Lindeque P, Fileman E, Halsband C, Goodhead R, Moger J, Galloway TS: Microplastic ingestion by zooplankton. *Environ Sci Technol* 2013, **47**:6646–6655.

32. Lee KW, Shim WJ, Kwon OY, Kang JH: Size-dependent effects of micro polystyrene particles in the marine copepod *Tigriopus japonicus*. *Environ Sci Technol* 2013, **47**:11278–11283.

33. Setälä O, Fleming-Lehtinen V, Lehtiniemi M: Ingestion and transfer of microplastics in the planktonic food web. *Environ Pollut* 2014, **185**:77–83.

34. Besseling E, Wegner A, Foekema EM, van den Heuvel-Greve MJ, Koelmans AA: Effects of microplastic on fitness and PCB bioaccumulation by the lugworm *Arenicola marina* (L.). *Environ Sci Technol* 2013, **47**:593–600.

35. Browne MA, Dissanayake A, Galloway TS, Lowe DM, Thompson RC: Ingested microscopic plastic translocates to the circulatory system of the mussel, *Mytilus edulis* (L.). *Environ Sci Technol* 2008, **42**:5026–5031.

36. von Moos N, Burkhardt-Holm P, Kohler A: Uptake and effects of microplastics on cells and tissue of the blue mussel *Mytilus edulis* L. after an experimental exposure. *Environ Sci Technol* 2012, **46**:11327–11335.

37. Wegner A, Besseling E, Foekema EM, Kamermans P, Koelmans AA: Effects of nanopolystyrene on the feeding behavior of the blue mussel (*Mytilus edulis* L.). *Environ Toxicol Chem* 2012, **31**:2490–2497.

38. Ugolini A, Ungherese G, Ciofini M, Lapucci A, Camaiti M: Microplastic debris in sandhoppers. *Estuar Coast Shelf Sci* 2013, **129**:19–22.

39. Oliveira M, Ribeiro A, Hylland K, Guilhermino L: Single and combined effects of microplastics and pyrene on juveniles (0+ group) of the common goby *Pomatoschistus microps* (Teleostei, Gobiidae). *Ecol Indic* 2013, **34**:641–647.

40. Farrell P, Nelson K: Trophic level transfer of microplastic: *Mytilus edulis* (L.) to *Carcinus maenas* (L.). *Environ Pollut* 2013, **177**:1–3.

41. Eriksen M, Mason S, Wilson S, Box C, Zellers A, Edwards W, Farley H, Amato S: Microplastic pollution in the surface waters of the Laurentian Great Lakes. *Mar Pollut Bull* 2013, **77**:177–182.

42. Zbyszewski M, Corcoran PL: Distribution and degradation of fresh water plastic particles along the beaches of Lake Huron, Canada. *Water Air Soil Pollut* 2011, **220**:365–372.

43. Zbyszewski M, Corcoran PL, Hockin A: Comparison of the distribution and degradation of plastic debris along shorelines of the Great Lakes, North America. *J Great Lakes Res* 2014, **40**:288–299.

44. Imhof HK, Ivleva NP, Schmid J, Niessner R, Laforsch C: Contamination of beach sediments of a subalpine lake with microplastic particles. *Curr Biol* 2013, **23**:R867–R868.

45. Moore CJ, Lattin GL, Zellers AF: *Working Our Way Upstream: A Snapshot of Land-Based Contributions of Plastic and Other Trash to Coastal Waters and Beaches of Southern California*. Long Beach: Algalita Marine Research Foundation; 2005.

46. Lechner A, Keckeis H, Lumesberger-Loisl F, Zens B, Krusch R, Tritthart M, Glas M, Schludermann E: The Danube so colourful: a potpourri of plastic litter outnumbers fish larvae in Europe's second largest river. *Environ Pollut* 2014, **188**:177–181.

47. Law KL, Moret-Ferguson S, Maximenko NA, Proskurowski G, Peacock EE, Hafner J, Reddy CM: Plastic accumulation in the North Atlantic subtropical gyre. *Science* 2010, **329**:1185–1188.

48. Zubris KA, Richards BK: Synthetic fibers as an indicator of land application of sludge. *Environ Pollut* 2005, **138**:201–211.

49. Leslie HA, Moester M, de Kreuk M, Vethaak AD: Verkennende studie naar lozing van microplastics door rwzi's. (Pilot study on emissions of microplastics from wastewater treatment plants). *H2O* 2012, **14/15**:45–47 (in Dutch).

50. Fendall LS, Sewell MA: Contributing to marine pollution by washing your face: microplastics in facial cleansers. *Mar Pollut Bull* 2009, **58**:1225–1228.

51. Sanchez W, Bender C, Porcher JM: Wild gudgeons (*Gobio gobio*) from French rivers are contaminated by microplastics: preliminary study and first evidence. *Environ Res* 2014, **128**:98–100.

52. Rosenkranz P, Chaudhry Q, Stone V, Fernandes TF: A comparison of nanoparticle and fine particle uptake by *Daphnia magna*. *Environ Toxicol Chem* 2009, **28**:2142–2149.

53. Ashton K, Holmes L, Turner A: Association of metals with plastic production pellets in the marine environment. *Mar Pollut Bull* 2010, **60**:2050–2055.

54. Koelmans AA, Besseling E, Wegner A, Foekema EM: Plastic as a carrier of POPs to aquatic organisms: a model analysis. *Environ Sci Technol* 2013, **47**:7812–7820.

55. Engler RE: The complex interaction between marine debris and toxic chemicals in the ocean. *Environ Sci Technol* 2012, **46**:12302–12315.

56. Antunes JC, Frias JGL, Micaelo AC, Sobral P: Resin pellets from beaches of the Portuguese coast and adsorbed persistent organic pollutants. *Estuar Coast Shelf S* 2013, **130**:62–69.

57. Bakir A, Rowland SJ, Thompson RC: Enhanced desorption of persistent organic pollutants from microplastics under simulated physiological conditions. *Environ Pollut* 2013, **185C**:16–23.

58. Fisner M, Taniguchi S, Majer AP, Bicego MC, Turra A: Concentration and composition of polycyclic aromatic hydrocarbons (PAHs) in plastic pellets: implications for small-scale diagnostic and environmental monitoring. *Mar Pollut Bull* 2013, **76**:349–354.

59. Zarfl C, Fleet D, Fries E, Galgani F, Gerdts G, Hanke G, Matthies M: Microplastics in oceans. *Mar Pollut Bull* 2011, **62**:1589–1591.

60. Fries E, Zarfl C: Sorption of polycyclic aromatic hydrocarbons (PAHs) to low and high density polyethylene (PE). *Environ Sci Pollut Res Int* 2012, **19**:1296–1304.

61. Rios LM, Moore C, Jones PR: Persistent organic pollutants carried by synthetic polymers in the ocean environment. *Mar Pollut Bull* 2007, **54**:1230–1237.

62. Mato Y, Isobe T, Takada H, Kanehiro H, Ohtake C, Kaminuma T: Plastic resin pellets as a transport medium for toxic chemicals in the marine environment. *Environ Sci Technol* 2001, **35**:318–324.

63. Hirai H, Takada H, Ogata Y, Yamashita R, Mizukawa K, Saha M, Kwan C, Moore C, Gray H, Laursen D, Zettler ER, Farrington JW, Reddy CM, Peacock EE, Ward MW: Organic micropollutants in marine plastics debris from the open ocean and remote and urban beaches. *Mar Pollut Bull* 2011, **62**:1683–1692.

64. Wagner M, Oehlmann J: Endocrine disruptors in bottled mineral water: total estrogenic burden and migration from plastic bottles. *Environ Sci Pollut Res* 2009, **16**:278–286.

65. Wagner M, Oehlmann J: Endocrine disruptors in bottled mineral water: estrogenic activity in the E-Screen. *J Steroid Biochem Mol Biol* 2011, **127**:128–135.

66. Teuten EL, Rowland SJ, Galloway TS, Thompson RC: Potential for plastics to transport hydrophobic contaminants. *Environ Sci Technol* 2007, **41**:7759–7764.

67. Gouin T, Roche N, Lohmann R, Hodges G: A thermodynamic approach for assessing the environmental exposure of chemicals absorbed to microplastic. *Environ Sci Technol* 2011, **45**:1466–1472.

68. Rochman CM, Hoh E, Kurobe T, Teh SJ: Ingested plastic transfers hazardous chemicals to fish and induces hepatic stress. *Sci Rep* 2013, **3**:3263.

69. Carson HS, Nerheim MS, Carroll KA, Eriksen M: The plastic-associated microorganisms of the North Pacific Gyre. *Mar Pollut Bull* 2013, **75**:126–132.

70. Harrison JP, Sapp M, Schratzberger M, Osborn AM: Interactions between microorganisms and marine microplastics: a call for research. *Mar Technol Soc J* 2011, **45**:12–20.

71. Harrison JP: *The Spectroscopic Detection and Bacterial Colonisation of Synthetic Microplastics in Coastal Marine Sediments*. Sheffield: University of Sheffield; 2012.

72. Zettler ER, Mincer TJ, Amaral-Zettler LA: Life in the "plastisphere": microbial communities on plastic marine debris. *Environ Sci Technol* 2013, **47**:7137–7146.

73. Galgani F, Hanke G, Werner S, De Vrees L: **Marine litter within the European Marine Strategy Framework Directive.** *ICES J Mar Sci* 2013, **70:**1055–1064.

74. European Commission: *Green Paper on a European Strategy on Plastic Waste in the Environment.* Brussels: European Commission; 2013.

75. Rochman CM: **Plastics and priority pollutants: a multiple stressor in aquatic habitats.** *Environ Sci Technol* 2013, **47:**2439–2440.

76. MSFD GES Technical Subgroup on Marine Litter (TSG-ML): *Monitoring Guidance for Marine Litter in European Seas*, Draft report. Brussels: European Commission; 2013.

15

Comparative assessment of plant protection products: how many cases will regulatory authorities have to answer?

Michael Faust[1], Carolina Vogs[2], Stefanie Rotter[2], Janina Wöltjen[3], Andreas Höllrigl-Rosta[3], Thomas Backhaus[1] and Rolf Altenburger[2*]

Abstract

Background: The substitution principle has been included in the EU pesticides legislation as a new element. Comparative assessments will have to be conducted for all uses of plant protection products (PPPs) that contain active substances with certain hazardous properties, the so-called candidates for substitution (CFS). This study investigated the resulting workload in terms of the number of cases for comparative assessments that regulatory authorities may have to face. The analysis was carried out for Germany as an example.

Main text: In Germany, the requirement for comparative assessments may affect up to 25% of all PPPs and around 50% of all uses of PPPs. In absolute terms, these are around 350 candidate products with 1,850 different uses. Alternative products without CFS may be available for around 40% of these uses. On average, a candidate product is authorised for around 18 different uses. For 11 of these uses, no alternatives are authorised. For the remaining seven uses, slightly more than seven alternatives are available on average. Multiplication of these factors gives an indicative figure of around 18,500 possible pairwise comparisons of candidate products with alternative products for every common use.

Conclusions: The high number of expectable cases poses a formidable challenge for the efficient conduct of the new task of comparative assessments by competent Member States authorities. To this end, new data handling systems, assessment procedures, and decision rules need to be established.

Keywords: Substitution principle; Comparative risk assessment; Pesticides; Plant protection products; Active substances; Candidates for substitution; PBT

Background

The substitution principle in the EU pesticides legislation

The substitution principle is a new element of the legislation on plant protection products (PPPs) in the European Union (EU). It was introduced with the new Regulation (EC) No 1107/2009 [1], in the following shortly denoted as the PPP Regulation. This replaced the old Directive 91/414/EEC on PPPs [2] in June 2011. In parallel, the substitution principle was also included in the new Regulation (EU) No 528/2012 on biocidal products [3], which came into force in September 2013. PPPs and biocidal products are collectively denoted as 'pesticides' under EU law, as has been defined in Article 3 of Directive 2009/128/EC on the sustainable use of pesticides in the European Community (EC) [4]. As a common rule, pesticides shall not be placed on the market or used unless they have been authorised in accordance with the applicable regulations. In general, 'substitution' of pesticides means that an authorisation is refused or withdrawn in favour of an alternative product or a non-chemical control or prevention method which presents a 'significantly lower risk,' according to Annex IV of the PPP Regulation (EC) No 1107/2009 and Article 23 of the biocidal products Regulation (EU) No 528/2012, respectively. In detail, the conditions, rules, and criteria for applying the substitution principle

* Correspondence: rolf.altenburger@ufz.de
[2]Department Bioanalytical Ecotoxicology, UFZ - Helmholtz Centre for Environmental Research, Permoser Str. 15, Leipzig 04318, Germany
Full list of author information is available at the end of the article

differ for PPPs and for biocidal products. In this paper, we focus on substitution under the Regulation for PPPs.

The inclusion of the substitution aspect in the EU pesticides legislation is an outcome of a broader and long-lasting discussion about the guiding principles of chemicals regulation under EU law. As a generic policy principle, substitution means the replacement of hazardous chemical substances and products by less hazardous alternatives [5]. Whether this idea should be established as a legal demand for actors in the field has been subject to heated debates. Opponents, such as the German chemical industries for instance, argued that substitution was superfluous if safe use of a hazardous chemical could be ensured by appropriate risk management measures [6]. In 2001, during the preparation of the REACH legislation, the Commission of the European Communities (COM) considered the substitution of hazardous chemicals as one of the 'key elements' of the proposed 'Strategy for a future Chemicals Policy' [7]. Five years later, in the final REACH legislation [8], legal requirements for feasibility analyses for substitution were, however, confined to substances of very high concern (SVHC) that are subject to authorisation (Article 55 of Regulation (EC) No 1907/2006). In all three pieces of legislation, where substitution has now been included as an element of authorisation procedures (REACH, biocidal products and PPPs), hazardous properties of chemicals serve only as a trigger for considerations for substitution, but are considered insufficient for decision making. Instead, comparative risk assessments of products have to be conducted as the basis for substitution decisions, which is novel and challenging.

Conventional risk assessments for individual PPPs, as they have been established under the old Directive 91/414/EEC, aim to ensure that regulatory acceptable exposure levels are not exceeded, but they do not provide incentives for reducing risks any further. This is changed by the complementary instrument of comparative risk assessment which supports a process of continuous improvement by identifying those PPPs that allow to achieve a desired purpose with minimal risks at a given point in time. This is particularly favourable for environmental risks, where the authorisation requirements still allow to tolerate temporary adverse effects as acceptable. Moreover, acceptable exposure levels for many pesticides on the market are only achievable by applying risk mitigation measures, such as protective equipment for workers or buffer zones between sprayed agricultural land and surface waters. Such measures may fail accidentally or may be disregarded negligently. Substitution of such products by alternatives that require less risk mitigation measures is therefore desirable and shall be supported by the new instrument of comparative assessments.

While the intended improvements are clear, the detailed procedures and methodologies for applying the substitution principle are not. Only in the Nordic countries, particularly in Sweden, the principle has been included in the national chemicals legislation since the beginning of the 1990s [9]. Other EU Member States (MS) have no comparable legislative tradition. Against this background, there is high uncertainty about potential impacts of this new element of EU pesticides legislation and the best way towards its efficient implementation.

Candidates for substitution

Plant protection products contain one or more active substances. Under EU law, PPPs are authorised on the Member States level, while active substances are approved on the Community level. Approved active substances are included in a positive list established by the European Commission. Member States shall not authorise PPPs that contain active substances other than those on the positive list. Authorisations are only granted for specified uses, usually defined by a combination of a protected crop and a targeted pest.

The revised legislation now requires that certain active substances shall be approved by the European Commission only as 'candidates for substitution' (CFS) and listed separately from other approved active substances. Member States shall not grant authorisation to PPPs that contain such CFS, if a comparative assessment reveals that a significantly safer alternative is available for the same use.

CFS are active substances that have one or more of the hazardous properties listed in Table 1. As laid down in the PPP Regulation, their identification constitutes one task within the regular assessment of active substances on Community level. In order to speed up the process for the already approved active substances, an obligation for the European Commission (COM) was included in the PPP Regulation to establish an initial list of CFS until the end of 2013. However, completion of this task is now expected to be delayed by a few months [10]. At the time of writing of this manuscript (March 2014), the official list was not yet available.

As a support for the preparation of the initial list, COM commissioned a contract study to the Food Chain Evaluation Consortium (FCEC). The FCEC delivered their study report in July 2013 [11]. Subsequently, the report was presented to the competent authorities of the Member States, and it was made available to all stakeholders via the Commission's CIRCA platform [12]. Thus, although not formerly published, the report is in the public domain and was accessible for the purposes of this paper via the German Federal Environment Agency (UBA).

It was not the task of the FCEC to set up the initial list of CFS - this is the privilege of COM - but to do the necessary preparatory work, which was

Table 1 Criteria for the identification of active substances as candidates for substitution (CFS)

Number	Legal text (Regulation (EC) No 1107/2009, Annex II, point 4) [1][a]
1	Its ADI, ARfD, or AOEL is significantly lower than those of the majority of the approved active substances within groups of substances/use categories
2	It meets two of the criteria to be considered as a PBT substance
3	There are reasons for concern linked to the nature of the critical effects (such as developmental neurotoxic or immunotoxic effects) which, in combination with the use/exposure patterns, amount to situations of use that could still cause concern, for example, high potential of risk to groundwater; even with very restrictive risk management measures (such as extensive personal protective equipment or very large buffer zones)
4	It contains a significant proportion of non-active isomers
5	It is or is to be classified, in accordance with the provisions of Regulation (EC) No 1272/2008, as carcinogen category 1A or 1B, if the substance has not been excluded in accordance with the criteria laid down in point 3.6.3[b]
6	It is or is to be classified, in accordance with the provisions of Regulation (EC) No 1272/2008, as toxic for reproduction category 1A or 1B if the substance has not been excluded in accordance with the criteria laid down in point 3.6.4[b]
7	If, on the basis of the assessment of Community or internationally agreed test guidelines or other available data and information, reviewed by the Authority, it is considered to have endocrine disrupting properties that may cause adverse effects in humans if the substance has not been excluded in accordance with the criteria laid down in point 3.6.5[b,c]

[a]The criteria apply independently, i.e. CFS meet one or more of them. [b]Points 3.6.2 to 3.6.5 of Annex II of Regulation (EC) No 1107/2009 [1] define hazard-based criteria for substances that must not be approved, so-called cut-off criteria. [c]For endocrine disrupters, currently the interim criteria laid down under point 3.6.5 of Annex II of Regulation (EC) No 1107/2009 [1] apply, i.e. substances classified as carcinogenic category 2 and toxic for reproduction category 2.

(i) to compile the data needed for decision-making from the legally relevant documents, i.e. those documents on which the original decisions for approval of active substances have been based, such as the official Review Reports, EFSA Conclusions, and Draft Assessment Reports, and

(ii) to explore options for the interpretation and operationalisation of the criteria for the identification of CFS (Table 1), where the legal text and the available data leave room for judgments and where corresponding rules for data assessments had not already been fixed in a corresponding Commission Working Document on 'Evidence Needed to Identify POP, PBT and vPvB Properties of Pesticides' [13].

As a consequence, the FCEC report did not directly provide a list of CFS, but it included separate lists of active substances that were considered to fulfil individual CFS criteria or sub-criteria, for example persistence in soil, in sediments, and in water. Where applicable, the report provided various versions of these lists, each representing the outcome of a different interpretation of a legal criterion, such as different measures and trigger values for a 'significantly lower ADI'. In each of these cases the report provided arguments for the option that the authors considered to be the most appropriate one.

Thus, by combining the information from these individual lists, it is possible to obtain a list of potential CFS that have a high chance for becoming actually included in the initial list of CFS that COM is going to establish. This opportunity was used for the purpose of this study.

Comparative assessments

In the future, EU Member States shall perform a comparative assessment whenever they evaluate any application for authorisation of a PPP that contains a CFS, in the following shortly denoted as candidate product. A comparative assessment may be initiated by an application for the authorisation of a new candidate product, for the renewal of an existing authorisation, or for the amendment of an authorisation for new uses of a candidate product. Comparative assessments must be performed for each use of a candidate product. A candidate product shall not be authorised for a use for which an alternative chemical product or a non-chemical control method is available, if the following requirements are fulfilled (Article 50 in conjunction with Annex IV of the PPP Regulation):

(i) Experience from practical use of the alternative is available.

(ii) The alternative has a comparable efficacy against target pests.

(iii) The alternative can be used without significant economic or practical disadvantages, including impacts on the so-called minor uses.

(iv) The substitution does not compromise resistance management and the minimisation of the occurrence of resistance.

(v) The alternative product or method is 'significantly safer for human or animal health or the environment'.

Thus, the comparative assessment can be divided into two major parts: a comparative agronomic assessment covering points (i) to (iv) and a comparative safety assessment as required by point (v). This paper focuses on the safety assessment part. In addition, it does not discuss comparisons of chemical PPPs with non-chemical protection methods (e.g. mechanical methods or bio-pesticides such as viruses and bacterial strains).

Annex IV to the PPP Regulation clarifies that the increase in safety that is achieved by a substitution shall be demonstrated in terms of a 'significantly lower risk'. In general, competent authorities shall identify such significant differences in risk 'on a case-by-case basis'. A significance level is not specified for the comparative human health risk assessment but for the environmental risk assessment: 'if relevant, a factor of at least 10 for the toxicity/exposure ratio (TER) of different plant protection products is considered a significant difference in risk' [1].

For conducting the agronomic part of the comparative assessments, guidance has been developed by EPPO, the European and Mediterranean Plant Protection Organization [14]. For the risk assessment part, COM is currently working on a guidance document, based on a proposal by Sweden [15]. This guidance aims to support the Member States, but it will not establish detailed and legally binding rules. Basically, it will be up to the decision of the Member States on how they actually conduct comparative risk assessments.

Regulatory impact

In order to prepare for the new task of conducting comparative product assessments, competent authorities need to obtain an overview about the dimension of the additional workload they may be confronted with. Earlier assessments of the potential impacts of the new PPP legislation estimated that about 15% to 25% of the approved active substances could become candidates for substitution [16,17], but these studies were based on limited and uncertain databases and they did not further explore the consequences in terms of numbers of potentially affected products and uses of products, and in terms of numbers of pairs or groups of products that could become subject to a comparative risk assessment. However, with the data from the recent FCEC report, a solid basis is now available for such calculations.

Aims and approach

This study aimed to estimate the potential number of cases for which a comparative assessment will need to be carried out, in terms of (i) the number of CFS-containing candidate products, (ii) the number of different uses of such candidate products, (iii) the number of uses of candidate products for which alternative products are available, and (iv) the possible number of pairwise comparisons of candidate and alternative products for all common uses. For the purpose of this study, we assumed that candidate products are only compared to alternative products that

do not contain any CFS. In principle, neither the legal text nor the available draft of the EU Guidance Document excludes the possibility of a substitution of a candidate product by another candidate product, if a significant risk reduction would result. However, we assumed that this would be a rare situation and excluded it from further consideration.

The exercise did not cover the whole European Union but was conducted for the German PPP market as an example. The calculations were based on the simplistic assumption of a static market share of candidate products. This means that authorisations for all uses of all CFS-containing PPPs that are currently on the market would become subject to renewal in the future or that they will be replaced by an equal number of new candidate products with an equal number of uses. In this way we assumed to obtain upper limit estimates of the potential number of cases for comparative assessments that may accumulate over a period of time equal to the average duration of authorisations (typically granted for 10 years).

The identification of CFS results from hazardous properties for humans or for the environment or for both. For human and environmental risk assessments of PPPs, different aims, methods, and procedures apply and they are often performed separately by different institutions. In Germany, for instance, the Federal Institute for Risk Assessment (BfR), the Julius Kühn Institute (JKI), and the Federal Environment Agency (UBA) separately assess risks for humans, for honey bees, and for the environment, respectively. Therefore, we were also interested to see whether the number of cases for comparative risk assessments would be significantly reduced, if comparative environmental assessments would be conducted only for those candidate products that contain CFS as identified by hazardous environmental properties. To this end, we performed all calculations twice: once for all candidate products and once only for those products that contain CFS that have been identified for reasons of environmental hazards (exclusively or in addition to human health hazards). The way by which we achieved this discrimination is explained in the 'Main text' section.

The list of potential CFS resulting from the FCEC report is currently subject to final revisions by the Commission and the Member States. In order to avoid any false discrimination of substances or products in this interim situation, this paper is exclusively focused on the quantitative aspects of the subject, providing numbers of substances, products, and uses, but no names. As a result of the ongoing revisions, the forthcoming official CFS list can be expected to be slightly but not substantially shorter than assumed in this paper [10]. Consequently, the figures provided in this paper for the number of candidate products can be expected to represent a slight overestimation of the findings that will be obtainable by performing the same analyses on the basis of the final CFS list.

Main text

How many CFS?

Almost 100 active substances are potential candidates for substitution, which is roughly a quarter of all approved active substances. This derives from an examination of the information provided in the FCEC report to the Commission [11], as detailed in the 'Methods' section. The number refers to the list of approved active substances laid down in the Annex to Commission Implementing Regulation (EU) No 540/2011 [18], as amended until 31 January 2013. Where the legal definition of an approved active substance includes different varieties of a parent compound or where the approval applies to a defined group of compounds, these were counted as a single entity. Earlier impact studies expected CFS proportions between 15% and 25% [16,17]. Our findings show that these prognoses indicated the dimension of the problem quite accurately.

Potential CFS are spread across all major use categories, as defined in the EU Pesticides database [19]. Between 10% and 25% of substances used as fungicides, insecticides, plant growth regulators, or multi-purpose pesticides are potential CFS (Table 2). For acaricides and herbicides, the proportions are higher, with around 40% of the substances being potentially affected. From the small group of rodenticides, even 60% are potential CFS.

The so-called PBT properties have an outstanding importance for the number of substances identified as a potential CFS (Table 3). Almost 80% of the potential CFS fulfil the second CFS criterion (see Table 1), which means that they meet two of the criteria to be considered as persistent (P), bioaccumulative (B), and/or toxic (T), as defined under point 3.7.2 of Annex II to the PPP Regulation. Low ADI, ARfD, or AOEL values; reproductive toxicity class 1A or 1B; or endocrine disrupting properties (criteria 1, 6, and 7 in Table 1) have a lower impact on the number of potential CFS, i.e. 22%, 9%, and 7% of the candidates, respectively. Even less influence on the number of CFS has criterion 4 (Table 1), i.e. a significant proportion of non-active isomers: only 2% of the potential CFS meet this criterion. The remaining two criteria 3 and 5 ('nature of critical effects' and carcinogens class 1A/1B) were not found to apply to any approved active substances in the FCEC report. Sixteen percent of all potential CFS were found to meet two or more of the seven identification criteria. There is no obvious association between applicable CFS criteria and the use categories of substances (Table 3). Substances meeting two PBT criteria, for instance, come from all major use categories, whereupon the proportions basically reflect the different sizes of the use groups.

The seven legal CFS criteria include both human health hazards and environmental hazards, but both aspects are separable, as explained in the following. CFS criteria 1, 5, 6, and 7 (low ADI/ARfD/AOEL, carcinogenicity, reproductive toxicity, and endocrine disrupting properties that may cause adverse effects in humans as defined in the legislation) are exclusively based on tests used for human toxicity assessments. Criterion 3 (nature of critical effects) could be interpreted to include both aspects but was not found to be relevant for the initial list of CFS. Criterion 4 (non-active isomers) clearly has a relevance for both human and environmental hazard and risk assessments. Criterion 2 (two PBT criteria) theoretically comprises three different situations: P + B, P + T, and B + T. Practically, however, the combination P + B has apparently no relevance for the initial CFS list. Hence, criterion 2 always combines an exposure indicator (P or B) with a toxicity indicator (T). As laid down under point 3.7.2.3 of Annex II of the PPP Regulation, two different types of toxicity indicators are applicable: indicators of human toxicity (CMR and STOT RE classifications) and indicators of aquatic toxicity (long-term NOEC), in the following denoted as T_{HUMAN} and T_{AQUA}, respectively. As a consequence, it is possible to distinguish between CFS that meet criterion 2 for reasons of human health protection (T_{HUMAN} in combination with P or B) or for reasons of environmental protection (T_{AQUA} in combination with P or B), or both. In summary of these considerations, CFS identified by environmental hazard criteria are those that meet the T criterion for water organisms in addition to the P or the B criterion, and those that contain a significant proportion of non-active isomers.

Table 2 Proportion of active substances identified as potential CFS, broken down by use categories

Use category[a]	Total number of approved active substances[b]	Proportion of potential CFS[c] (%)
AC	8	38
FU	103	25
HB	110	41
IN	55	18
PG	25	12
RE	17	0
RO	5	60
Other	14	0
Multi	35	23
Not assigned	6	0
All	378	26

[a]Assignment to use categories as given in the EU Pesticides database [19]; AC, acaricides; FU, fungicides; HB, herbicides; IN, insecticides; PG, plant growth regulators; RE, repellents; RO, rodenticides; Other, attractants, bactericides, elicitors, molluscicides, nematicides, and plant activators; Multi, multiple use categories apply to the same substance, also including uses as dessicant in addition to one or more of the other listed categories. [b]Reference date: 31 January 2013; counting of approved active substances refers to the legal definitions listed in Commission Implementing Regulation (EU) 540/2011 [18]; from a chemical perspective, parts of these are groups of substances or mixtures of substances; in addition to chemicals, the legal substance definition also includes viruses and bacteria, the so-called bio-pesticides. [c]Percentages rounded to integer values; potential CFS are those identified in the FCEC report [11], as detailed in the text.

Table 3 Breakdown of the number of potential CFS by use categories and identification criteria

CFS criterion[a]		Use category[e]							Sum
		AC	FU	HB	IN	PG	RO	Multi	
1. Low ADI/ARfD/AOEL		0	6	8	2	1	0	5	22
2. Two PBT criteria	T_{AQUA}[b]	3	20	32	8	2	2	3	70
	T_{HUMAN}[c]	0	8	8	1	0	0	3	20
	All	3	23	35	8	2	2	4	77
3. Nature of critical effects		0	0	0	0	0	0	0	0
4. Non-active isomers		0	1	1	0	0	0	0	2
5. Carcinogen 1A/1B		0	0	0	0	0	0	0	0
6. Toxic for reproduction 1A/1B		0	3	5	0	0	1	0	9
7. Endocrine disrupting properties[d]		0	2	4	1	0	0	0	7
All		3	26	45	10	3	3	8	98

[a]The numbering refers to the full legal definition of criteria as given in Table 1; corresponding short descriptions are those used in the EU pesticide database [19]; different criteria for identification as a CFS may apply to one and the same substance; therefore figures in lines 'All' do not equal sums of values given in the columns. [b]T_{AQUA}: substances fulfil the toxicity criterion for water organisms laid down under point 3.7.2.3 of Annex II to Regulation (EC) No 1107/2009 [1], in addition to the P or the B criterion. [c]T_{HUMAN}: substances fulfil any of the human toxicity criteria defined under point 3.7.2.3 of Annex II to Regulation (EC) No 1107/2009 [1] in terms of CMR or STOT RE classifications, in addition to the P or the B criterion. [d]For endocrine disrupters, the interim criteria laid down under point 3.6.5 of Annex II of Regulation (EC) No 1107/2009 [1] were applied, i.e. substances classified as carcinogenic category 2 and toxic for reproduction category 2. [e]Assignment to use categories as given in the EU Pesticides database [19]; AC, acaricides; FU, fungicides; HB, herbicides; IN, insecticides; PG, plant growth regulators; RO, rodenticides; Multi, multiple use categories apply, also including uses as dessicant, nematicide, or repellent in addition to one or more of the categories AC, FU, HB, IN, PG, and RO.

All other CFS are identified for reasons of human health hazards.

By applying these categorisations, we found that only 19% of the potential CFS are exclusively identified by human health criteria; 81% meet environmental hazard criteria, whereby 27% meet both human and environmental hazard criteria and 54% are exclusively identified for reason of environmental hazards.

How many candidate products?

The analysis of the German register of authorised PPPs revealed that 25% of the products contain one or more potential CFS and may hence be considered as potential candidate products (Table 4). This proportion is almost identical to the share of CFS in the number of approved active substances. Interestingly, however, only 67 out of 98 potential CFS were actually found in any PPP on the German market, i.e. around 30% of the CFS currently have no relevance for comparative assessments in Germany. In absolute figures, 351 out of 1,378 authorised PPPs are potential candidate products. Two hundred thirty-seven products contain a potential CFS that has been identified for reason of environmental hazards, which is a share of 17% of all products and 67% of all potential candidate products. Thus, on the level of products on the German market, the importance of environmental hazard criteria for the number of candidates is only slightly lower than seen on the level of active substances, where the corresponding fraction is 81% of all potential CFS.

The potential candidate PPPs are spread across seven major use categories, with shares ranging between 2%

and 45%. This is essentially the same situation as observed on the level of CFS. Interestingly, however, the proportions can be quite different for the same use category. In the small group of rodenticides, for instance, 60% of the active substances were found to be potential CFS, but no more than 2% of the rodenticidal products authorised in Germany actually contain any of these CFS. The opposite situation does also occur: In the large group of fungicides, for instance, 25% of the active substances are potential CFS, but 45% of the products on the German market contain one or more of these CFS. The high proportion of 45% potential candidate products amongst fungicides is exceptional; for all other use categories, the fractions are at or below the average of 25%.

How many uses of candidate products?

On average, potential candidate products are authorised for a broader spectrum of uses than other PPPs. As a consequence, they may necessitate comparative assessments for a disproportionally high number of uses; they account for no more than 25% of all products (see above), but they are authorised for around 50% of all uses. This is the essential outcome of an analysis detailed in the following.

Authorised uses of PPPs are defined in terms of a combination of a crop and a pest from which the crop shall be protected. A small number of other treatment objects (such as food storage rooms) and treatment aims (such as plant growth regulation) are subsumed here also under the terms crop and pest, respectively. Using the status of May 2013, a total of 1,378 PPPs was authorised in Germany for a total of 3,606 uses, defined as different

Table 4 Proportion of candidate products containing potential CFS authorised in Germany

Use category[a]	Total number of authorised PPPs[b]	Proportion of candidate PPPs containing one or more potential CFS[c]	
		Containing any potential CFS (%)	Containing CFS identified by environmental hazard criteria[d] (%)
Herbicides	567	25	21
Fungicides	308	45	32
Insecticides	281	25	6
Acaricides	109	12	2
Plant growth regulators	59	14	7
Molluscicides	58	0	0
Rodenticides	41	2	2
Glue, sealing wax	31	0	0
Repellents	29	0	0
Sprout inhibitors	18	0	0
Bactericides	9	11	11
Pheromones	3	0	0
Viricides	1	0	0
Nematicides	1	0	0
All[b]	1,378	25	17

As of May 2013. [a]Categorisation as given in the BVL database on plant protection products [20]; this categorisation of products is largely but not entirely consistent with the categorisation of active substances used in the EU pesticide database [19] (see preceding tables). [b]The given categorisation of products is non-exclusive, i.e. multiple use categories may apply to a single product; this is the case for around 10% of all PPP; as a consequence, the overall number of 1,378 PPPs is smaller than the sum of values in the column. [c]Percentages rounded to integer values; potential CFS are those identified in the FCEC report [11], as detailed in the text. [d]Active substances that fulfil the toxicity criterion for water organisms laid down under point 3.7.2.3 of Annex II to Regulation (EC) No 1107/2009 in addition to the P or the B criterion as defined under points 3.7.2.1 and 3.7.2.1 of the same Annex II [1], and active substances that contain a significant proportion of non-active isomers; human health criteria may apply additionally.

combinations from of a total of 309 crops and 477 pests (Table 5). The fraction of 351 potential candidate products (25% of all products) was authorised for a total of 1,863 uses (52% of all uses), defined as different combinations from 209 crops (68% of all crops) and 264 pests (55% of all pests). The sub-fraction of 237 potential candidate products that contain CFS identified by environmental hazard criteria (17% of all products) was authorised for a total of 1,501 uses (42% of all uses), defined as different combinations from 186 crops (60% of all crops) and 228 pests (48% of all pests).

On average, every PPP is authorised for use against five different pests in seven different crops. For potential candidate products, the same mean values apply. On the level of uses, however, the situation is different: while the mean number of authorised uses is 13 for all PPPs, potential candidate products have an average of 18 different authorised uses.

How many alternatives are available?
For 767 different uses of potential candidate products, at least one CFS-free alternative product is authorised (Table 6). This is slightly more than 40% of all 1,863 uses of potential candidate products. For around 9%, even more than 10 different alternative products are authorised. For the

sub-fraction of potential candidate products that contain CFS identified by environmental hazard criteria, the corresponding proportions are slightly higher: at least one alternative is available for 687 of 1,501 uses, i.e. almost 46%. For almost 12%, even more than 10 different alternative products are authorised.

On average, one or more alternative products are authorised for around 7 out of 18 different uses of a potential candidate product. For each of these seven uses, the mean number of available alternative products is also around seven. For the sub-fraction of potential candidate products that contain CFS identified by environmental hazard criteria, these mean values differ only marginally.

How many cases for comparative assessments?
In principle, every application for authorisation of a candidate product constitutes a case for comparative assessments, as required by the PPP Regulation. While the assessments may be stopped in an early phase, if alternatives are not available for any of the uses of the candidate product, our analysis revealed that only for 6 out of 351 potential candidate products no alternative products are authorised for any of their uses. The remaining 345 potential candidate products (around 98%) may trigger more detailed comparative assessments, at least of the

Table 5 Number of plant protection products, protected crops, controlled pests, and authorised uses in Germany

Parameter			All PPPs	Candidate PPPs containing one or more potential CFS	
				Containing any potential CFS	Containing CFS identified by environmental hazard criteria[a]
Products	Total number		1,378	351	237
Crops[b]	Total number of different crops for which use of PPPs has been authorised		309	209	186
	Number of different crops for which use of an individual PPP has been authorised	Min	1	1	1
		Max	85	85	85
		Median	4	4	4
		Mean[e]	7	7	7
Pests[c]	Total number of different pests against which use of PPPs has been authorised		477	264	228
	Number of different pests against which use of an individual PPP has been authorised	Min	1	1	1
		Max	64	39	39
		Median	3	4	4
		Mean[e]	5	5	5
Uses[d]	Total number of different uses for which PPPs have been authorised		3,606	1,863	1,501
	Number of different uses for which an individual PPP has been authorised	Min	1	1	1
		Max	337	337	337
		Median	6	8	9
		Mean[e]	13	18	18

As of May 2013. Broken down by potential CFS content of products. [a]As defined in the corresponding footnote to Table 4. [b]Including other authorised treatment objects such as food storage rooms for instance. [c]Including other authorised treatment aims such plant growth regulation for instance. [d]Defined by a combination of a crop (or another treatment object) and a pest from which the crop shall be protected (or another treatment aim). [e]Arithmetic mean rounded to integer values.

agronomic aspects, and where appropriate also of the safety aspects. For the sub-fraction of potential candidate products that contain CFS identified by environmental hazard criteria, the proportion is basically the same (Table 7).

In terms of affected uses, more or less detailed comparative assessments may be required for the 767 different uses of potential candidate products for which at least one alternative product is currently authorised. However, even for the same use, not all candidate products will become subject to authorisation or re-authorisation at the same time. Whenever an authorisation is requested for a single candidate product, comparative assessments will have to be conducted for all its uses. This means that comparative assessments for the same use may have to be conducted repeatedly over time. Therefore, as an additional indicator of the potential workload for competent authorities, the number of potential candidate products can be multiplied with the average number of uses for which alternatives are potentially available. This yields an indicative figure of roughly 2,500 cases of use-specific assessments of candidate products that may become necessary during the next years. For the sub-fraction of potential candidate products that contain CFS identified by environmental hazard criteria, the figure is roughly 25% smaller.

It must be noted that our analysis of uses was based on the exact wording of the use descriptions in the German PPP register. Hence, we did not explore the potential for substituting a candidate product that is only used for a specific pest and/or crop by an alternative product that is authorised for broad-spectrum use against a large group of pests and/or in multiple crops. This could further increase the number of comparative product assessments.

In addition to the number of uses, the number of available alternative products must be taken into account. The broader the spectrum, the higher will be the workload for comparatively assessing all products that are available for a given use. As a further workload indicator, we therefore derived the product of all three factors: The number of potential candidate products × the average number of uses for which alternatives are potentially available × the average number of alternative products that is available for each of these uses. This results in roughly 18,500 cases. This figure is the number of all possible pairwise comparisons of candidate products with alternative products for all common uses of two products. For the sub-fraction of potential candidate products that contain CFS identified by environmental hazard criteria, the figure is roughly 20% smaller.

Table 6 Potential availability of alternative products for authorised uses of CFS-containing candidate products in Germany

Parameter	All potential candidate products ($n = 351$)	Potential candidate products containing CFS identified by environmental hazard criteria[a] ($n = 237$)
Availability of alternative products for uses of candidate products	Number of uses	
No alternatives available	1,096	813
Any alternatives available	767	687
1 alternative available	220	162
2 alternatives available	126	112
3 to 5 alternatives available	172	163
6 to 10 alternatives available	82	77
11 to 20 alternatives available	112	118
21 to 50 alternatives available	42	44
51 to 100 alternatives available	12	10
More than 100 alternatives available	1	1
Average spectrum of uses of candidate products	Mean number of uses[b]	
Including all uses of candidate products	18	18
Including only uses of candidate products for which alternatives are available	7	8
Average availability of alternatives	Mean number of alternative products[b]	
For all uses of candidate products	3	4
For all uses of candidate products for which alternatives are available	7	8

Scenario based on the status of authorised products in May 2013. Candidate products are products containing one or more potential CFS; alternatives are products containing no potential CFS. [a]As defined in the corresponding footnote to Table 4. [b]Arithmetic mean rounded to integer values.

Discussion

Estimated case numbers in this paper are based on a draft CFS list and on a simplistic static scenario, reflecting neither fluctuations in the number of authorised products nor any potential future trends. Therefore, the figures should not be taken as precise estimates but rather as indications of the expectable dimension of the upcoming demands on comparative assessments. In any case, the results of the present study indicate that several thousands of cases might need to be evaluated. Hence,

Table 7 Potential number of cases for comparative risk assessment of plant protection products in Germany

Cases defined in terms of...		Counts	
		For products containing any potential CFS	For products containing CFS identified by environmental hazard criteria[a]
Products[b]	Number of all candidate products	351	237
	Number of candidate products for which alternatives are available for one or more of their uses	345	232
Uses	Number of all authorised uses of candidate products	1,863	1,501
	Number of authorised uses of candidate products for which alternative products are available	767	687
Products × Uses	Number of all candidate products times average number of uses[c]	6,232	4,175
	Number of all candidate products times average number of uses for which alternatives are potentially available[c]	2,569	1,910
Products × Uses × Alternatives	Number of all possible pairwise risk comparisons of candidate products with alternative products for all common uses[d]	18,479	15,287

Scenario based on the status of authorised products in May 2013. [a]As defined in the corresponding footnote to Table 4. [b]Candidate products are products containing one or more potential CFS; alternatives are products containing no potential CFS. [c]Calculated with non-rounded values for mean numbers of uses. [d]Calculated by multiplying the number of all candidate products with the average number of uses for which alternatives are potentially available and the mean number of such alternatives for every use; non-rounded mean values were used for the calculations.

the new task of performing comparative assessments of PPPs may pose a formidable challenge to competent authorities. Not only are the quantitative demands considerable, the established assessment methodologies for PPPs and the existing organisational arrangements are also only designed for checking compliance of individual PPPs with legal acceptability criteria. They were not developed for comparing products with each other. New data handling systems, assessment procedures, and decision rules might hence be required for efficiently dealing with the new task.

The comparative agronomic assessment and the comparative safety assessment may be performed either subsequently or in parallel. The EPPO guidance on comparative assessment proposed that the agronomic assessment should be carried out first [14], and the latest draft of the forthcoming Commission guidance document suggests the same [15]. With the aim of optimal efficiency of the overall procedure, this appears to be a rational and self-evident approach, as the agronomic assessment can be expected to provide a strong filter. Thus, the novel and demanding task of comparing the human and environmental risks of different products would be limited to the lowest possible number of cases. On the other hand, however, it must be taken into consideration that the agronomic part of the assessment includes criteria that need further interpretation and specification, in particular the requirement for no 'significant' economic or practical disadvantages. Corresponding definitions of significance may be based on isolated agronomic considerations, but it is also imaginable, that the assessment of the significance of economic disadvantages should take account of potential advantages for human health and the environment. Thus, the number of cases for comparative human and environmental risk assessments will depend on the detailed criteria and procedural arrangements of the overall process.

The workload for performing a complete set of comparative human and environmental risk assessments for all the uses of a candidate product will certainly be quite different, depending on the respective product. These different assessment situations result from the different numbers of uses of potential candidate products, which vary between one and a few hundreds, and also from the different numbers of potentially available alternatives, which range between zero and more than a hundred (Tables 5 and 6). In addition, the initial comparative assessments of substances available for a certain common use may be more demanding than a repeated assessment that is triggered by an application for authorisation of another product for the same use. Furthermore, the risks resulting from different uses of the same product may be quite similar or even identical, and most likely there will also be situations where the available

spectrum of candidate products and alternative products is identical for different uses. Thus, a more detailed assessment of the workload from comparative risks assessments and the development of an optimal strategy for efficiently dealing with the issue would need to consider these different types of assessment situations. Such analyses will become possible as soon as the final official CFS list becomes available.

Besides the need to further clarify these details, the high number of expectable cases calls for the development of an electronic decision support instrument which would enable competent authorities to rapidly carry out initial comparative risk assessments. Such a tool should allow to filter out clear-cut cases in a semi-automatic manner and to separate them from borderline cases which require in-depth expert judgment. To this end, the legal requirement for a significant risk reduction needs to be translated into well-defined programmable decision rules. In addition, an efficient application of such a computer-based decision support tool depends on the availability of a continuously updated database with all regulatory relevant information on the risks of individual products that is required for the comparative assessments. In Germany, regulatory risk assessment reports for individual PPPs currently exist in the form of text files only, and the same may presumably apply to most other EU Member States. For performing a comparative assessment, all the relevant risk indicators, such as TER values for the various ecotoxicological endpoints, must therefore be compiled manually from the individual assessment reports. Improving this situation is essential for a less time-consuming future practice. In addition, efforts should be made to identify representative endpoints or to define suitable indicators for facilitating decision-making within the process, e.g. for sorting products according to their risk. Within a broader scope, cooperation between Member States authorities may be considered as a further means for efficiently dealing with the large numbers of comparative risk assessments needed. Comparative risk assessment of PPPs is clearly a national task, but a zonal authorisation of PPPs across different Member States has been introduced with the new PPP Regulation. In this situation, sharing the workload for comparative risk assessments between Member States in the same zone seems to be a self-suggesting option that deserves further exploration.

Conclusions

The expectable number of cases for comparative assessments of PPPs is high. In Germany, it may comprise up to a quarter of all products and half of all uses of products. This puts regulatory authorities under considerable pressure to develop appropriate strategies for efficient handling of the task.

Methods

Data presented in this study were generated by systematically compiling and interlinking information from three different sources:

- The FCEC report to the European Commission [11] for information on CFS properties of approved active substances (approval status as of 31 January 2013)
- The EU Pesticides database [19] for allocating approved active substances to use categories
- The database on Authorised Plant Protection Products of the German Federal Office of Consumer Protection and Food Safety (BVL) [20] for information on the number, nature, active ingredients, uses, and use categories of authorised PPP in Germany (authorisation status as of May 2013)

A list of potential CFS was derived from the FCEC report by aggregating relevant information as follows:

- CFS criterion 1 (low ADI/ARfD/AOEL): Separate assessments of ADI, ARfD, and AOEL values (Tables A1, A2, and A3 of the FCEC report) were merged under the assumption that the following decision rule applies: values are considered to fulfil the criterion when they are below the 5% percentile of a use group as defined in the EU pesticides database.
- CFS criterion 2 (two PBT criteria): Separate assessments of half-life in water, sediments, and soil (Tables A5, A7, and A9 of the FCEC report) were merged for assessments of persistence (P). Assessments of bioaccumulation (B) were directly retrieved (from Table A10 of the FCEC report). Assessments of aquatic toxicity (T_{AQUA}) were generated by merging separate assessments of the toxicity to fish, algae, daphnids, and other aquatic species (as provided in Tables A11, A12, A13, and A14 of the FCEC report). Assessments of human toxicity (T_{HUMAN}) were obtained by merging the separate information on CMR and STOT RE classifications (provided in Tables A16, A20, A22, A24, A26, and A 27 of the FCEC report).
- CFS criterion 4 (non-active isomers): Assessments were directly abstracted (from Table A15 of the FCEC report).
- CFS criterion 6 toxic for (reproduction 1A/1B): Separate information on existing and forthcoming classifications of reproductive toxicity was merged (from Tables A22 and A23 of the FCEC report).

- CFS criterion 7 (endocrine disrupting properties): For assessments of endocrine disrupting properties according to the interim criteria laid down in Regulation (EC) No 1272/2008, i.e. substances that are classified as both carcinogenic category 2 and toxic for reproduction category 2, separate information on these two properties (in Tables A21 and A25 of the FCEC report) were combined accordingly.
- CFS criteria 3 and 5 ('nature of critical effects' and carcinogen 1A/1B): No substances fulfilling any of these criteria were identified in the FCEC report.

Abbreviations

ADI: acceptable daily intake; AOEL: acceptable operator exposure level; ARfD: acute reference dose; BVL: Bundesamt für Verbraucherschutz und Lebensmittelsicherheit (German Federal Office of Consumer Protection and Food Safety); BfR: Bundesinstitut für Risikobewertung (German Federal Institute for Risk Assessment); CFS: candidate for substitution; CMR: carcinogenic, mutagenic or toxic for reproduction; COM: European Commission, formerly Commission of the European Communities; EC: European Community; EEC: European Economic Community; EU: European Union; EFSA: European Food Safety Authority; FCEC: Food Chain Evaluation Consortium; JKI: Julius Kühn-Institut (German Federal Research Centre for Cultivated Plants); MS: Member States of the EU; NOEC: no observed effect concentration; PBT: persistent, bioaccumulative, and toxic; PPP: plant protection product; REACH: registration, evaluation, authorisation and restriction of chemicals; STOT RE: specific target organ toxicity - repeated exposure; TER: toxicity/exposure ratio; UBA: Umweltbundesamt (German Federal Environment Agency).

Competing interests

The authors declare that they have no competing interests. They are personally and financially independent from any commercial actors in the field of plant protection products. Stefanie Rotter, Carolina Vogs, and Rolf Altenburger are employees of the Helmholtz Centre for Environmental Research (UFZ), which is part of Germany's largest state-funded research organisation. Michael Faust and Thomas Backhaus are co-owners of a small and independent environmental consulting enterprise, and Thomas Backhaus also holds a professorship at the University of Gothenburg, Sweden. Janina Wöltjen and Andreas Höllrigl-Rosta are employees of the UBA, Germany's main environmental protection agency; they are actively involved in the regulatory risk assessment of plant protection products.

Authors' contributions

All authors made substantial contributions to the conception and design of the study and the acquisition of data. SR and CV performed the database analyses. MF drafted the manuscript. All authors commented and agreed on the final version.

Acknowledgements

Funding by the German Federal Environment Agency (FKZ 3712 67 406) is gratefully acknowledged.

Disclaimer

The findings and conclusions in this paper are those of the authors and do not necessarily represent the view of the German Federal Environment Agency (UBA).

Author details

[1]Faust & Backhaus Environmental Consulting, Fahrenheitstr. 1, Bremen 28359, Germany. [2]Department Bioanalytical Ecotoxicology, UFZ - Helmholtz Centre for Environmental Research, Permoser Str. 15, Leipzig 04318, Germany. [3]Federal Environment Agency, Wörlitzer Platz 1, Dessau-Roßlau 06844, Germany.

References

1. European Union: **Regulation (EC) No 1107/2009 of the European Parliament and of the Council of 21 October 2009 concerning the placing of plant protection products on the market and repealing Council Directives 79/117/EEC and 91/414/EEC.** *Off J Eur Union* 2009, **L** 309:1–50.

2. European Communities: **Council Directive of 15 July 1991 concerning the placing of plant protection products on the market (91/414/EEC).** *Off J Eur Communities* 1991, **L** 230:1–32.

3. European Union: **Regulation (EU) No 528/2012 of the European Parliament and of the Council of 22 May 2012 concerning the making available on the market and use of biocidal products.** *Off J Eur Union* 2012, **L** 167:1–123.

4. European Union: **Directive 2009/128/EC of the European Parliament and of the Council of 21 October 2009 establishing a framework for Community action to achieve the sustainable use of pesticides.** *Off J Eur Union* 2009, **L** 309:71–86.

5. KEMI (Swedish Chemicals Agency): **The substitution principle. KEMI Report Nr. 8/07.** 2007. Available online at the agencies website: www.kemi.se.

6. VCI (Verband der Chemischen Industrie) (Association of chemical industries): **Reform des europäischen Chemikalienrechts - Fakten, Meinungen, Hintergründe.** 2005. (Reform of the European chemicals legislation – facts, opinions, background; in German). 2005. Available online at https://www.vci. de/Downloads/Publikation/28_Reform-europaeischen-Chemikalienrechts.pdf.

7. Commission of the European Communities: **White Paper strategy for a future chemicals policy.** 2001. Brussels, 27.2.2001, COM(2001) 88 final.

8. European Union: **Regulation (EC) No 1907/2006 of the European Parliament and of the Council of 18 December 2006 concerning the Registration, Evaluation, Authorisation and Restriction of Chemicals (REACH), establishing a European Chemicals Agency, amending Directive 1999/45/EC and repealing Council Regulation (EEC) No 793/93 and Commission Regulation (EC) No 1488/94 as well as Council Directive 76/769/EEC and Commission Directives 91/155/EEC, 93/67/EEC, 93/105/EC and 2000/21/EC.** *Off J Eur Union* 2006, **L** 396:1–849.

9. KEMI (Swedish Chemicals Agency): **Management of chemicals and pesticides in Sweden - national profile.** 2000. Available online at: http://www2.unitar.org/cwm/publications/cw/np/np_pdf/Sweden_National_Profile.pdf.

10. HSE (Health and Safety Executive): **Regulation (EC) No 1107/2009: progress on endocrine disrupters and candidates for substitution, Information Update: 01/2014.** 2014. Issued: 17th January 2014 online at http://www.pesticides.gov.uk/guidance/industries/pesticides/topics/pesticide-approvals/eu/european-regulation/Regulation+_EC_No_1107_2009-progress_on_endocrine_disrupters_and_candidates_for+substitution.

11. FCEC (Food Chain Evaluation Consortium): **Ad-hoc study to support the initial establishment of the list of candidates for substitution as required in Article 80(7) of Regulation (EC) No 1107/2009.** 2013. Framework Contract for evaluation and evaluation related services - Lot 3: Food Chain, Final report submitted to the European Commission, Directorate General for Health and Consumers, European Commission DG SANCO, Rue de la Loi 200, 1049 Brussels, 09.07.2013. Available online to Member States authorities and registered stakeholders via the Commission's CIRCA platform.

12. European Union: **CIRCA (Communication & Information Resource Centre for Administrations, Businesses and Citizens).** 2014. https://circabc.europa.eu.

13. COM (European Commission, Health and Consumers Directorate General): **DG SANCO working document on "Evidence needed to identify POP, PBT and vPvB properties of pesticides".** 2012. Brussels, 25.09.2012 – rev. 3. Available online at http://ec.europa.eu/food/plant/pesticides/approval_active_substances/docs/wd_evidence_needed_to_identify_pop_pbt_vpvb_properties_rev3_en.pdf.

14. OEPP/EPPO (Organisation Européenne et Méditerranéenne pour la Protection des Plantes/European and Mediterranean Plant Protection Organization): **Guidance on comparative assessment.** *Bull OEPP/EPPO Bull* 2011, **41**:256–259.

15. COM (European Commission, Health and Consumers Directorate General): **Draft Guidance document on comparative assessment and substitution of plant protection products in accordance with Regulation (EC) No** 1107/2009. 2014. SANCO/11507/2013 rev. 8, March 2014. Available to stakeholders and competent Member States authorities.

16. PSD (Pesticide Safety Directorate, United Kingdom): **Assessment of the impact on crop protection in the UK of the 'cut-off criteria' and substitution provisions in the proposed Regulation of the European Parliament and of the Council concerning the placing of plant protection products in the market.** 2008. Online available at http://www.nefyto.nl/6008/Impact-report-final-PSD-UK-%28May-2008%29.pdf.

17. Rapagani MR, Maglulio M, Piccolo M, Nencini L, Galassi T, Mazzini F: **Future availability of pesticides in the integrated pest management agricultural programme in Italy in accordance with the application of the new European Regulation No. 1107/2009 concerning the placing of plant protection products on the market: impact of the application of cut-off criteria and selection criteria for substances that are candidate for substitution.** 2011. ENEA - Agenzia Nazionale per le Nuove Tecnologie, L'Energia e lo Sviluppo Economico Sostenibile Future, Report RT/2011/8/ENEA, Roma, Italy, 2011. Available at http://www.dinamica-fp.it/centri/ra/docs/allegatoENEA.pdf.

18. European Union: **Commission Implementing Regulation (EU) No 540/2011 of 25 May 2011 implementing Regulation (EC) No 1107/2009 of the European Parliament and of the Council as regards the list of approved active substances.** *Basic Legal Act published in Off J Eur Union L* 2011, **L** 153:1–186. Regulation as amended until 31 January 2013.

19. European Union: **EU Pesticides database.** 2014. http://ec.europa.eu/sanco_pesticides/public/.

20. BVL (Bundesamt für Verbraucherschutz und Lebensmittelsicherheit) (German Federal Office of Consumer Protection and Food Safety): **Database on authorised plant protection products.** 2014. Available upon subscription as detailed on the agencies website at http://www.bvl.bund.de.

16

Industrial sludge containing pharmaceutical residues and explosives alters inherent toxic properties when co-digested with oat and post-treated in reed beds

Lillemor K Gustavsson[1*], Sebastian Heger[2], Jörgen Ejlertsson[3], Veronica Ribé[4], Henner Hollert[2] and Steffen H Keiter[2]

Abstract

Background: Methane production as biofuels is a fast and strong growing technique for renewable energy. Substrates like waste (e.g. food, sludge from waste water treatment plants (WWTP), industrial wastes) can be used as a suitable resource for methane gas production, but in some cases, with elevated toxicity in the digestion residue. Former investigations have shown that co-digesting of contaminated waste such as sludge together with other substrates can produce a less toxic residue. In addition, wetlands and reed beds demonstrated good results in dewatering and detoxifying of sludge. The aim of the present study was to investigate if the toxicity may alter in industrial sludge co-digested with oat and post-treatment in reed beds. In this study, digestion of sludge from Bjorkborn industrial area in Karlskoga (reactor D6) and co-digestion of the same sludge mixed with oat (reactor D5) and post-treatment in reed beds were investigated in parallel. Methane production as well as changes in cytotoxicity (Microtox(R); ISO 11348–3), genotoxicity (Umu-C assay; ISO/13829) and AhR-mediated toxicity (7-ethoxyresorufin-O-deethylase (EROD) assay using RTW cells) were measured.

Results: The result showed good methane production of industrial sludge (D6) although the digested residue was more toxic than the ingoing material measured using microtox$_{30min}$ and Umu-C. Co-digestion of toxic industrial sludge and oat (D5) showed higher methane production and significantly less toxic sludge residue than reactor D6. Furthermore, dewatering and treatment in reed beds showed low and non-detectable toxicity in reed bed material and outgoing water as well as reduced nutrients.

Conclusions: Co-digestion of sludge and oat followed by dewatering and treatment of sludge residue in reed beds can be a sustainable waste management and energy production. We recommend that future studies should involve co-digestion of decontaminated waste mixed with different non-toxic material to find a substrate mixture that produce the highest biogas yield and lowest toxicity within the sludge residue.

Keywords: Biogas; Co-digestion; EROD; Industrial sludge; Microtox; Oat; Reed beds; Sustainable waste management; Toxicity; Umu-C

* Correspondence: lillemor.gustavsson@karlskogaenergi.se
[1]Karlskoga Environment and Energy Company, Box 42, Karlskoga 69121, Sweden
Full list of author information is available at the end of the article

Background

Methane production as biofuels is a fast and strong growing technique for renewable energy. According to Eurostat [1], total production of biogas in Europe was 100 million tonnes of oil equivalents (toe), corresponding to 9% of the total biofuel production in the European Union [1]. In Sweden, total renewable energy production was 9,993 ktonnes [1], corresponding to 1,363 GWh. Biogas from landfill, waste water treatment plants (WWTP), co-digestion plants and industrial plants accounted for 22%, 44%, 25% and 8%, respectively [2]. The aim of EU council [3] states that 20% of final energy consumption should be provided by renewable sources by 2020. Although the production of biofuels is growing, this goal is probably not realistic [4].

The part of 8% industrial waste (food waste, sludge from WWTP and industrial wastes) has potential to increase since some of these type of waste possess good methane gas potential comparable with common substrates for biogas production like grass, wheat and straw [5]. However, digested WWTP sludge is highly questioned for use as fertilizer in agriculture [6-9], and industrial waste can also possess inherent toxicity inhibiting digestion process and resulting in a digested residue needing post-treatment. Co-digesting of hardly degradable and toxic material together with some other substrate has also been demonstrated to be a realistic option [10-15].

The main alternatives for a digested residue with toxic or environmental hazardous properties are combustion, composting and/or use in less sensitive land applications such as covering of landfills. One alternative cost-efficient and low-technology demanding method, efficient in reducing nutrient content, carbons and toxicity, is the dewatering and treatment of sludge in constructed wetlands (CW) and reed beds [16-18]. More than 7,000 CW is operating in Europe and North America with increasing number in South America, Australia, New Zeeland as well as Africa and Asia [16-24]. The removal efficiency of nutrients and pollutants by CW and reed beds can be explained by the rhizosphere providing a large attachment area for both aerobic and anaerobic microorganisms [16,17,23,25-28] and as well as dewatering capacity by evapotranspiration and mechanical impact of shoots, roots and rhizome growth [17,22-24,29]. The capacity of planted beds in treating sludge from the same industrial area as in the present study, in comparison to filter beds without vegetation, has been demonstrated earlier [29,30]. Results showed that reed-planted beds were more efficient than unplanted at retaining toxicants, reducing the water-soluble toxicity [30], total organic carbon (TOC), biological oxygen demand (BOD) and chemical oxygen demand (COD) in the outgoing water [29,30].

Common reed has also demonstrated a high adaptive capacity to sewage sludge environment, and a doubling of shoot density compared to natural stands has been observed [20,24,29].

In this study, digestion of industrial sludge from Björkborn industrial area in Karlskoga containing nitroaromatic compounds, explosives and pharmaceutical residue and co-digestion of the same sludge mixed with oat was studied in parallel as well as post-treatment and dewatering of the digested sludge in reed beds. Earlier studies of the sludge used in this study industrial sludge from Björkborn industrial area in Karlskoga, showed good methane production potential during mesophilic conditions [31] but increased toxicity in the digested sludge [32-34]. Dewatering and treatment of this particular sludge demonstrated high dewatering capacity and reduced nutrient levels in outgoing water and sludge residue [29] as well as significantly reduced toxicity in outgoing water and bed material of reed beds measured with DR-CALUX, Umu-C assay and fish embryo toxicity test using *Danio rerio* [30].

The aim of the present study was to investigate co-digestion of industrial sludge from Björkborn industrial area in Karlskoga and oat. We wanted to investigate if co-digestion of substrate, considered as waste, together with common crop is suitable for biogas production. Moreover, we wanted to check if oat could alter the methane yield and toxic properties of industrial sludge by measuring methane production as well as change of cytotoxicity, genotoxicity and dioxin-like activity. In addition, we also wanted to investigate if co-digestion produced a less toxic and post-treatment demanding residue when dewatered through reed beds.

Results and discussion

Biogas production

The results of the biogas measurements (Table 1) showed that the mixed reactor with sludge and oat (D5) produced methane gas at a level below the control reactor with oat (D4) but possessed a methane gas potentially higher than the sludge reactor (D6). However, gas production in reactor D6 (Table 1) was almost twice as high compared to methane production of the same sludge in a former study where a gas production of 2,000 ml/day at 37°C (60% methane) was achieved using the same organic loading rate (OLR) of 3 g VS/L reactor/day [31].

The measured toluene, benzene, ethylbenzene and xylen concentrations during routine controls (ALS; Table 2) exceeded the limits for land application and land use in Sweden [35]. These compounds are mainly degradation products of 2,4,6-trinitrotoluene (TNT), di-nitrotoluens, nitrobenzoic acids and a range of other compounds used for the manufacturing of explosives, pharmaceutical and chemical intermediates [32-34].

Table 1 Biogas and methane yield

Reactor	Substrate	OLR (g VS/L/day)	CH4 (%)	Gas production (ml/day at 37°C)	Specific gas production (ml methane/g VS at 0°C)
D4	Oat	6	51	17,500	300
D5	Oat + sludge	3 + 3	53	14,000	270
D6	sludge	3	63	4,000	180

However, the OLR of 3 gVS/L used in the present study did not inhibited the digestion process and resulted in gas production comparable with ordinary municipal sewage sludge in Sweden with an average gas production of 160 to 350 m^3 CH$_4$/tonne VS [36,37]. OLR in D6 is half of OLR in D5, resulting in twice as high hydraulic retention time (HRT) which may give the microorganisms time for adaptation but can also explain the lower methane yield. The lower HRT in D5 (Table 2) may also explain lower toxicity. A previous study [38] found that decreasing HRTs results in higher feeding and outgoing flow rates and, consequently, rapid withdrawal of toxic intermediates and less accumulation of inhibiting intermediates [38]. Intermediates can originate from the degradation of aromatic amino acids [39].

Table 2 Content of organics and metals in a month sample of undigested sludge

Organic compounds		Metals	
Parameter	(mg/kg TS)	Parameter	(mg/kg TS)
TS (%)	20	As	<3
VS (% of TS)	68.4	Ba	79.8
AOX	84	Be	0.264
Benzene	0.76	Ca	25,600
Toluene	260	Cd	<0.1
Etylbenzene	0.27	Co	4.17
Xylene	13	Cr	74.3
Di-etylftalat	4.1	Cu	25.2
Di-n-butylftalat	0.19	Fe	63,000
Di-n-pentylftalat	32	Hg	<1
Di-(2etylhexyl)ftalat	3.1	Li	0.517
PAH (sum)	<4.3	Mn	85
4-Nonylphenol	<0.25	Mo	4.9
RDX	0.43	Na	2,050
HMX	0.45	Ni	4.08
TNT	0.82	P	24,600
		Pb	29.2
		S	17,300
		Sr	34.3
		V	21
		Zn	99.8

From Björkborn industrial area collected during routine sampling.

A decreased HRT can prevent the accumulation of toxic substances and inhibition of the digesting process [10,38], but as a consequence, the methane yield would be reduced [38]. Instead, co-digestion could be a promising alternative option. Olive mill waste (OMW) possesses a high energy potential (45 to 220 g of COD/L) but also a low pH, alkalinity and nitrogen content; additionally, a lipophilic fraction and phenolic compounds are present. These characteristics make this substrate toxic and complex to degrade during anaerobic conditions [14]. However, it is increasing the methane yield when co-digested with manure (Table 3). Earlier studies (Table 3) of co-digesting different substrates revealed promising biogas production and showed high methane yield.

Co-digesting of starch-rich and ammonia strong wastes obtained gas yields comparable with yields obtained in the present study when co-digesting sewage sludge and potato processing industrial waste [13]. Co-digesting of manure, slaughterhouse and agricultural waste revealed higher gas yield (Table 3) with higher diversity of substrate [13]. This is consistent with the study performed by Chan et al. [11] who tested co-digestion of sewage sludge and marine dredgings mixed with municipal refuse at 13 different ratios [11]. Additionally, results in this study are strengthened by other studies which also found that co-digesting enhanced biogas production compared to digestion of single material such as high-strength COD substrate [10,13,14].

The methane yield in this study is lower than in many other co-digestion studies (Table 3). The explanation can be the presence of hardly degradable and toxic nitro-aromatic compounds (Table 2). The present study of 4,000 ml biogas/day and 300 ml CH$_4$/g VS added (Table 1) are confirmed by earlier studies demonstrating digestion of nitroaromatic compounds where methane gas production of 2,300 ml/day was achieved using a nitro-benzene loading rate of 30 mg/L/day [40]. In another study, methane yields between 116 and 209 ml CH$_4$/gVS L^{-1} were obtained by adding p-nitrophenol [38].

A low methane yield can also be explained by low carbon/nitrogen (C/N) ratio. The C/N ratio of 1.7 (Table 4) makes the industrial sludge unfavourable for digestion since a ratio of 16 is required to balance the anaerobic degradation between accumulation of volatile fatty acid (VFA) during digestion with high C/N ratio or accumulation of ammonia with low C/N ratio [10,13]. C/N ratio in oat mixed with sludge was higher (3.6) but still

Table 3 Examples of previously performed studies of co-digestion

Substrate	(Mixed ratios v/v)	HRT (days)	OLR (g VS/L/day)	Methane yield (ml CH$_4$/g VS/day)	References
OMW + cattle manure	75% + 25%	13	3,4	700 to 1,000	Angelidaki and Ahring [10]
OMW + pig manure	69% + 31%	6	2.9 + 2.6	2,700	Sampaio et a.l [14]
Sewage sludge + potato waste	44% + 56%	20	2.7	600	Murto et al. [13]
Industrial waste + pig manure	17% + 83%	30	2.6	800	Murto et al. [13]
Industrial waste + pig manure + slaughterhouse waste	17% + 71% + 12%	28	3.1	900	Murto et al. [13]
Industrial waste + pig manure + slaughterhouse waste	17% + 66% + 12% + 5%	36	2.6	1,000	Murto et al. [13]
Sewage sludge + marine dredgings + municipal refuse	20 + 5% + 75%	36 (batch)		900 to 1,200	Chan et al. [11]
OMW + piggery effluent	83% + 17%	6 to 7	3.5	1,300	Marques [12]
Oat + sludge	50% + 50%	28	6	500	Present study

unfavourable according to [13]. On the other hand, if the activity of methanogenic bacteria is low, less of the proteins will be degraded to free ammonia ions, inhibiting the digestion process [5].

Post-treatment in reed beds

Reduction of nutrients and carbon throughout the three months of digestion and dewatering in reed beds is shown in Table 4. The results show a high reduction of ammonia for both reed bed lines (Table 4) despite the short treatment time. This result is consistent with the findings of an earlier study of dewatering the sludge from Björkborn industrial area [29]. The authors found that reed beds were able to reduce COD and TOC to more than 90%, BOD and total nitrogen (N$_{tot}$) to more

Table 4 Nutrient and organic change in the sludge and sludge + oat reactor

(mg/L)	N$_{tot}$	TOC	P$_{tot}$	NH$_4$	C/N
Ingoing sludge D5	103	369	4.2	46.7	3.6
Digested D5	110	215	1.5	89.0	2.0
Sludge residue D5	28	275	1.4	0.4	9.9
Bed material D5	49	8	1.3	17.2	0.2
Outgoing water D5	Nm	118	1.5	2.7	0.2
Red (%)		68	65.0	94.2	*-0.2*
Ingoing sludge D6	811	1,340	10.7	800	1.7
Digested D6	777	1,290	13.4	746	1.7
Sludge residue D6	439	350	3,4	0,31	0,8
Bed material D6	123	373	32.3	24.3	3
Outgoing water D6	347	151	10.1	164	0.4
Red (%)	57	89	5.6	79.5	1.6

Before and after digestion and dewatering through reed beds after 3 months. Reduction (%) of nutrients and organics is calculated between ingoing sludge and outgoing water from reed beds. Italic number means increase. nm, not measured.

than 80% and total phosphorous (P$_{tot}$) to over 85% during the growth period (April to October). During the resting period (November to March), reduction of COD, BOD, N$_{tot}$ and P$_{tot}$ decreased to 66%, 28%, 35% and 55%, respectively [29].

Additionally, several other studies have shown a high post-treatment capacity of reed beds. BOD removal efficiency of 63% to 79% independent from the season or age of the system was reported [41]. Moreover, other studies demonstrated removal efficiency of nutrients, nitrogen, phosphorus, BOD and total suspended solids (TSS) using constructed wetlands [21,42].

Toxicity tests
Microtox

The result of the present study showed an increased toxicity and an accumulation potential of non-reduced nitroaromatic compounds of the industrial sludge in reactor D6 during digestion. This is consistent with earlier studies investigating the same sludge where an increased acute toxicity and decreased cell vitality were measured after exposure to extracts of digested sludge using the microtox [31] and neutral red assay [32], respectively. Reed beds containing industrial sludge (D6) showed a decreased toxicity in sludge residual and bed material, suggesting that some of the compounds causing toxicity were transformed to water-soluble compounds and rapidly transported through the reed beds, ending up in the outgoing water. A large portion of those compounds, which are not trapped in the bed material or flushed out with the outgoing water, were probably degraded in the reed beds as shown in earlier studies of dewatering sludge from Björborn industrial area [30].

A contrary result was observed by testing a mixture of sludge and oat from D5. Ingoing sludge of D5 was less toxic compared to ingoing sludge of D6 (Figure 1). The

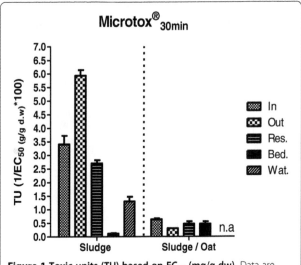

Figure 1 Toxic units (TU) based on EC$_{50}$ (mg/g dw). Data are given as mean and 95% confidence interval. Each sample was tested in two independent replicates.

differences between the reactors (Figure 1) are larger than the dilution effect of oat by a factor of 2 (Table 1). Additionally, sludge from D5 showed decreased toxicity after digestion in opposite to digested D6 sludge demonstrating increased toxicity. The comparison of both reactors after digestion demonstrates a larger difference in toxicity that cannot be explained by dilution effect alone. Acute toxicity could not be detected in outgoing water from reed beds of D5. Sludge residue and bed material trapped toxic compounds, demonstrated by slightly increased TU values in D5 (Figure 1, Table 5).

Umu-C

In the Umu-C assay (ISO 13829), a genotoxic effect is significant if the induction factor is above 1.5 compared to the negative control. Figure 2 shows the used concentrations of the different samples reaching an induction factor of 1.5. Genotoxicity above 1.5 was detected in the undigested and digested D6 sludge with LID values of 82.5 and 41.25 μl/ml, respectively (Figure 2, Table 5). This result clearly demonstrates that the toxicity increased after digestion with only half the concentration needed to be genotoxic compared to undigested sludge. Outgoing water from reed beds treating D6 sludge showed higher LID values indicating that genotoxic compounds may have been adsorbed in bed material or degraded.

The mixed material with sludge and oat from reactor D5 showed unchanged genotoxic activity before and after digestion but decreased LID values in bed material which points at adsorption of genotoxic compounds in the bed material (Figure 2).

The genotoxic properties of this industrial sludge have been demonstrated before. A previous study [32] showed significant genotoxic potential in the digested sludge compared to undigested, tested in the comet assay with RTL-W1-cells. A former study of large scale anaerobic treatment of sludge from Björkborn industrial area demonstrated an increased genotoxic activity during treatment from induction factor 1.5 to induction factor 2.8 in the Umu-C assay [34]. This is higher as the genotoxic potential of D6 in the present study, although the pattern is the same, an increasing induction factor during anaerobic treatment (Figure 2, Table 5).

Different publications describe the genotoxic properties of nitro-aromatic compounds such as TNT, nitrobenzoic acids, nitrobenzenes and degradation products [43-46] and increasing toxicity with increasing number of nitro-groups [34,47]. Additionally, literature has also shown higher toxicity with nitro-substituted aromatics compared with their corresponding amines [38,48-50]. This may explain the increased genotoxicity, although weak, by the presence of unreduced nitro-aromatics within the sludge used in this study. 7-Ethoxyresorufin-O-deethylase (EROD) EROD activity could only be detected in three samples (Figure 3, Table 6). The EC$_5$ was comparably low in D5 Res (36.8 mg/ml), D6 Res (20.4 mg/ml) and in D6 Bed (68.5 mg/ml). All other samples did not induced EROD activity.

After 3 months of loading, detectable levels of EROD inducers could only be found in sludge residue on top of the reed beds from both D5 and D6 and in the bed material from D6. However, the levels are very low, and it was impossible to calculate Bio-TEQ values. Mesophilic digestion can increase EROD activity compared to undigested material. In a former study by [33], the same sludge obtained from the same manufacturing area as in this study, demonstrated three to six times higher levels of EROD activity in the digested sludge than in the incoming flux [33]. Additionally, earlier studies of methanogenic digestion of household waste showed that acid anaerobic conditions [51,52].

The identity of these EROD inducers was neither clarified in the former study by [33] nor in this present study. Additionally, it has been shown that a variety of different compounds apart from the well-known dioxins and polychlorinated biphenyls (PCBs) can induce EROD activity. For instance, conversion of proteins like tryptophan to indole acetic acid (IAA) and transformed compounds like indole-3-carbinole (I3C) and indolo-3.2-β-carbazole (ICZ) demonstrated 1×10^2 and 1×10^5 times higher AhR binding affinity than the parent compound, respectively [53,54]. Additionally, several phytochemicals including caffeic acid, chlorogenic acid, diosmin, ferulic acid and resveratrol showed both inhibition and induction of EROD [55].

Conclusions

Digestion of sludge from Björkborn industrial area (D6) resulted in a methane production of 180 ml CH$_4$/g VS,

Table 5 Summary table of the result of ecotoxicity test

Sample	EROD		Umu-C induction > 1.5	Microtox TU
	EC$_5$	EC$_{10}$	LID (µl/ml)	(mg/g dw)
D5 undigested	n.a		165	0.6
D5 digested	n.a		165	0.3
D5 sludge residue	36.8	100.8	165	0.5
D5 bed material	n.a		165	0.11
D5 outgoing water	n.a		n.m	n.a
D6 undigested	n.a		82.5	3.3
D6 digested	n.a		41.25	5.9
D6 Sludge residue	20.4	37.8	n.m	2.7
D6 Bed material	68.5		n.m	0.04
D6 Outgoing water	n.a		82.5	1.3

n.a, not available; n.m, not measured.

comparable with methane production of WWTP sludge. However, the digested residue was more toxic than the ingoing material measured using microtox$_{30min}$ and Umu-C.

Co-digestion of toxic industrial sludge and oat (D5) showed higher methane production (270 ml CH$_4$/g VS) despite the fact that just half of the HRT was used. Moreover, the digested residue was significantly less toxic than the sludge residue of D6. The differences in toxicity (i.e. Microtox) cannot be explained by dilution effects (OLR and HRT) as discussed in the 'Results and discussion' section and in this section. This clearly demonstrates the benefits of co-digestion of industrial sludge together with oat. Furthermore, dewatering and treatment in reed beds showed low and non-detectable toxicity in reed bed material and outgoing water as well as reduced ammonium (NH$_4^+$), N$_{tot}$ and TOC. Moreover, toxicity of the dewatered D5 sludge on top of the reed

beds were significantly lower than corresponding D6 for all three bioanalytical tests used in this study.

A less-contaminated waste stream demands less energy and monitoring during treatment. Therefore, digestion of sludge resulting in a less toxic residue, with a shorter and less complex post-treatment is the most cost-efficient option. We have demonstrated that co-digestion of industrial sludge with oat fulfilled that requirement. Additionally, dewatering and treatment of sludge in reed beds can be recommended as a post-treatment method of digested sludge.

Future studies should involve co-digestion of this industrial sludge or other waste mixed with different straw and grass in different proportions in order to find a substrate mixture that produces the highest biogas yield and lowest toxicity within the sludge residue. Using waste as a substrate in a sustainable way can also increase the possibilities to reach the aim of the EU council [3],

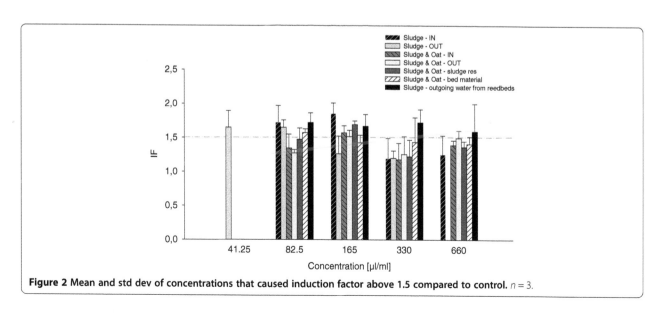

Figure 2 Mean and std dev of concentrations that caused induction factor above 1.5 compared to control. $n = 3$.

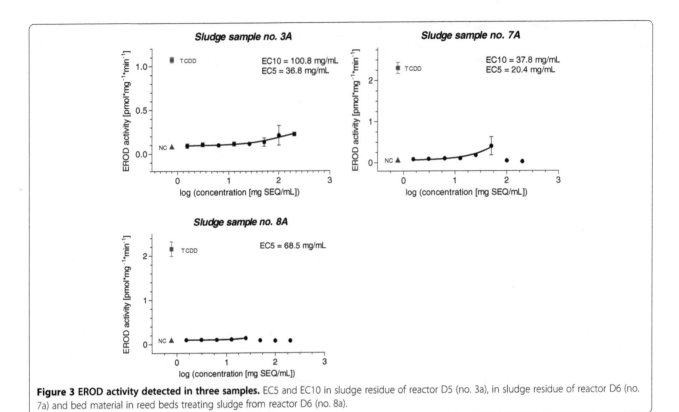

Figure 3 EROD activity detected in three samples. EC5 and EC10 in sludge residue of reactor D5 (no. 3a), in sludge residue of reactor D6 (no. 7a) and bed material in reed beds treating sludge from reactor D6 (no. 8a).

stating that 20% of final energy consumption should be provided by renewable sources by the year 2020.

Methods

Substrates and inoculums used in the study

Industrial sludge used as feeding material to the bioreactors was collected from the wastewater treatment plant of Björkborn industrial area, (Karlskoga, Sweden). Approximately 25 kg of dehydrated sludge was collected and carefully mixed and aliquots were stored in 1 L polyethylene bottles at –20°C. Total solid content (TS) of the sludge mounted 15.8, and volatile solid (VS) was 65.8% of TS. The oat was received from Söderslätts Spannmålsgrupp and milled to a grain size of 1 mm prior storage at room temperature in 1 L polyethylene cans during the experimental period. TS of the mounted

96%, and the VS of TS was 97.3%. Inoculum for the laboratory digesters consisted of digested sewage sludge from Reningsverket Nykvarn (Linköping, Sweden) and cow manure from Swedish dairy farm (Hags gård, Rimforsa, Sweden).

Digestion of sludge and oat in bioreactors

In the study, two digesters and one control were operated at 37°C with 20 days of HRT for the co-digestion of industrial sludge, milled oat (D5) and industrial sludge (D6), in parallel with a control reactor fed with milled oat (D4). The control reactor was operated according to the same protocol as the experiment reactors. Each digester contained an active liquid volume of 4 L and was equipped with a tube for feeding substrate/withdrawal of reactor material, a gas outlet and a central placed

Table 6 Samples, extraction solvents and analysis used in this study

Sampling point	Sample	Analysis			
		TOC, NH$_4^+$, Tot N, Tot P	*Microtox*	*Umu-C assay*	*EROD assay*
Ingoing sludge	IN	Water phase	Water phase	Water phase	Toluene (Soxhlet)
Digested sludge	Out	Water extract	Water extract	Water extract	Toluene (Soxhlet)
Sludge residual	Res	Water phase	Water phase	Water phase	Toluene (Soxhlet)
Bed material	Bed	Water extract	Water extract	Water extract	Toluene (Soxhlet)
Outgoing water	Wat	Native water	Native water	Native water	Toluene

The material used in the analysis is in bold in the first column. The first row in bold and italics describes the analysis used in the study, and the cells from the second row and column describes the extracts used.

impeller (Ø:70 mm) for mixing. Mixing was performed in 15 min intervals four times a day and for about 10 min in connection to feeding by use of a servomotor (MAC050-A1; All motion technology, New York, NY, USA) at 500 rpm.

Digester D5 was inoculated with 2.5 L digested sewage sludge and 500 g of cow manure followed by the addition of 1 L deionized water. The following day, 200 g of digester liquid was withdrawn followed by the feeding of 200 g of a substrate blend consisting of the industrial sludge (2.0 g VS/L/day), oat 0.25 (g VS/L/day) and deionized water. The same feeding procedure was done for 28 days. To start reactor D6, digested sludge (200 g per day) from reactor D5 was collected for the last 10 days and transferred to a digester. At an active volume of 2 L in D6, 1 L digester liquid was transferred from D5 to D6.

Digester D5 was fed with the same substrate blend until an active volume of 4 L was resumed. The loading rate was allowed to increase with 0.5 kg VS/L/day every 5 days until 3 g VS/L/day was reached first with the industrial sludge and then 3 g VS/L/day with oat. Digester D5 was then fed with this substrate blend for 60 days when the experiment was terminated. Digester D6 was fed with the industrial sludge (2 g VS/L) and deionized water until an active volume of 4 L was resumed. The loading rate was then allowed to increase with 0.5 kg VS/L/day every 5 days until 3 g VS/L/day was reached. This loading rate was kept for the remaining experimental period of 60 days.

Gas production was recorded on daily basis. The methane content of the produced gas was measured once a week. The produced gas was collected in a balloon during 24 h, and the gas composition was determined using a portable gas detector (Gas data, GFM series, Whitley, Coventry, UK). Analyzed gases, besides CH_4, were CO_2, O_2 and H_2S. Samples were also taken from the reactor liquid; concentrations of individual VFAs (acetic, propionic, butyric, isobutyric, capronic, isocapronic, valeric and isovaleric acid) were determined twice a week by GC-FID [56], pH at least twice a week and TS/VS once a week following the protocols from Swedish Standard SS-EN 12176 and SS 028113, respectively.

Post-treatment in reed beds

Two lines of reed beds treating sludge from D5 and D6 were constructed. Three beds with 32 cm of sand and gravel, from top to bottom 10 cm of sand, 9 cm of coarse sand, 7 cm of gravel and 6 cm of small stones (Figure 4) with vertical flow, were constructed indoors with a volume of 25 L and a upper surface area of approximately 700 cm^2 (Figure 1). The beds were planted with common reed (*Phragmites australis*) and kept indoors under controlled conditions with 300 mmol

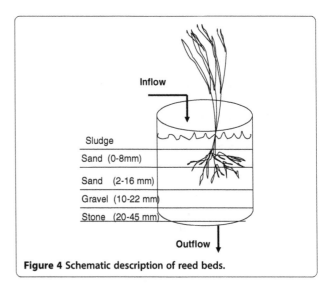

Figure 4 Schematic description of reed beds.

photons/m^2 s^{-1} and loaded with sludge (diluted to 1.2% dry weight, 1.0 L/day); retention time was 2 h. The loading was continued within 3 months.

Sampling, preparation, extraction and cleanup

All tests in this study were performed on the undigested sludge, digested sludge and outgoing water (Figure 5) of the last month sample after 90 days of digesting. Outgoing water was collected daily and pooled into a monthly sample and stored at −20°C prior to analysis. Sludge residue on top of the reed beds and bed material (Figure 5) was collected at the end of the study.

Approximately 500 g of the bed material and sludge residual was collected from the upper part of the beds consisting of sand. Only the upper layer was collected

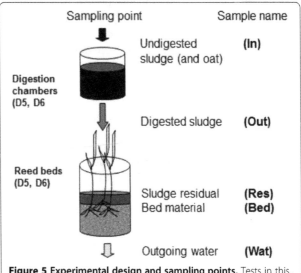

Figure 5 Experimental design and sampling points. Tests in this study performed on the undigested sludge, digested sludge and outgoing water.

for the experiments based on the assumption that most of the compounds will be trapped in the upper part of the beds due to the large specific area of this bed material compared to the coarse material below the upper part. In brief (Figure 5), ingoing sludge (In), digested sludge (Out), sludge residue (Res), bed material (Bed) and outgoing water (Wat) were extracted with water and toluene (Table 6) according to a former study in order to estimate both the bioavailable toxicity and the total toxic potential, respectively [30].

For toxicity testing of sludge and bed material, a Soxhlet extraction was conducted (24 h, 3 cycles/h) by using toluene (Riedel-de Haën, >99.8% (GC)), Envisolv according to [57]. The toluene extracts for the EROD assay were cleaned up using a multi-layer silica column in order to isolate the persistent lipophilic compounds according to the protocol shown in [58] and [59]. The silica column consisted from the bottom of 1 cm^3 copper powder to precipitate the remaining sulphate and 5.3 cm^3 KOH, 0.88 cm^3 neutral silica, 5.3 cm^3 40% H_2SO_4, 2.65 cm^3 20% H_2SO_4, 1.76 cm^3 neutral silica and 1.76 cm^3 $NaSO_4$ (monohydrate). The remaining fraction contained persistent dioxin-like compounds and included, e.g. PCDDs/PCDFs and PCBs and were eluted with n-hexane. The solvents were evaporated under a nitrogen stream, and the sample was transferred to DMSO (Sigma assay (GC) minimum 99.5%, Sigma-Aldrich, St. Louis, MO, USA) for the subsequent EROD assay.

The ingoing sludge, bed material and sludge residue were also prepared for analysis of toxicity, TOC, N_{tot}, NH_4^+ and P_{tot} by shaking with 1:5 proportion of deionized water for 24 h followed by centrifugation at 5,700 × g. The water-phase supernatant was used for testing. The outgoing water from the beds was collected, and 100 ml from each time point was pooled to a monthly sample and stored at −20°C until analysis. The water was centrifuged and tested undiluted.

Analysis of organics and metals were performed of the ingoing sludge of reactor D6 (Table 2) at the commercial laboratory, ALS Scandinavia AB, Luleå, Sweden (ALS).

Toxicity tests
Microtox
Toxicity to *Vibrio fischeri* of water extracted samples was assessed according to the Microtox® ISO 11348–3 test protocol (2007) by using Microtox Omni™ Software (Azur Environmentals, Newark, DE, USA). The samples and two controls consisting of deionized water were adjusted to a salinity of 2 ppt. Light inhibition in the sample compared to the control was measured after 30 min of incubation. Samples were diluted by a 1:2 series, and each dilution step was prepared in duplicates. The sample concentrations tested were 80%, 50%, 33.33%, 25%, 16.67%, 12.50%, 8.33% and 6.25%. EC_{50} values (30 min)

were determined from concentration-response curves. Toxic units TU (*g/g* dw) was calculated using the formula: TU = $1/(EC_{50} × 100)$.

Umu-C
Genotoxicity of water extracts from the sludge and bed material as well as outgoing water from the beds was detected using the Umu-C test with *Salmonella typhimurium* TA1535/pSK1002 according to the standard protocol ISO 13829 (2000). The bacteria were cultured in tryptone/glucose medium in 96-well plates (Labdesign, TCT, Lake Charles, LA, USA). All concentrations were tested in triplicates. As positive and negative control, 50 µg 4-nitroquinoline-1-oxide (4-NQO)/L and pure medium was used, respectively. Induction of genotoxicity, expressed as β-galactosidase activity was measured as the absorbance at 420 nm after 2 h of exposure followed by 2 h post-incubation. Growth was measured as the absorbance at 600 nm. Absorbance was measured with a microplate reader (Expert 96, MikroWin 2000, Asys/Hitech, Eugendorf, Austria). The result was calculated as an induction ratio related to growth in Equation 1.

$$\text{Induction ratio} = (1/\text{Growth}_{\text{Abs600 nm}} × \quad (1)$$
$$(\text{Samples}_{\text{Abs 420 nm}}/\text{Control}_{\text{Abs420 nm}}))$$

The test was considered valid if the growth factor at a wavelength of 600 nm of exposed bacteria versus negative control was not below 0.5 and the induction ratio measured at 405 nm of the positive control was at least twice compared to the negative control. The samples were considered genotoxic if the induction factor exceeded 1.5 (exposed bacteria versus negative control) measured at 405 nm.

EROD assay
Induction of 7-ethoxyresorufin-O-deethylase was measured in the CYP 1A expressing permanent fish cell line RTL-W1 (rainbow trout liver, *Oncorhynchus mykiss*). Cells were obtained from Dr. Niels C. Bols (University of Waterloo, Canada) [60] and maintained at 20°C in 75 cm^2 plastic culture flasks (TPP, Trasadingen, Switzerland) without additional gassing in Leibovitz medium (L15) supplemented with 9% foetal bovine serum (Th. Geyer, Renningen, FRG), 1% penicillin/streptomycin (Sigma-Aldrich). Induction of EROD was measured in confluent cell monolayers in 96-well microtiter plates (TPP) with 3 to 4 × 10^5 cells/ml according to a previously published method [61,62]. Before exposure to the sludge extracts, cells were seeded in 96-well microtiter plates at a density of 3 to 4 × 10^5 cells/ml and allowed to grow at 20°C to confluency for 72 h. Subsequently, the medium was removed and the cells were exposed for 72 h with the sludge extracts water dilutions in L15 medium, negative control (L15 medium) and positive

control using 100 to 3.125 pM 2,3,7,8-tetrachlorodibenzo-p-dioxin (TCDD, Sigma-Aldrich, Deisenhofen, FRG). After exposure, all plates were shock-frozen and stored at −80°C for at least 1 h until EROD measurement.

For measurement of the EROD activity, the plates were thawed for 10 min, the protein standard solution (10 to 1.25 mg/ml bovine serum albumin (Sigma-Aldrich) in phosphate buffer (0.1 M Na_2HPO_4-solution; Malinckrodt Baker, Deventer, Netherlands) adjusted to pH 7.8 with 0.1 M NaH_2PO_4-solution (Merck, Darmstadt, FRG) was added in triplicates and 100 to 3.125 nM resorufin standard (Sigma-Aldrich) in phosphate buffer was added in duplicates. The 7-ethoxyresorufin solution (100 μl, 1.2 μM, Sigma-Aldrich) was added to each well except the wells containing either the protein standard or the resorufin standard. The plates were incubated for 10 min. NADPH (50 μl, 0.09 mM, Sigma-Aldrich) was added to all wells and the plates were incubated 10 min at room temperature. The deethylation reaction was stopped by adding 100 μl of 0.54 mM fluorescamine (in acetonitrile) to each well. The production of resorufin was measured in a fluorescence plate reader (TECANinfiniteM200, Tecan Austria GmbH, Grödig, Austria; excitation 544 nm, emission 590 nm) after 15 min. The EROD activity was expressed as picomole resorufin produced per milligramme protein per minute (pmol/(mg protein/min)). Protein was determined fluorometrically using the fluorescamine method (excitation 355 nm, emission 465 nm) [57,63]. Concentration-response curves and EC_5, EC_{10} and EC_{25} values were calculated using non-linear regression analyses of GraphPad Prism 5.0 (GraphPad, San Diego, USA).

Competing interests
The authors declare that they have no competing interests.

Authors' contributions
LG designed and conceived of the study, collected and prepared the samples, carried out the microtox, participated in the EROD assay and the Umu-C assay and drafted the manuscript. SH carried out the EROD assay. JE carried out the digestion part, and VR performed the Umu-C assay. HH participated in its design and coordination and helped to draft the manuscript. SK participated in the design of the study, supervised the study in Aachen and performed the statistical analysis. All authors read and approved the final manuscript.

Acknowledgements
We would like to thank Karlskoga Biogas for the financial support to this study. The authors would like to express their thanks to Drs. Niels C. Bols and Lucy Lee (University of Waterloo, Canada) for providing RTL-W1 cells.

Author details
[1]Karlskoga Environment and Energy Company, Box 42, Karlskoga 69121, Sweden. [2]Department of Ecosystem Analysis, Institute for Environmental Research, RWTH Aachen University, Aachen 52062, Germany. [3]Scandinavian Biogas Fuels AB, Holländaregatan 21A, Stockholm 11160, Sweden. [4]School of Sustainable Development of Society and Technology, Mälardalen University, Västerås 72123, Sweden.

References

1. Corselli-Nordblad Louise: Renewable energy in the EU28. *Eurostat Press Office* 2014, 37. http://epp.eurostat.ec.europa.eu/cache/ITY_PUBLIC/8-10032014-AP/EN/8-10032014-AP-EN.PDF.
2. Energimyndigheten (Swedish authority for energy use) Produktion och användning av biogas år 2010 (Production and use of biogas during 2010): Report ES 2011:07. http://www.biogasvast.se/upload/Milj%C3%B6/Biogasproduktion%20Energimyndigheten%202011.pdf.
3. EU Council Directive 2009/28/ EC on the Promotion of the Use of Energy from Renewable Sources of the European Parliament and of the Council. http://europa.eu/legislation_summaries/energy/renewable_energy/en0009_en.htm.
4. EU Council: COMMISSION STAFF WORKING DOCUMENT Recent Progress in Developing Renewable Energy Sources and Technical Evaluation of the use of Biofuels and Other Renewable Fuels in Transport in Accordance with Article 3 of Directive 2001/77/EC and Article 4(2) of Directive 2003/30/EC. 2011, http://europa.eu/legislation_summaries/energy/renewable_energy/l27065_en.htm.
5. Jarvis Å, Schnürer A: Mikrobiologisk handbok för biogasanläggningar. *Swed Gas Technol Cent, Rapp SGC* 2009, 207:1102–7371.
6. Engwall M, Hjelm K: Uptake of dioxin-like compounds from sewage sludge into various plant species–assessment of levels using a sensitive bioassay. *Chemosphere* 2000, 40:1189–1195.
7. Hernández T, Moreno JI, Costa F: Influence of sewage sludge application on crop yield and heavy metal availability. *Soil Sci Plant Nutr* 1991, 37:201–10.
8. Hooda PS, McNulty D, Alloway BJ, Aitken MN: Plant availability of heavy metals in soils previously amended with heavy applications of sewage sludge. *J Sci Food Agricul* 1997, 73:446–54.
9. Corey RB, King LD, Lue-Hing C, Flanning DS: Effects of sludge properties on accumulation of trace elements by crops. In *Land Application of Sludge – Food Chain Implications*. Edited by Page AL, Logan TJ, Ryan JA. Chelsea MI (US): Lewis Publisher; 1987.
10. Angelidaki I, Ahring BK: Codigestion of olive oil mill wastewaters with manure, household waste or sewage sludge. *Biodegradation* 1997, 8:221–226.
11. Chan YSG, Chu LM, Wong MH: Codisposal of municipal refuse, sewage sludge and marine dredgings for methane production. *Environ Pollut* 1999, 106:123–128.
12. Marques IP: Anaerobic digestion treatment of olive mill wastewater for effluent re-use in irrigation. *Desalination* 2001, 137:233–239.
13. Murto M, Björnsson L, Mattiasson B: Impact of food industrial waste on anaerobic co-digestion of sewage sludge and pig manure. *J Environ Manage* 2004, 70:101–107.
14. Sampaio MA, Gonçalves MR, Marques IP: Anaerobic digestion challenge of raw olive mill wastewater. *Bioresour technol* 2011, 102:10810–10818.
15. Heger S, Bluhm K, Agler MT, Maletz S, Schaeffer A, Seiler T-B, Angenent LT, Hollert H: Biotests for hazard assessment of biofuel fermentation. *Energ & Environ Sci* 2012, 5:9778–9788.
16. Kadlec RH, Knight RL: *Treatment Wetlands*. London, New York: CRC Press; 1996.
17. Nielsen S, Willoughby N: Sludge treatment and recycling of sludge and environmental impact. Topic in 10th European Biosolids and Biowaste Conference UK, November 2005. *Water Environ J* 2005, 19:285–296.
18. Ugetti E, Ferrer I, Molist J, García J: Technical, economic and environmental assessment of sludge treatment wetlands. *Water Res* 2011, 45:573–582.
19. Bialowiec A, Randerson PF: Phytotoxicity of landfill leachate on willow - Salix amygdalina L. *Waste Manage* 2010, 30:1587–1593.
20. Kengne IM, Dodane P-H, Akoa A, Kone D: Vertical-flow constructed wetlands as sustainable sanitation approach for faecal sludge dewatering in developing countries. *Desalination* 2010, 251:291–297.
21. Merlin G, Pajean J-L, Lissolo T: Performance of constructed wetlands for municipal wastewater treatment in rural mountainous area. *Hydrobiologia* 2002, 469:87–98.
22. Nielsen S: Sludge drying reedbeds. *Wat Sci and Tec* 2003, 48:101–109.
23. Troesch S, Liénard A, Molle P, Merlin G, Esser D: Sludge drying reed beds: s full and pilot-scales study for activated sludge treatment. *Wat Sci Tech* 2009, 60:1145–1154.
24. Ugetti E, Ferrer I, Llorens E, García J: Sludge treatment wetlands: a review on the state of the art. *Bioresour Tech* 2010, 101:2905–2912.

25. Coleman J, Hench K, Garbutt K, Sexstone A, Bissonette G, Skousen J: Treatment of domestic wastewater by three plant species in constructed wetlands. *Water Air Soil Poll* 2001, **128:**283–285.

26. Decamp O, Warren A: **Bacteriovory in ciliates isolated from constructed wetlands (reed beds) used for wastewater treatment.** *Water Res* 1998, **12:**1989–1996.

27. Decamp O, Warren A: **Investigation of Escherichia coli removal in various designs of subsurface flow wetlands used for wastewater treatment.** *Ecol Eng* 2000, **14:**293–299.

28. Moshiri GA: *Constructed wetlands for water quality improvement.* Michigan, USA: Lewis Publishers; 1993.

29. Gustavsson L, Engwall M: **Treatment of sludge containing nitro-aromatic compounds in reed-bed mesocosms - water, BOD, carbon and nutrient removal.** *Waste Manage* 2012, **32:**104–109.

30. Gustavsson L, Hollert H, Jönsson S, van Bavel B, Engwall M: **Reed beds receiving industrial sludge containing nitroaromatic compounds - effects of outgoing water and bed material extracts in the umu-C genotoxicity assay, DR-CALUX assay and on early life stage development in zebrafish (danio rerio).** *Environ Sci Pollut Res* 2007, **14:**202–211.

31. Gustavsson L, Engwall M, Jönsson S, Van Bavel B: **Biological treatment of sludge containing explosives and pharmaceutical residues – effects on toxicity and chemical concentrations.** In *Third Conference of Disposal of Energetic Material.* Karlskoga: KCEM (Knowledge Centre for Energetic Materials) - Section of Detonation and Combustion; 2003.

32. Klee N, Gustavsson L, Kosmehl T, Engwall M, Erdinger L, Braunbeck T, Hollert H: **Changes in toxicity and genotoxicity of industrial sewage sludge samples containing nitro- and amino-aromatic compounds following treatment in bioreactors with different oxygen regimes.** *ESPR, Environ Sci Pollut Res* 2004, **11:**313–320.

33. Gustavsson L, Klee N, Olsman H, Hollert H, Engwall M: **Fate of Ah receptor agonists during biological treatment of an industrial sludge containing explosives and pharmaceutical residues.** *ESPR, Environ Sci Pollut Res* 2004, **11:**379–386.

34. Gustavsson L, Engwall M: **Genotoxic activity of nitroarene- contaminated industrial sludge following large-scale treatment in aerated and non-aerated sacs.** *Sci Tot Env* 2006, **367:**694–703.

35. Swedish EPA: **Riktvärden för förorenad mark.** *NV Rapport* 2009, 5976.

36. Swedish Waste Association: **Den Svenska Biogaspotentialen Från Inhemska Varor.** 2008, **2.** Report 2008.

37. Swedish Waste Association (Avfall Sverige Utveckling): **Substrathandbok för Biogasproduktion.** 2009, 14 . Report U2009.

38. Kuşçu ÖS, Sponza DT: **Kinetics of para-nitrophenol and chemical oxygen demand removal from synthetic wastewater in an anaerobic migrating blanket reactor.** *J Hazard Mater* 2009, **161:**787–799.

39. Hecht C, Griehl C: **Investigation of the accumulation of aromatic compounds during biogas production from kitchen waste.** *Bioresour Tech* 2009, **100:**654–658.

40. Kuşçu ÖS, Sponza DT: **Effect of increasing nitrobenzene loading rates on the performance of anaerobic migrating blanket reactor and sequential anaerobic migrating blanket reactor/completely stirred tank reactor system.** *J Hazard Mater* 2009, **168:**390–399.

41. Karathanasis AD, Potter CL, Coyone MS: **Vegetation effects on fecal bacteria, BOD and suspended solid removal in constructed wetlands treating domestic wastewater.** *Ecol Eng* 2003, **20:**157–169.

42. Luederitz V, Eckert E, Lange-Weber M, Lange A, Gersberg RM: **Nutrient removal efficiency and resource economics of vertical flow and horizontal flow constructed wetlands.** *Ecol Eng* 2001, **18:**157–171.

43. Boopathy R, Kulpa C: **Anaerobic biodegradation of explosives and related compounds by sulfate-reducing and methanogenic bacteria; a review.** *Bioresour Tech* 1997, **63:**81–89.

44. Padda RS, Wang CY: **Mutagenicity of trinitrotoluene and metabolites formed during anaerobic degradation by Clostridium acetobutylicum ATCC 824.** *Environ Toxicol Chem* 2000, **19:**2871–2875.

45. Rogers JD, Bunce NJ: **Treatment methods for the remediation of nitroaromatic explosives.** *Water Res* 2001, **35:**2101–2111.

46. Sundvall A, Marklund H, Rannug U: **The mutagenicity on Salmonella typhimurium of nitrobenzoic acids and other wastewater components generated in the production of nitrobenzoic acids and nitrotoluenes.** *Mutat Res* 1984, **137:**71–78.

47. Boopathy R, Manning JF, Kupla CF: **A laboratory study of the bioremediation of 2,4,6-trinitrotoluene-contaminated soil using aerobic/anoxic soil slurry reactor.** *Water Environ Res* 1998, **70:**80–86.

48. Razo-Flores E, Donlon B, Lettinga G, Field JA: **Biotransformation and biodegradation of N-substituted aromatics in methanogenic granular sludge.** *FEMS Microbiol Rev* 1997, **20:**525–538.

49. Toze S, Zappia L: **Microbial degradation of munition compounds in production wastewater.** *Water Res* 1999, **33:**3040–3045.

50. Umbuzeiro GA, Franco A, Martins MH, Kummrow F, Carvalho L, Schmeiser HH, Leykauf J, Stiborova M, Claxton LD: **Mutagenicity and DNA adduct formation of PAH, nitro-PAH, and oxy-PAH fractions of atmospheric particulate matter from São Paulo, Brazil.** *Mutat Res* 2008, **652:**172–80.

51. Engwall M, Schnürer A: **Fate of Ah-receptor agonists in organic household waste during anaerobic degradation-estimation of levels using EROD induction in organic cultures of chick embryo livers.** *Sci Tot Environ* 2002, **297:**105–108.

52. Olsman H, Björnfoth H, van Bavel B, Lindström G, Schnürer A, Engwall M: **Characterisation of dioxin-like compounds in anaerobically digested organic material by bioassay-directed fractionation.** *Organohalogen Compd* 2002, **58:**345–348.

53. Chen I, Safe S, Bjeldanes L: **Indole-3-carbinol and diindolmethane as aryl hydrocarbon (Ah) receptor agonist and antagonist in T47D human breast cancer cells.** *Biochem Pharmacol* 1996, **51:**1069–1076.

54. Naur P, Hansen HC, Bak S, Hansen GB, Jensen NB, Nielsen HL, Halkier BA: **CYP79B1 from Sinapsis alba coverts tryptophan to indole-3-acetaldoxime.** *Arch Biochem Biophys* 2003, **409:**235–241.

55. Teel RW, Huynh H: **Modulation by phytochemicals of cytochrome P450 linked enzyme activity.** *Cancer Lett* 1998, **133:**135–141.

56. Jonsson S, Borén H: **Analysis of mono- and diesters of o-phthalic acid by solid phase extractions with polystyrene-divinylbenzene-based polymers.** *J Chromatogr A* 2002, **963:**393–400.

57. Hollert H, Dürr M, Olsman H, Halldin K, Bavel B v, Brack W, Tysklind M, Engwall M, Braunbeck T: **Biological and chemical determination of dioxin-like compounds in sediments by means of a sediment triad approach in the catchment area of the Neckar River.** *Ecotoxicology* 2002, **11:**323–336.

58. Keiter S, Grund S, van Bavel B, Hagberg J, Engwall M, Kammann U, Klempt M, Manz W, Olsman H, Braunbeck T, Hollert H: **Activities and identification of aryl hydrocarbon receptor agonists in sediments from the Danube river.** *Anal Bioanal Chem* 2008, **390:**2009–2019.

59. Olsman H, Hagberg J, Kalbin G, Julander A, van Bavel B, Strid Å, Tysklind M, Engwall M: **Ah receptor agonists in UV-exposed toluene solutions of decabromodiphenyl ether (decaBDE) and in soils contaminated with polybrominated diphenyl ethers (PBDEs).** *ESPR - Environ Sci Pollut Res* 2005, **13:**161–169.

60. Lee LE, Clemons JH, Bechtel DG, Caldwell SJ, Han KB, Pasitschniak-Arts M, Mosser D, Bols NC: **Development and characterization of a rainbow trout liver cell line expressing cytochrome p450-dependent monooxygenase activity.** *Cell Bio Toxicol* 1993, **9:**279–294.

61. Behrens A, Schirmer K, Bols NC, Segner H: **Polycyclic aromatic hydrocarbons as inducers of cytochrome P4501A enzyme activity in the rainbow trout liver cell line, RTL-W1, and in primary cultures of rainbow trout hepatocytes.** *Environ Toxicol Chem* 2001, **20:**632–643.

62. Seiler TB, Rastall AC, Leist E, Erdinger L, Braunbeck T, Hollert H: **Membrane dialysis extraction (MDE): a novel approach for extracting toxicologically relevant hydrophobic organic compounds from soils and sediments for assessment in biotests.** *J Soils Sediments* 2006, **6:**20–29.

63. Brunström B, Halldin K: **EROD induction by environmental contaminants in avian embryo livers.** *Comp Biochem Physiol C Pharmacol Toxicol Endocrinol* 1998, **121:**213–219.

17

Embryotoxic and proteotoxic effects of water and sediment from the Neckar River (Southern Germany) to zebrafish (*Danio rerio*) embryos

Krisztina Vincze[*], Katharina Graf, Volker Scheil, Heinz-R Köhler and Rita Triebskorn

Abstract

Background: The Neckar River (Southern Germany) represents an aquatic system strongly affected by numerous anthropogenic activities. Thus, it is an excellent model for ecotoxicological investigations. The present study aims to assess time and spatial variations of embryo- and proteotoxic effects in surface water and sediment. For this end, embryos of zebrafish (*Danio rerio*) were exposed to Neckar River samples collected in the Tübingen region in different seasons over 2 years. Additionally, quantification of the heat shock (stress) protein Hsp70 was carried out in newly hatched larvae; furthermore, physico-chemical water parameters were measured in order to gain baseline information about the limnologic conditions.

Results: Nearly all of the investigated Neckar River sites caused elevated mortality, developmental retardation and failures, modified heart rate and reduced hatching success in zebrafish embryos and larvae. Additionally, exposure to Neckar River water and sediment led to changes in larval Hsp70 level. During the 2 years of investigation, seasonal differences of embryo- and proteotoxic effects occurred. Along these lines, physico-chemical measurements delivered basic information for the interpretation of *in vivo* test data.

Conclusions: Our study suggests a changing toxic burden in the whole investigated study area. Consequently, for ecotoxicological field studies, time and spatial variations on small scale must be dealt with. The lethal and sublethal endpoints of the fish embryo test combined with Hsp70 level measurements proved to be effective tools for toxicity assessment of environmental samples.

Keywords: Fish embryo test; Hsp70; Biomarker; Time and spatial variations

Background

Anthropogenic chemicals entering freshwater via sewage effluents and other sources may pose a potential threat to aquatic ecosystems; therefore, environmental monitoring is a key issue in order to draw unbiased conclusions about the toxic effects on living organisms. Since it is hard to predict how organisms of different systematic levels may react to the presence of multiple pollutants, there is an obvious need for investigating their physiological responses to certain exposure situations. In this context, biological effect studies including biomarkers provide a way to carry out an overall health assessment at various organization levels [1,2]. Biomarkers are defined as biological responses of organisms that give a measure of exposure and/or of toxic effect [3]. Biomarker responses, such as gene expression alterations, increased biotransformation enzyme levels, immune responses, histological impairments etc. can provide an early warning on environmental and ecological effects of chemicals [4].

Heat shock proteins (Hsps), also named as stress proteins, are one of the molecular biomarkers of effect which can be used as indicators for cellular and proteotoxic hazard [5]. Among the various stress response pathways, the heat shock response is one of the major ones [6]. The Hsp70 family of stress proteins is an essential class of highly conserved molecular chaperones which are present in different cell types and cellular compartments [7]. Hsp70s serve multiple roles in the prevention of protein aggregation; they are also involved in nascent protein folding, correct refolding of damaged

---
* Correspondence: krisztina.vincze@student.uni-tuebingen.de
Animal Physiological Ecology, Institute for Evolution and Ecology, University of Tübingen, Konrad-Adenauer-Strasse 20, Tübingen 72072, Germany

polypeptides and assisting cellular protein transloca-
tion [8]. A large number of studies [9-11] have demonstrated that Hsp70 expression is closely linked to a
variety of biotic and abiotic stress factors showing a
high sensitivity even to minor impacts. For example,
Hallare and colleagues [12] recorded an up-regulation
of Hsp70 levels in zebrafish (*Danio rerio*) embryos exposed to contaminated sediments and organic extracts
collected along the Laguna Lake, Philippines. Their
study also confirmed that developing zebrafish embryos are able to detect chemical stressors in their immediate environment [13-16].

Zebrafish are frequently used organisms in biological
as well as in ecotoxicological research. The short and
well-characterized embryonic ontogenesis (for details,
see e.g. [17]) and the transparency of eggs offer a unique
opportunity for developmental observations. The fish
embryo test (FET) with the zebrafish is a reliable alternative to existing *in vivo* ecotoxicity test methods [18] for
investigating the adverse effects of aquatic pollutants
[19-22]. Compared to other toxicity screening assays,
the FET is in most cases more sensitive than the acute
toxicity test with adult zebrafish or than certain cell line
tests [23].

The current work is part of an extended monitoring
project focussing on the Neckar River system in the
Tübingen area (Southern Germany). The Neckar River is
an excellent model for ecotoxicological investigations,
since it is still recovering from a former heavy pollution
caused by rapid industrial development in the last five
decades. Today, the water quality is considered as 'moderate' according to the European Water Framework Directive [24]; however, there are still several wastewater
loaded sections. Consequently, Braunbeck and colleagues [25] reported the genotoxic effects and strong
histological impairments in roach (*Rutilus rutilus*) and
gudgeon (*Gobio gobio*) from the lower Neckar region.
Also, sediment extracts from a Neckar River basin creek
receiving treated hospital wastewater were found to ex-
hibit strong aromatic hydrocarbon receptor-mediated
effects in a rainbow trout liver cell line (RTL-W1) as
well as high mutagenicity in the *Salmonella* microsome
assay [26].

The present study investigates embryo- and proteotoxic effects of native Neckar River water and sediment
from the Tübingen area collected in different seasons
over 2 years. For this end, zebrafish embryo tests combined with an analysis of the stress protein Hsp70 in
newly hatched larvae were conducted. Additionally,
measurements of physico-chemical parameters were carried out in order to provide basic information about the
limnological conditions at the river sections. Our goals
are, on the one hand, the evaluation of early developmental responses of zebrafish to native environmental

samples at the whole organism level as well as on the
cellular level and, on the other hand, the assessment of
the ecotoxicological consequences of small-scale time
and spatial variation in a model aquatic system.

Results

Physico-chemical water properties

The Neckar River revealed good dissolved oxygen conditions and a normal pH and hardness range at all sampling sites over both seasons and years (Figures 1 and 2).
According to the LAWA water classification, nitrite-N
values varied between the classes I and I-II, nitrate-N
concentrations were between class III and subclass III-
IV and ammonium-N concentrates lay overall below
0.04 mg/L (class I) during the four sampling events.
Chloride ions were detected between subclass I-II and
class II; the highest values occurred generally at the first
sampling site (S1; downstream of the sewage treatment
plant (STP) lead-in). Extremely high phosphate values
(subclass III-IV and class IV) were found in May 2011 at
S3 (upstream of the Ammer lead-in) and S4 (Tübingen-
Hirschau), while in 2012, the average range was in class
II and subclass II-III.

Fish embryo test

Exposure to Neckar River water and sediment induced
numerous effects during zebrafish ontogenesis. In 2011,
a significantly higher mortality was detected in embryos
exposed to S4 water and sediment compared to the
negative control (Figure 3A,B). Only a minor mortality
could be observed in the following spring (Figure 3C),
while in autumn 2012, all Neckar River samples induced
an elevated mortality (Figure 3D). In the first 60 h of
ontogenesis, strong developmental retardations (delay of
spontaneous contractions, low pigmentation, under-
developed circulation system) were noted at S1 (May
2011, May and September 2012), S2 (May and September 2012), S3 (Mai 2011) and S4 (May and September in
both years) (Figure 4). According to this, the above-
mentioned Neckar River samples also led to modified or
even to lack of heartbeat of 48 hpf embryos (Figure 5).
During further ontogenesis, developmental failures (yolk
and pericardial oedema, pigment failure and tail deform-
ation) occurred in embryos exposed to water and sediment collected at S1 (May 2011, May and September
2012), S2 (May and September 2012), S3 (May and September in both years) and S4 (May and September in
both years) (Figure 6). Hatching occurred predominantly
between 60 and 96 hpf. The main hatching times (50%
of the surviving individuals reached hatching) are shown
in Table 1. Generally, negative control embryos showed
a mean hatching time between 66 and 72 hpf; furthermore, 100% of the surviving individuals reached the larval stage at the age of 96 hpf. In contrast, hatching

Physicochemical features	Site 1		Site 3		Site 4	
Year 2011	May	September	May	September	May	September
Air temperature (°C)	8.9	15.3	11.7	18.7	11.6	19.9
Water temperature (°C)	13.5	17.2	13.3	17.9	14.4	18.2
O_2 (mg/L)	9.5	7.8	11.7	9.2	11.6	9.4
O_2 (%)	96	83	106	99	123	99
Conductivity (µS)	-	901	-	930	-	904
pH	8.12	8.05	8.21	8.34	8.39	8.39
Nitrite-N (mg/L)	0.03	0.01	0.04	0.02	0.04	0.02
Nitrate-N (mg/L)	5.6	4.2	2.1	3.3	5.3	3.0
Ammonium-N (mg/L)	<0.04	<0.04	<0.04	<0.04	<0.04	<0.04
Chloride (mg/L)	88.7	58	54.81	60	58.02	50
Ortho-Phosphate-P (mg/L)	0.23	0.06	0.74	0.08	1.11	0.05
Overall hardness (°dH)	29	26	30	27	30	28
Carbonate hardness (°dH)	17	13	16	14	16	16

Figure 1 Basic physical and chemical conditions at three sampling sites along the Neckar River in 2011. Water classification was based on the directives of the German Working Group for Water Issues (LAWA): class I (dark blue) - unpolluted to very slightly polluted, class I-II (light blue) - slightly polluted, class II (dark green) - moderately polluted, class II-III (light green) - critically polluted, class III (yellow) - heavily polluted, class III-IV (orange) - very heavily polluted and class IV (red) - excessively polluted [27]. In May 2011, conductivity measurements could not be carried out because of equipment failure. Ammonium-N values are given in a range due to detecting limitations.

success was significantly lower at S1 (May 2011), S2 (May and September 2012), S3 (September 2011, May 2012) and S4 (both years and seasons) (Figure 7). The strongest effects on hatching were induced by S2 samples with a mean hatching time of 93 hpf in May 2012, while in September 2012, only 26% of the individuals reached the larval stage at the end of the FET.

Hsp70 analysis

Neckar River water and sediment from S1 collected in May 2011 led to changes in Hsp70 level of newly hatched zebrafish larvae when compared to the negative control (Figure 8A). Autumn samples from the same year, however, did not reveal any proteotoxic effects (Figure 8B). In the following spring, S1 embryos showed a significantly lower Hsp70 level, while S3 samples led to

a stress protein induction compared to the negative control (Figure 8C). S2 embryos of the above-mentioned sampling event did not manage to reach the larval stage; thus, heat shock protein measurements could not be carried out. In contrast, S2 samples from autumn 2012 led to a significantly reduced Hsp70 level in comparison to the negative control (Figure 8D).

Discussion

The present work reports spatial and temporal variations of embryo- and proteotoxic effects during the early development of zebrafish exposed to native Neckar River samples. A large number of studies revealed that hydrological events such as high/low water discharges and sediment re-mobilization through heavy rainfall could lead to short-term toxicity changes and to a seasonal

Physicochemical features	Site 1		Site 2		Site 3		Site 4	
Year 2012	May	Sept.	May	Sept.	May	Sept.	May	Sept.
Air temperature (°C)	13.3	17.4	15.3	21.0	24.1	27.6	25.1	22.7
Water temperature (°C)	14.5	15.9	15.1	15.9	16.2	15.7	16.0	16.1
O_2 (mg/L)	9.7	9.56	9.55	9.21	11.82	10.6	11.8	13.24
O_2 (%)	99.1	101.2	97.3	95.1	112	107.2	108	134
Conductivity (µS)	924	1035	901	953	889	938	849	937
pH	7.74	8.25	8.12	8.05	8.28	8.08	8.48	8.22
Nitrite-N (mg/L)	0.03	0.01	0.03	0.02	0.04	0.02	0.04	0.01
Nitrate-N (mg/L)	5.0	5.2	3.4	3.6	3.8	3.4	3.4	3.8
Ammonium-N (mg/L)	<0.04	<0.04	<0.04	<0.04	<0.04	<0.04	<0.04	<0.04
Chloride (mg/L)	58	70	51	55	50	55	46	55
Ortho-Phosphate-P (mg/L)	0.20	0.16	0.16	0.10	0.16	0.10	0.16	0.10
Overall hardness (°dH)	26	29	26	28	28	28	27	27
Carbonate hardness (°dH)	15	14	16	16	17	15	16	15

Figure 2 Basic physical and chemical conditions at four sampling sites along the Neckar River in 2012. Water classification was based on the directives of the German Working Group for Water Issues (LAWA): class I (dark blue) - unpolluted to very slightly polluted, class I-II (light blue) - slightly polluted, class II (dark green) - moderately polluted, class II-III (light green) - critically polluted, class III (yellow) - heavily polluted, class III-IV (orange) - very heavily polluted and class IV (red) - excessively polluted [27]. Ammonium-N values are given in a range because of detecting limitations.

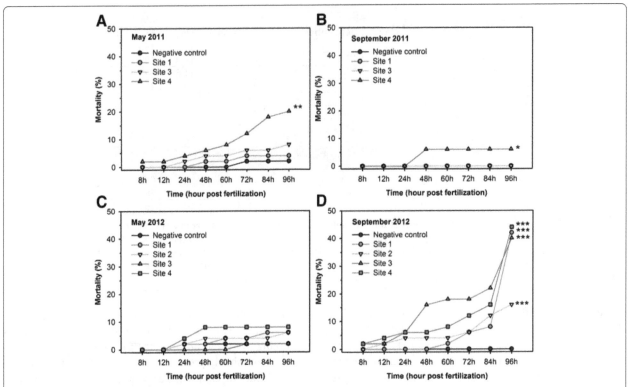

Figure 3 Mortality during the fish embryo test. Zebrafish embryos were exposed to the Neckar River samples collected in May 2011 **(A)**, September 2011 **(B)**, May 2012 **(C)** and September 2012 **(D)**. Asterisks show significant differences (*$p < 0.05$, **$p < 0.01$, ***$p < 0.001$, $a = 0.05$) between sampling site exposure and negative control. Combined data of two replicates was assessed by the Cox proportional hazards survival model.

burden [28-32]. There were strong water level alterations at the Neckar River and tributaries during the 2 years of investigation. An average gauge was recorded in the late spring 2011, while autumn rains led to an elevated water level. The Neckar River revealed in the following May a normal water level, while in September 2012, heavy rainfall events occurred again (stronger than in 2011).

S1 exposure led to developmental deficits during the zebrafish ontogenesis. Since the site mentioned was situated downstream of the Tübingen STP, a contaminant input through the municipal wastewater discharge has to be considered. As a further consequence of the STP presence [33], increased chloride and nitrate values could be detected at S1. In 2011, a slower development and higher number of malformations were observed in spring compared to autumn. There are several studies reporting a lower contaminant level in the rainy period [32,34-36]; thus, increased dilution of discharged contaminants may be a possible interpretation for lower embryotoxicity by autumn samples. Unexpectedly, in September 2012, an extraordinary high mortality was observed at all sites, especially after 84 hpf. Since heavy rainfalls occurred just shortly before this sampling event, there is a possibility of toxic input through untreated effluents of the STP storm water overflow and through

urban runoff as well [37]. The noticeable mortality increase after hatching can be elucidated through the barrier function of the embryonic chorion, which has a structure pierced by pore canals. Polymers and higher molecular weight surfactants, for instance, are suspected to be blocked by the chorion [38].

Native samples from S2 induced numerous negative effects during zebrafish ontogenesis. The close proximity of the Ammer River lead-in may explain the extremely slow development and malformations of the embryos; thus as mentioned before, a series of organic contaminants were reported in this Neckar River tributary. Liu and colleagues highlighted that high flow events by the Ammer River may result in the redistribution of sediments [39], which may clarify why autumn S2 samples induced stronger teratogenic impairments compared to the spring ones. A similar process was reported at the lower Neckar area as well: there were persistent organic compounds (polychlorinated dibenzo-p-dioxins, dibenzofurans, polychlorinated biphenyls, polycyclic aromatic hydrocarbons) detected in settling particulate matter (SPM) during a flood event in the Heidelberg region [40]. Hollert and co-workers [41] investigated the cyto- and genotoxic potentials of the above-mentioned river samples: SPM taken during the period of flood rise

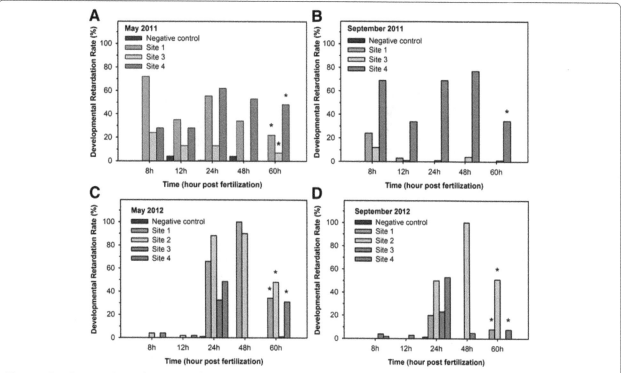

Figure 4 Developmental retardation rate of zebrafish embryos in the first 60 hpf. Fish embryo test with water and sediment from the Neckar River collected in May 2011 **(A)**, September 2011 **(B)**, May 2012 **(C)** and September 2012 **(D)**. Developmental retardation rate is given as percentage of the observed retardations divided by the total possible ones at appropriate time points. Vertical bars show the mean of two replicates. Asterisks show significant differences (*$p < a$ when adjusted according to Holm-Bonferroni's method) between sampling site exposure and negative control at 60 hpf assessed by Fisher's exact test.

showed the highest cytotoxic activities in the neutral red retention assay, while the Ames test delivered no evidence for any genotoxic activity. Therefore, as a consequence of flood events, runoff and remobilized sediments may cause an increase of ecotoxicologically relevant effects from contaminant reservoirs [40].

S3 was located 800 m upstream of the Ammer River, with the Steinlach creek as the only lead-in. Our study confirmed the fact of Steinlach having a good water quality, since during the most sampling events only minor developmental abnormalities could be observed by the FET. Interestingly, in 2012, S3 embryos showed an increased heart rate compared to the negative control. This phenomenon may be a sign for an elevated metabolism due to toxicant elimination, since heartbeat frequency in fish can be strongly affected by the metabolic rate [42]. In autumn 2012, an elevated mortality was detected at the whole investigated Neckar section, including S3. Here, the potential pollution source is unclear as of jet; however, substance drifting from S4 due to the increased water level may be a reasonable possibility. As an evidence, in spring 2011, extraordinarily high phosphate concentrations were detected at S3, while S4 showed an even higher value.

S4 was considered as the most nature-close sampling site, since it was located outside the city area with no

discharges in close proximity; however, an agricultural activity was still present. Surprisingly, this site caused the highest mortality and heartbeat reduction during all four sampling events. Numerous developmental retardations and failures and lower hatching rates were recorded as well. One possible contaminant source could be the agricultural runoff, which may explain the strong embryotoxic effects and the elevated nitrate and phosphate levels in spring. Abandoned discharges and past substance burden would deliver further explanations; unfortunately, there is no information available about these.

Parallel to the present study, Hollert and colleagues [43] observed a broad range of embryotoxic effects of native water and sediments from the Heidelberg Neckar region through the zebrafish sediment contact assay. Nevertheless, they also proved the suitability of the zebrafish embryo test for the analysis of complex environmental samples including the whole sediments. According to their FET results, sediment extracts from the above-mentioned area exhibited dioxin-like potentials when tested with the 7-ethoxyresorufin-O-deethylase (EROD) assay on a permanent cell line [44].

The present study is the first work reporting proteotoxic effects in the Neckar River. The results of the

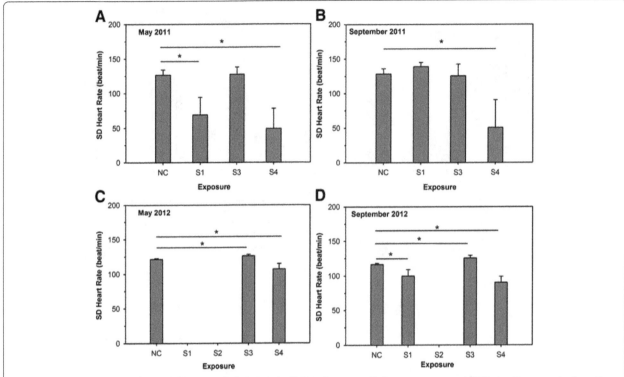

Figure 5 Heart rate of zebrafish embryos at 48 hpf during the fish embryo test. Embryos were exposed to Neckar River samples from May 2011 **(A)**, September 2011 **(B)**, May 2012 **(C)** and September 2012 **(D)**. Vertical bars show standard deviations. Asterisks indicate significant differences (*$p < 0.05$, $a = 0.05$) between river sample treatment and negative control assessed by Tukey-Kramer test. Results of two independent replicates are shown.

Hsp70 quantification in newly hatched zebrafish larvae provided additional information about the sublethal effects of environmental samples on the molecular level. There was a decreased heat shock protein level detected at S1 (May 2011 and 2012) and S2 (September 2012), while S3 (May 2012) exposure led to a stress protein induction when compared to the negative control. Elevated stress protein values indicate proteotoxic conditions, since heat shock proteins play an essential role in protein integrity maintenance and prevent aggregation [45]. Decreasing cellular heat shock protein amounts, however, can be interpreted as signs of a very intense stress response [46] in which case the organism is heading towards physiological breakdown and destruction. For this reason, the molecular biomarker Hsp70 should be combined with additional methods such as histopathology [47], biotransformation enzyme activity measurements [48,49] etc. in order to provide reliable information on toxic effects in different organisms. For example, in the frame of an extended monitoring project at Lake Constance tributaries, Triebskorn and colleagues integrated Hsp70 measurements in a battery of chemical and biological methods also including the fish embryo test [50]. The combination of the various techniques offered a way to establish plausible connections between

the presence of micropollutants and the respective effects on fish health.

Conclusions

The current work is a field survey not only of local but also of general relevance. The Neckar River proved to be an excellent model for ecotoxicological investigations due to the impact of variable anthropogenic pollutant sources. Toxicity alterations along the investigated river section could be connected to the potential pollution sources, sampling seasons and raining events. Our study highlighted that a seasonal burden has to be taken into account during environmental monitoring. The *in vivo* assays applied in the present work focussed on diverse toxic aspects on different organization levels of zebrafish; thus, the fish embryo test revealed developmental toxicity on the organismal level, while the analysis of the stress protein Hsp70 assessed proteotoxicity on the subcellular level. The basic physico-chemical measurements contributed only in a limited way to the interpretation of the FET results. Therefore, little is known about the actual toxic loads of the Neckar River in the Tübingen area. The authors also want to highlight that the present work focussed only on the effects of a short-term exposure in a laboratory model organism to environmental

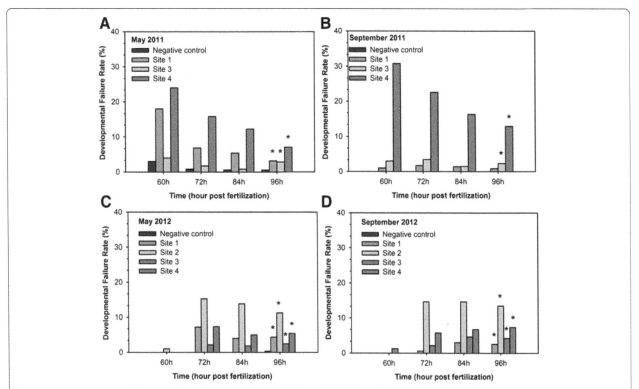

Figure 6 Developmental failure rate between 60 and 96 hpf during the fish embryo test. Zebrafish embryos were exposed to water and sediment collected at the Neckar River in May 2011 **(A)**, September 2011 **(B)**, May 2012 **(C)** and September 2012 **(D)**. Developmental failure rate is given as percentage of the observed failures divided by the total possible ones at appropriate time points. Vertical bars show the mean of two replicates. Asterisks indicate significant differences (*$p < a$ when adjusted according to Holm-Bonferroni's method) between sampling site exposure and negative control at 96 hpf assessed by Fisher's exact test.

samples; the physiological responses of feral animals, long-term consequences and cause-effect relationships (biological data combined with chemical analysis results) will be discussed in the frame of further publications.

Methods

Ethics statement

Zebrafish were maintained according to the European Union Animals Ethics Directive [51]. Fish embryo tests were carried out on the non-protected embryonal and larval stages. Experiments with extended exposure times (168 h) were approved by the Animal Ethics Committee of the Regional Council Tübingen (permission number: ZP 2/11). Zebrafish embryos and larvae were sacrificed

through an anaesthetic overdose of 40 mg/mL benzocaine solution.

Sampling sites

For the current project, four sites were selected along the Neckar River in the Tübingen region (Figure 9). S1 was located 150 m downstream of the local STP discharger (9°6′41.08″N, 48°32′16.44″E). The Tübingen STP is a conventional municipal treatment plant with no additional upgrades and a daily load between 40,000 and 50,000 m³ [52]. At S1, the Neckar River is about 20 m wide, 2 m deep with a flow velocity between 0.3 and 0.5 m/s (depending on raining events). The second site (S2) was situated 1 km upstream of the local STP and 100 m

Table 1 Mean hatching times during the zebrafish embryo test

Mean hatching time	Negative control	Site 1	Site 2	Site 3	Site 4
May 2011	66 hpf	69 hpf	n.a.	58 hpf	79 hpf
September 2011	67 hpf	63 hpf	n.a.	69 hpf	87 hpf
May 2012	72 hpf	81 hpf	93 hpf	81 hpf	85 hpf
September 2012	68 hpf	59 hpf	26% at 96 hpf	63 hpf	68 hpf

Mean hatching time is defined as the time point when 50% of the surviving fish embryos hatched. Mean hatching times were calculated in SigmaPlot 10.0 by graphic analysis of the combined hatching curves of two independent replicates. n.a. not available.

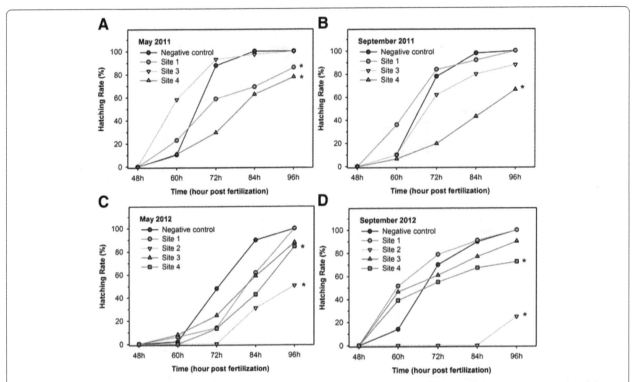

Figure 7 Hatching rate of zebrafish embryos during the fish embryo test. Embryos were exposed to water and sediment samples of the Neckar River from May 2011 **(A)**, September 2011 **(B)**, May 2012 **(C)** and September 2012 **(D)**. Hatching rate shows the percentage of hatched larvae in relation to the surviving individuals. The combined data of two replicates is shown. Asterisks indicate significant differences (*$p < a$ when adjusted according to Holm-Bonferroni's method) between sampling site exposure and negative control at 96 hpf assessed by Fisher's exact test.

downstream of the influent of the Ammer River (9°5′ 19.97″N, 48°31′34.32″E). It reveals a flow velocity of 0.2 m/s, a width of 36 m and a depth of 2 m. In the past few years, there were organic contaminants detected in the Ammer River: polycyclic aromatic hydrocarbons were found in the sediment with a concentration ranging from 112 to 22.900 ng/g dry weight [39], and an accumulation of polychlorinated biphenyls in the tissues of brown trouts was observed as well [53]. On this account, a third site (S3) was assigned located 800 m upstream of the Ammer lead in (9°5′1.28″N, 48°31′23.05″E) with a flow velocity of 0.3 to 0.5 m/s, a width of 40 m and an overall depth of approximately 1 m. The Steinlach River leads into the Neckar River about 300 m upstream of S3, its water quality is considered good and showed recently a strong improvement [54] contrary to the Ammer. The fourth sampling site (S4) was located about 4.2 km upstream, outside the city area (9°1′8.76″N, 48°29′59.24″ E) where no STP effluents are registered in close proximity. The flow velocity of the Neckar River at this site ranges between 0.1 and 0.3 m/s, its overall width 30 m and depth 1.2 m.

Sampling events
Monitoring was conducted over 2 years: during 2011 and 2012. Sampling was performed two times per year:

in May and September. At this point, the authors want to highlight that S2 was subsequently added to the study in 2012; thus in 2011, only S1, S3 and S4 were investigated.

For the fish embryo test and Hsp70 analysis, 2 L of water and 200 g of fine-grained sediment from the upper aerobic layers were taken near the riverbank in sterile glass flasks (Schott Duran, Mainz, Germany) at each sampling site. Samples were transported in a cooling box at approximately 4°C to the laboratory facility of the Animal Physiological Ecology Group of the University of Tübingen. Water samples were divided into 0.5-L glass flasks (Schott Duran) while the sediment was distributed into 50-g packages wrapped in aluminium foil (Roth, Karlsruhe, Germany). Neckar River water and sediment were frozen immediately and stored at −20°C until further use.

Physico-chemical water parameter
On the day of sampling, air and water temperatures were measured close to the riverbank with a multi-thermometer (Voltcraft, Hirschau, Germany) at all field sites. Oxygen, pH and conductivity levels were captured through sensors (Oxi 340-A/SET, pH 330/SET TRW, LF 330/SET, WTW, Weilheim, Germany) on the same spots where water and sediment were obtained for the biotests. Additionally, 1 L of water was collected in a sterile

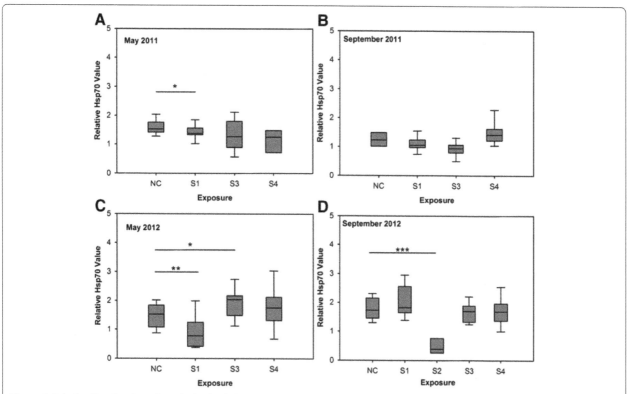

Figure 8 Relative Hsp70 value of 168 hpf zebrafish larvae. Larvae were exposed to Neckar River water and sediment collected in May 2011 (**A**), September 2011 (**B**), May 2012 (**C**) and September 2012 (**D**). Relative Hsp70 value is given in relation to the mean of two internal standards. Box plots represent medians and 5th/95th percentiles. Where $n < 10$, no plot whiskers are shown. Asterisks indicate significant (*$p < 0.05$, **$p < 0.01$, ***$p < 0.001$, $a = 0.05$) differences between river sample treatment and negative control assessed by Steel-Dwass (nonparametric) or Tukey-Kramer (parametric) test.

glass flask (Schott Duran) at all Neckar River sites and transported to the laboratory facility of the Animal Physiological Ecology Group of the Tübingen University in a cooling box (4°C). Chloride, nitrate, nitrite, ammonium, phosphate (test kits from Macherey-Nagel, Düren, Germany), carbonate and overall hardness (test kits from Merck, Darmstadt, Germany) levels of the water samples were determined immediately. Classification of the water quality was carried out according to the directives of the German Working Group for Water Issues (LAWA) including seven assessment groups: class I - unpolluted to very slightly polluted, class I-II - slightly polluted, class II - moderately polluted, class II-III - critically polluted, class III - heavily polluted, class III-IV - very heavily polluted and class IV - excessively polluted [27].

Maintenance and breeding of zebrafish

The Animal Physiological Ecology Group of the Tübingen University reared several stocks of zebrafish based on the West Aquarium strain. Fish were kept at 26°C ± 1°C in 100- to 200-L tanks in filtered (AE-2 L water filter equipped with an ABL-0240-29 activated carbon filter (0.3 μm), Reiser, Seligenstadt, Germany) tap water under semi-static conditions, with 30% of water volume

being exchanged every 14 days. The room was light-isolated, and an artificial dark-light cycle of 12:12 h was maintained. Animals were fed three times daily with dry flake food (TetraMin™, Tetra, Melle, Germany) and additionally with freshly hatched Artemia larvae (Sanders, Mt. Green, MT, USA) on the day before spawning. For egg production, a stainless steel grid box with a mesh size of 1.5 mm in a plastic basin was positioned at the bottom of the aquaria in the evening before spawning. The grid allowed the passage of eggs into the separate spawning tray, thus preventing predation by adult zebrafish, while green plastic wire material adjusted to the boxes served as spawning stimulus. Spawning took place in the early morning period after the onset of light. Eggs were collected 30 to 60 min after spawning.

The fish embryo test

For each sampling event, two independent replicates (two tests at different time points) of the fish embryo test were conducted according to the work of Hallare and colleagues [12]. In the evening before the test onset, five glass Petri dishes (30-mm diameter with cover, Schott Duran) per each field site were saturated with approximately 2.5 g of the corresponding Neckar River

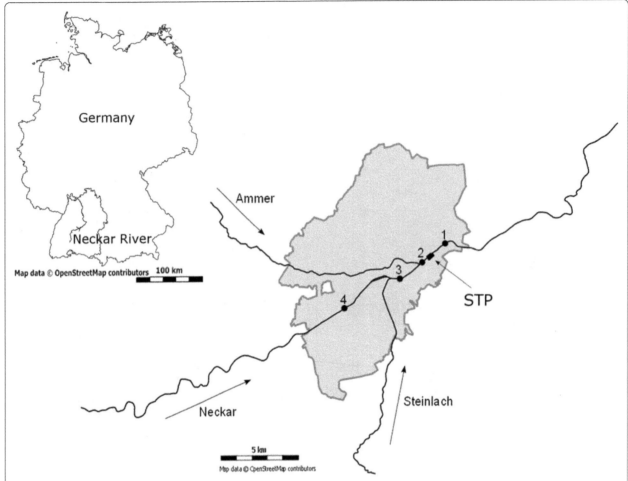

Figure 9 Neckar River overview and sampling sites in the Tübingen area. S1: downstream of the local STP; S2: upstream of the STP and downstream of Ammer lead-in; S3: upstream of the Ammer lead-in; S4: Tübingen-Hirschau. Arrows indicate the stream direction. Map source: Open Street Maps.

sediment and filled with the appropriate water sample to the top. For negative control, additional five Petri dishes were saturated with reconstituted soft water according to ISO 7346/3 [55]. Test dishes were incubated in a heating cabinet (Aqualytic, Dortmund, Germany) at 26°C ± 1°C overnight.

In the morning of the test onset, saturation medium was replaced with 2.5 g of the corresponding sediment sample, and Petri dishes were filled with the appropriate Neckar water to the top. There were identical water and sediment batches used for the saturation step and for the actual test. Reconstituted soft water of the negative control was changed as well. Freshly laid eggs (max 60 min post-fertilization) were collected from the aquaria and were immediately distributed into sterile 90-mm-diameter glass Petri dishes (Schott Duran) filled with Neckar water samples and ISO water, respectively (one dish per treatment). Thus, exposure circumstances were ensured from the very beginning of the development.

After 2 h of incubation at 26°C, five fertilized embryos were transferred subsequently into each test Petri glass from the corresponding pre-exposure dish. Embryos were incubated in a heating cabinet at 26°C ± 1°C for 96 h at a 12:12-h dark-light cycle. Coagulated embryos were removed daily. If necessary, the amount of evaporated water was substituted from the corresponding sample batch. The development of embryos was observed through a binocular (Stemi 2000-C, Zeiss, Oberkochen, Germany) at several time points (6, 12, 24, 48, 60, 72, 84 and 96 h post-fertilization (hpf)). Mortality was recorded at all observation events. Heartbeat was assessed at the age of 48 hpf: for each embryo, heartbeats were counted for 20 s, and test dishes were evaluated in a random order to avoid temperature effects. Developmental retardation and failures were recorded at relevant time points (Table 2). Retardation and failure rates for each time point were estimated as percentage of the observed retardations or failures divided by the total possible

Table 2 Endpoints of the fish embryo test with the zebrafish (based on Nagel [22])

Observed endpoints	Hours post-fertilization							
	8	12	24	48	60	72	84	96
Mortality	•	•	•	•	•	•	•	•
Hatching						•	•	•
Slowed ontogenesis/Retardation								
Epiboly	•							
Gastrulation		•						
Exogastrulation		•						
Formation of somites			•					
Tail detachment			•					
Spontaneous movements			•					
Development of eyes			•					
Number of heartbeats (beats/min)				•				
Blood circulation				•				
Sacculi/Otolith formation					•			
Presence of melanocytes					•			
Developmental failures								
Oedema						•	•	•
Malformation of head						•	•	•
Tail deformation							•	•
Pigmentation failure							•	•
Helical body							•	•
Scoliosis							•	•

ones. The time of hatching was recorded between 60 and 96 hpf. Mean hatching times (50% of the individuals hatched) were estimated as well.

Hsp70 quantification

For each sampling site, three 90-mm-diameter glass Petri dishes (Schott Duran) were saturated with 12.5 g of Neckar River sediment and were filled with the corresponding water sample to the top the day before the test onset. Reconstituted soft water according to ISO 7346/3 served as negative control. In the next morning, saturation media was replaced with the same amount of water and sediment from the identical sample batch. In the following step, 50 freshly laid (30 to 60 min post-fertilization) zebrafish eggs were placed into each dish. Embryos were incubated in a heating cabinet for 168 h by the same conditions as described in the FET. Coagulated embryos were removed daily. The amount of evaporated water was replaced from the corresponding sample batch if necessary. At the end of the exposure, newly hatched larvae were collected by pouring the water carefully from the Petri dish through a cellulose paper filter (595½, 185-mm diameter, Schleicher & Schüll, Dassel, Germany). Eight larvae were pooled into

an Eppendorf tube (Eppendorf, Hamburg, Germany) and frozen immediately in liquid nitrogen. Samples were stored at −20°C until further processing.

Hsp70 quantification was carried out based on the work of Hallare and colleagues [20]. Pooled zebrafish larvae were ultrasonically homogenized in 20-μL extraction buffer (80 mM potassium acetate, 4 mM magnesium acetate, 20 mM Hepes pH 7.5 (Sigma Aldrich, Deisenhofen, Germany)) for 5 s and centrifuged for 12 min at 20,000 g. The total protein concentration for each supernatant was determined according to the method of Bradford [56]. Supernatants for the actual Hsp70 analysis were diluted 2:1 with 3% sodium dodecyl sulphate (SDS) in TRIS buffer (pH 7) (Sigma Aldrich) and heated at 96°C for 5 min. For each sample, 10 μg of total protein per lane were loaded on a minigel SDS page. Two lanes per gel were filled with 4 μL of a reference homogenate (standard) made of adult zebrafish in order to ensure comparability. Electrophoresis took place at 80 V for 15 min followed by 120 V for approximately 90 min at 360 mA. Proteins were transferred to a nitrocellulose membrane (Macherey-Nagel) by semi-dry blotting at 10 V and 90 mA/filter for 2 h. After blotting, filters were incubated for 2 h in a blocking solution (1:1 horse serum (Sigma Aldrich) - Tris-buffered saline (TBS) pH 5.7 (Roth)) at room temperature. Hsp70 bands were marked with a mouse anti-human Hsp70 monoclonal antibody (Dianova, Hamburg, Germany) (diluted 1:5,000 in 1:9 horse serum/TBS) and incubated overnight at room temperature. After rinsing the filter in TBS, a goat anti-mouse IgG (H + L) antibody coupled to peroxidase (Dianova) (diluted 1:1,000 in 1:9 horse serum/TBS) was applied. Following 2 h of incubation, an antibody complex was detected by chloronaphthol (Sigma Aldrich) and 0.015% hydrogen peroxide (Sigma Aldrich) resolved in Tris buffer (pH 8.5) containing 6% methanol. After digitalization of the filters, the grey value intensity of Hsp70 bands was measured by densitometric image analysis (Herolab E.A.S.Y., Wiesloch, Germany). Sample Hsp70 levels were normalized by the mean of the two internal standards (reference homogenate) of the corresponding filter.

Statistical analysis

Mortality during the whole FET time span was evaluated through Cox proportional hazards analysis, a survival model considering multiple observation events [57] used in clinical, epidemiologic and also in ecotoxicological [58] research. Fisher's exact tests (two-tailed) were carried out with the absolute numbers of developmental retardations at 60 hpf, developmental failures at 96 hpf and hatched individuals at 96 hpf. In Fisher's exact test, Neckar River exposure groups and negative control were always compared pairwise. According to this, significance levels were adjusted using the Holm-Bonferroni method. For the evaluation of embryonic heartbeat and larval Hsp70 level, data was first tested for normal distribution using the Shapiro-Wilks test; in the following step, variance homogeneity was assessed with the Levene test. Where parameter assumptions were met, differences between exposure groups were determined using an all pairs Tukey-Kramer test. Data with non-normal distributions or inhomogeneous variances were analyzed through the non-parametric Steel-Dwass method. Statistical analysis was carried out using SAS JMP version 9.0 (SAS Institute GmbH, Böblingen, Germany).

Abbreviations
FET: fish embryo test; hpf: hours post-fertilization; Hsp: heat shock protein; LAWA: German Working Group for Water Issues (Länderarbeitsgemeinschaft Wasser); NC: negative control; STP: sewage treatment plant; S1: site one (downstream of the Tübingen sewage treatment plant); S2: site two (upstream of the Tübingen sewage treatment plant); S3: site three (upstream of the Ammer lead-in); S4: site four (outside the Tübingen city area).

Competing interests
The authors declare that they have no competing interests.

Authors' contributions
The physico-chemical measurements, statistical analysis, FET and Hsp70 assessment of 2012 samples was carried out by KV. The manuscript was mainly drafted by KV. KG conducted the FET and Hsp70 assessment of Neckar River samples collected in 2011 and participated in the sampling events. VS took part in the experimental design, participated in the sampling events and gave advice in the data analysis and manuscript drafting. H-RK and RT supervised the current work, supported the publication of the results and provided technical and financial background for the study. All authors read and approved the final manuscript.

Acknowledgements
The authors acknowledge the Carl Zeiss Foundation, the Foundation of the Landesnaturschutzverband (LNV) Baden-Württemberg and the Deutsche Forschungsgemeinschaft (DFG) Open Access Publishing Fund of the Tübingen University for their financial support. We also thank the co-workers of the Animal Physiological Ecology Group of the Tübingen University, especially to Andreas Dieterich, Anja Henneberg, Stefanie Krais, Diana Maier, Katharina Peschke, Alexandra Scheil, Simon Schwarz and Paul Thellmann for their technical assistance and help in the field. The authors want to thank the three anonymous reviewers for their constructive critics and useful suggestions.

References
1. Moore MN, Depledge MH, Readman JW, Paul Leonard DR: An integrated biomarker-based strategy for ecotoxicological evaluation of risk in environmental management. Mutat Res/Fundam Mol Mech Mutagen 2004, 552:247–268.
2. de la Torre FR, Salibián A, Ferrari L: Assessment of the pollution impact on biomarkers of effect of a freshwater fish. Chemosphere 2007, 68:1582–1590.
3. Peakall DB: The role of biomarkers in environmental assessment (1). Introduction. Ecotoxicol Environ Saf 1994, 3:157–160.
4. Walker CH: Biochemical biomarkers in ecotoxicology—some recent developments. Sci Total Environ 1995, 171:189–195.
5. Gupta SC, Sharma A, Mishra M, Mishra RK, Chowdhuri DK: Heat shock proteins in toxicology: how close and how far? Life Sci 2010, 86:377–384.
6. Westerheide SD, Morimoto RI: Heat shock response modulators as therapeutic tools for diseases of protein conformation. J Biol Chem 2005, 280:33097–33100.
7. Bonomo J, Welsh JP, Manthiram K, Swartz JR: Comparing the functional properties of the Hsp70 chaperones, DnaK and BiP. Biophys Chem 2010, 149:58–66.
8. Frydman J: Folding of newly translated proteins in vivo: the role of molecular chaperones. Annu Rev Biochem 2001, 70:603.
9. Köhler H-R, Eckwert H, Triebskorn R, Bengtsson G: Interaction between tolerance and 70 kDa stress protein (hsp70) induction in collembolan populations exposed to long-term metal pollution. Appl Soil Ecol 1999, 11:43–52.
10. Scheil V, Zürn A, Köhler H-R, Triebskorn R: Embryo development, stress protein (Hsp70) responses, and histopathology in zebrafish (Danio rerio) following exposure to nickel chloride, chlorpyrifos, and binary mixtures of them. Environ Toxicol 2010, 25:83–93.
11. Hofmann GE: Ecologically relevant variation in induction and function of heat shock proteins in marine organisms. Am Zool 1999, 39:889–900.
12. Hallare AV, Kosmehl T, Schulze T, Hollert H, Köhler HR, Triebskorn R: Assessing contamination levels of Laguna Lake sediments (Philippines) using a contact assay with zebrafish (Danio rerio) embryos. Sci Total Environ 2005, 347:254–271.
13. Kosmehl T, Otte JC, Yang L, Legradi J, Bluhm K, Zinsmeister C, Keiter SH, Reifferscheid G, Manz W, Braunbeck T, Strähle U, Hollert H: A combined DNA-microarray and mechanism-specific toxicity approach with zebrafish embryos to investigate the pollution of river sediments. Reprod Toxicol 2012, 33:245–253.
14. Osterauer R, Köhler H-R: Temperature-dependent effects of the pesticides thiacloprid and diazinon on the embryonic development of zebrafish (Danio rerio). Aquat Toxicol 2008, 86:485–494.
15. Yang L, Ho NY, Alshut R, Legradi J, Weiss C, Reischl M, Mikut R, Liebel U, Müller F, Strähle U: Zebrafish embryos as models for embryotoxic and teratological effects of chemicals. Reprod Toxicol 2009, 28:245–253.
16. Weigt S, Huebler N, Strecker R, Braunbeck T, Broschard TH: Zebrafish (Danio rerio) embryos as a model for testing proteratogens. Toxicology 2011, 281:25–36.
17. Kimmel CB, Ballard WW, Kimmel SR, Ullmann B, Schilling TF: Stages of embryonic development of the zebrafish. Dev Dyn 1995, 203:253–310.
18. Embry MR, Belanger SE, Braunbeck TA, Galay-Burgos M, Halder M, Hinton DE, Léonard MA, Lillicrap A, Norberg-King T, Whale G: The fish embryo toxicity test as an animal alternative method in hazard and risk assessment and scientific research. Aquat Toxicol 2010, 97:79–87.
19. Braunbeck T, Böttcher M, Hollert H, Kosmehl T, Lammer E, Leist E, Rudolf M, Seitz N: Towards an alternative for the acute fish LC50 test in chemical assessment: the fish embryo toxicity test goes multi-species - an update. Altex 2005, 22:87–102.
20. Hallare AV, Köhler HR, Triebskorn R: Developmental toxicity and stress protein responses in zebrafish embryos after exposure to diclofenac and its solvent, DMSO. Chemosphere 2004, 56:659–666.
21. Lammer E, Carr GJ, Wendler K, Rawlings JM, Belanger SE, Braunbeck T: Is the fish embryo toxicity test (FET) with the zebrafish (Danio rerio) a potential alternative for the fish acute toxicity test? Comp Biochem Physiol C 2009, 149:196–209.
22. Nagel R: DarT: The embryo test with the zebrafish Danio rerio—a general model in ecotoxicology and toxicology. ALTEX 2002, 19:38–48.
23. Lange M, Gebauer W, Markl J, Nagel R: Comparison of testing acute toxicity on embryo of zebrafish, Brachydanio rerio and RTG-2 cytotoxicity as possible alternatives to the acute fish test. Chemosphere 1995, 30:2087–2102.
24. Haberbosch R, Hoffmann R, Wnuck H: Mittlerer Neckar im Wandel der Zeit. In Vom Wildfluss zur Wasserstraße - Fischfauna und Fischerei im Mittleren Neckar. Edited by Haberbosch R, Hoffmann R, Wnuck H. Suttgart: VFG Service und Verlags GmbH; 2012:1721.
25. Braunbeck T, Brauns A, Keiter S, Hollert H, Schwartz P: Fischpopulationen unter Stress – das Beispiel des Unteren Neckars. Umweltwissenschaften und Schadstoff-Forschung 2009, 21:197–211.
26. Brack W, Schirmer K, Erdinger L, Hollert H: Effect-directed analysis of mutagens and ethoxyresorufin-O-deethylase inducers in aquatic sediments. Environ Toxicol Chem 2005, 24:2445–2458.
27. Chemical water quality classification, Working Group of the Federal States on Water Issues (LAWA). http://www.umweltbundesamt.de/wasser-e/themen/fluesse-und-seen/fluesse/bewertung/chemische-gewaesserklassifikation.htm.
28. Camacho-Muñoz D, Martín J, Santos JL, Aparicio I, Alonso E: Occurrence of surfactants in wastewater: hourly and seasonal variations in urban and

industrial wastewaters from Seville (Southern Spain). *Sci Total Environ* 2014, **468–469**:977–984.

29. Gan W, Guo W, Mo J, He Y, Liu Y, Liu W, Liang Y, Yang X: **The occurrence of disinfection by-products in municipal drinking water in China's Pearl River Delta and a multipathway cancer risk assessment.** *Sci Total Environ* 2013, **447**:108–115.

30. Hsieh C-Y, Yang L, Kuo W-C, Zen Y-P: **Efficiencies of freshwater and estuarine constructed wetlands for phenolic endocrine disruptor removal in Taiwan.** *Sci Total Environ* 2013, **463–464**:182–191.

31. Hsu P, Matthäi A, Heise S, Ahlf W: **Seasonal variation of sediment toxicity in the Rivers Dommel and Elbe.** *Environ Pollut* 2007, **148**:817–823.

32. Moliner-Martínez Y, Herraez-Hernandez R, Verdú-Andres J, Campíns-Falcó P, Garrido-Palanca C, Molins-Legua C, Seco A: **Study of the influence of temperature and precipitations on the levels of BTEX in natural waters.** *J Hazard Mater* 2013, **263**(Part 1):131–138.

33. Giri S, Singh AK: **Risk assessment, statistical source identification and seasonal fluctuation of dissolved metals in the Subarnarekha River, India.** *J Hazard Mater*, **265**:305–314.

34. Katip A, Karaer F, Ileri S, Sarmasik S, Aydogan N, Zenginay S: **Analysis and assessment of trace elements pollution in sediments of Lake Uluabat, Turkey.** *J Environ Biol* 2012, **33**:961–968.

35. Palma P, Alvarenga P, Palma V, Matos C, Fernandes R, Soares A, Barbosa I: **Evaluation of surface water quality using an ecotoxicological approach: a case study of the Alqueva Reservoir (Portugal).** *Environ Sci Pollut Res* 2010, **17**:703–716.

36. Zhu S, Chen H, Li J: **Sources, distribution and potential risks of pharmaceuticals and personal care products in Qingshan Lake basin, Eastern China.** *Ecotoxicol Environ Saf* 2013, **96**:154–159.

37. VanLandeghem MM, Meyer MD, Cox SB, Sharma B, Patiño R: **Spatial and temporal patterns of surface water quality and ichthyotoxicity in urban and rural river basins in Texas.** *Water Res* 2012, **46**:6638–6651.

38. Henn K, Braunbeck T: **Dechorionation as a tool to improve the fish embryo toxicity test (FET) with the zebrafish (*Danio rerio*).** *Comp Biochem Physiol C* 2011, **153**:91–98.

39. Liu Y, Beckingham B, Ruegner H, Li Z, Ma L, Schwientek M, Xie H, Zhao J, Grathwohl P: **Comparison of sedimentary PAHs in the rivers of Ammer (Germany) and Liangtan (China): differences between early- and newly-industrialized countries.** *Environ Sci Technol* 2013, **47**:701–709.

40. Wolz J, Engwall M, Maletz S, Olsman Takner H, van Bavel B, Kammann U, Klempt M, Weber R, Braunbeck T, Hollert H: **Changes in toxicity and Ah receptor agonist activity of suspended particulate matter during flood events at the rivers Neckar and Rhine - a mass balance approach using in vitro methods and chemical analysis.** *Environ Sci Pollut Res* 2008, **15**:536–553.

41. Hollert H, Dürr M, Erdinger L, Braunbeck T: **Cytotoxicity of settling particulate matter and sediments of the Neckar River (Germany) during a winter flood.** *Environ Toxicol Chem* 2000, **19**:528–534.

42. Brown C, Gardner C, Braithwaite VA: **Differential stress responses in fish from areas of high- and low-predation pressure.** *J Comp Physiol B* 2005, **175**:305–312.

43. Hollert H, Keiter S, König N, Rudolf M, Ulrich M, Braunbeck T: **A new sediment contact assay to assess particle-bound pollutants using zebrafish (*Danio rerio*) embryos.** *J Soils Sediments* 2003, **3**:197–207.

44. Hollert H, Dürr M, Olsman H, Halldin K, van Bavel B, Brack W, Tysklind M, Engwall M, Braunbeck T: **Biological and chemical determination of dioxin-like compounds in sediments by means of a sediment triad approach in the catchment area of the River Neckar.** *Ecotoxicology* 2002, **11**:323–336.

45. Porter CM, Janz DM: **Treated municipal sewage discharge affects multiple levels of biological organization in fish.** *Ecotoxicol Environ Saf* 2003, **54**:199–206.

46. Eckwert H, Alberti G, Kohler H-R: **The induction of stress proteins (hsp) in *Oniscus asellus* (Isopoda) as a molecular marker of multiple heavy metal exposure: I. Principles and toxicological assessment.** *Ecotoxicology* 1997, **6**:249–262.

47. Rajeshkumar S, Mini J, Munuswamy N: **Effects of heavy metals on antioxidants and expression of HSP70 in different tissues of milk fish (*Chanos chanos*) of Kaattuppalli Island, Chennai, India.** *Ecotoxicol Environ Saf* 2013, **98**:8–18.

48. Mayon N, Bertrand A, Leroy D, Malbrouck C, Mandiki SNM, Silvestre F, Goffart A, Thomé J-P, Kestemont P: **Multiscale approach of fish responses to different types of environmental contaminations: a case study.** *Sci Total Environ* 2006, **367**:715–731.

49. Weber LP, Diamond SL, Bandiera SM, Janz DM: **Expression of HSP70 and CYP1A protein in ovary and liver of juvenile rainbow trout exposed to β-naphthoflavone.** *Comp Biochem Physiol C* 2002, **131**:387–394.

50. Triebskorn R, Blaha L, Engesser B, Güde H, Hetzenauer H, Henneberg A, Köhler H-R, Krais S, Maier D, Peschke K, Thellmann P, Vogel H-J, Kuch B, Oehlmann J, Rault M, Suchail S, Rey P, Rischter D, Sacher F, Weyhmüller M, Wurm K: **SchussenAktiv – Eine Modellstudie zur Effizienz der Reduktion der Gehalte an anthropogenen Spurenstoffen durch Aktivkohle in Kläranlagen.** *Korrespondenz Wasserwirtschaft* 2013, **8**:427–437.

51. European Union: **Directive 2010/63/EU of the European Parliament and of the Council of 22 September 2010 on the protection of animals used for scientific purposes.** *Off J Eur Union* 2010, **276**:33–77.

52. EBT – Entsorgungsbetriebe Tübingen Anerkennung Dienstleistung 2004. http://www.um.baden-wuerttemberg.de/servlet/is/11336/.

53. **Wieder Fische mit zu hohem PCB-Wert in der Ammer.** http://www.tagblatt.de/Home/nachrichten/kreis-tuebingen/ammerbuch_artikel,-Wieder-Fische-mit-zu-hohem-PCB-Wert-in-der-Ammer-_arid,98441.html.

54. Landesanstalt für Umwelt Messungen und Naturschutz Baden-Württemberg (LUBW): **Umweltdaten 2012 Baden-Württemberg.** In *LUBW Landesanstalt für Umwelt, Messungen und Naturschutz Baden-Württemberg (ed) Referat 21 – Nachhaltigkeit, Ressourcenschonung.* Stuttgart: ABT Print und Medien GmbH Weinheim; 2012:83.

55. ISO: **Water quality - determination of the acute lethal toxicity of substances to a freshwater fish [*Brachydanio rerio* Hamilton-Buchanan (Teleostei, Cyprinidae)] ISO 7346/3.** 1996. http://www.iso.org/iso/catalogue_detail.htm?csnumber=14030.

56. Bradford MM: **A rapid and sensitive method for the quantitation of microgram quantities of protein utilizing the principle of protein-dye binding.** *Anal Biochem* 1976, **72**:248–254.

57. Smith T, Smithh B: **Survival analysis using Cox proportional hazards modeling for single and multiple event time data.** In *SAS Conference Proceedings.* Seattle: SAS Users Group International; 2003:245–228. 30 March to 2 April 2003.

58. Newman MC, McCloskey JT: **Time-to-event analyses of ecotoxicity data.** *Ecotoxicology* 1996, **5**:187–196.

Permissions

All chapters in this book were first published in ESE, by Springer; hereby published with permission under the Creative Commons Attribution License or equivalent. Every chapter published in this book has been scrutinized by our experts. Their significance has been extensively debated. The topics covered herein carry significant findings which will fuel the growth of the discipline. They may even be implemented as practical applications or may be referred to as a beginning point for another development.

The contributors of this book come from diverse backgrounds, making this book a truly international effort. This book will bring forth new frontiers with its revolutionizing research information and detailed analysis of the nascent developments around the world.

We would like to thank all the contributing authors for lending their expertise to make the book truly unique. They have played a crucial role in the development of this book. Without their invaluable contributions this book wouldn't have been possible. They have made vital efforts to compile up to date information on the varied aspects of this subject to make this book a valuable addition to the collection of many professionals and students.

This book was conceptualized with the vision of imparting up-to-date information and advanced data in this field. To ensure the same, a matchless editorial board was set up. Every individual on the board went through rigorous rounds of assessment to prove their worth. After which they invested a large part of their time researching and compiling the most relevant data for our readers.

The editorial board has been involved in producing this book since its inception. They have spent rigorous hours researching and exploring the diverse topics which have resulted in the successful publishing of this book. They have passed on their knowledge of decades through this book. To expedite this challenging task, the publisher supported the team at every step. A small team of assistant editors was also appointed to further simplify the editing procedure and attain best results for the readers.

Apart from the editorial board, the designing team has also invested a significant amount of their time in understanding the subject and creating the most relevant covers. They scrutinized every image to scout for the most suitable representation of the subject and create an appropriate cover for the book.

The publishing team has been an ardent support to the editorial, designing and production team. Their endless efforts to recruit the best for this project, has resulted in the accomplishment of this book. They are a veteran in the field of academics and their pool of knowledge is as vast as their experience in printing. Their expertise and guidance has proved useful at every step. Their uncompromising quality standards have made this book an exceptional effort. Their encouragement from time to time has been an inspiration for everyone.

The publisher and the editorial board hope that this book will prove to be a valuable piece of knowledge for researchers, students, practitioners and scholars across the globe.

List of Contributors

Ove Bergersen
Norwegian Institute for Agricultural and Environmental Research (Bioforsk), Soil and Environment Division, Fredrik A Dahls vei 20, N-1432 Ås, Norway

Kine Østnes Hanssen and Terje Vasskog
Department of Pharmacy, Faculty of Health Sciences, University of Tromsø, N-9037 Tromsø, Norway

Terje Vasskog
Norut (Northern Research Institute), N-9294 Tromsø, Norway

Holger Schmidt, Moritz Thom, Silke Wieprecht and Sabine Ulrike Gerbersdorf
Institute for Modelling Hydraulic and Environmental Systems, University Stuttgart, Pfaffenwaldring 61, 70569 Stuttgart, Germany

Kerstin Matthies and Ursula Obst
Institute of Functional Interfaces, Karlsruhe Institute of Technology (KIT), Hermann-von-Helmholtz-Platz 1, 76344 Eggenstein-Leopoldshafen, Germany

Sebastian Behrens
Geomicrobiology/ Microbial Ecology, Centre for Applied Geosciences (ZAG), Eberhard-Karls-University Tübingen, Sigwartstrasse 10, 72076 Tübingen, Germany

Andreas Schoenborn
Institute of Natural Resource Sciences, Zurich University of Applied Sciences, P.O. Box CH-8200, Waedenswil, Switzerland

Petra Kunz
Ecotox Centre Eawag/EPFL, P.O. Box 611, CH-8600 Duebendorf, Switzerland

Margie Koster
Kanton Thurgau, Amt für Umwelt, Abteilung Gewaesserqualitaet und -nutzung, Bahnhofstrasse 55, CH-8510 Frauenfeld, Switzerland

Ali Kamran and Muhammad Nawaz Chaudhry
College of Earth and Environmental Sciences, University of the Punjab, Lahore, Pakistan

Syeda Adila Batool
Department of Space Science, University of the Punjab, Lahore, Pakistan

Carolin Floeter and Susanne Heise
Hamburg University of Applied Sciences (HAW-Hamburg), Ulmenliet 20, 21033 Hamburg, Germany

Ulrich Förstner
University of Technology Hamburg-Harburg, Insti- tute of Environmental Technology and Energy Economics, 21071 Hamburg, Germany

Monika Hammers-Wirtz, Henner Hollert, Martina Roß-Nickoll and Andreas Schäffer
Gaiac-Research Institute for Ecosystem Analysis and Assessment, Kackertstraße 10, 52072 Aachen, Germany

Fred Heimbach
Tier3 solutions GmbH, Kolberger Str. 61-63, 51381 Leverkusen, Germany

Sebastian Hoess
Ecossa, Giselastr. 6, 82319 Starnberg, Germany

Henner Hollert, Martina Roß-Nickoll and Thomas-Benjamin Seiler
RWTH Aachen University, Institute for Environmental Research (Biology V), Worringerweg 1, 52074 Aachen, Germany

Udo Noack
DR.U.NOACK-LABORA- TORIEN, Käthe-Paulus-Str. 1, 31157 Sarstedt, Germany

Stephan Solloch
Ecolab, Monheim, Germany

Nathan Pechacek, Bridget Peterson and Magdalena Osorio
Ecolab, St. Paul, MN, USA

Jeffrey Caudill
Ecolab, Naperville, IL, USA

S. Hoppe, M. Sundbom, H. Borg and M. Breitholtz
Department of Environmental Science and Analytical Chemistry (ACES), Stockholm University, 106 91 Stockholm, Sweden

Nikolai Svoboda, Maximilian Strer and Johannes Hufnagel
Institute of Land Use Systems, Leibniz Centre for Agricultural Landscape Research, Eberswalder Straße 84, 15374 Müncheberg, Germany

Anja Henneberg and Rita Triebskorn
Animal Physiological Ecology, University of Tübingen, Auf der Morgenstelle 5, 72076 Tübingen, Germany

John P Giesy
Department of Veterinary Biomedical Sciences and Toxicology Centre, University of Saskatchewan, Saskatoon, SK S7B 5B3, Canada

Keith R Solomon
Centre for Toxicology, School of Environmental Sciences,
University of Guelph, Guelph, ON N1G 2 W1, Canada

Don Mackay
Centre for Environmental Modelling and Chemistry,
Trent University, Peterborough, ON K9J 7B8, Canada

Julie Anderson
Stantec, 603-386 Broadway Ave, Winnipeg, MB R3C 3R6,
Canada

**Barbara Guhl, Franz-Josef Stürenberg and
Gerhard Santora**
North Rhine-Westphalian State Agency for Nature,
Environment and Consumer Protection, Leibnizstr 10,
45659 Recklinghausen, Germany

Alyson Rogério Ribeiro and Gisela de Aragão Umbuzeiro
School of Technology, State University of Campinas -
UNICAMP, Paschoal Marmo Street 1888, Limeira, SP
13484-332, Brazil

Jan Riediger, Broder Breckling and Winfried Schröder
Chair of Landscape Ecology, University of Vechta,
Driverstraße 22, PO Box 15 53, Vechta 49377, Germany

Robert S Nuske
Northwest German Forest Research Station, Grätzelstraße
2, Göttingen 37079, Germany

Dana Kühnel
Department of Bioanalytical Ecotoxicology, Helmholtz
Centre for Environmental Research - UFZ, 04318 Leipzig,
Germany

Clarissa Marquardt and Katja Nau
Institute for Applied Computer Science, Karlsruhe Institute
of Technology (KIT), 76344 Eggenstein-Leopoldshafen,
Germany

Harald F Krug
Research Focus Area Health and Performance,
Empa - Swiss Federal Laboratories for Materials Science
and Technology, 9014 St. Gallen, Switzerland

Björn Mathes and Christoph Steinbach
Society for Chemical Engineering and Biotechnology
(DECHEMA), 60486 Frankfurt am Main, Germany

Martin Wagner and Christian Scherer
Department of Aquatic Ecotoxicology, Goethe University
Frankfurt am Main, Max-von-Laue-Str. 13, Frankfurt
60438, Germany

Diana Alvarez-Muñoz and Sara Rodriguez-Mozaz
Catalan Institute for Water Research (ICRA), Girona
17003, Spain

**Nicole Brennholt, Sebastian Buchinger and Georg
Reifferscheid**
Department Biochemistry and Ecotoxicology, Federal
Institute of Hydrology, Koblenz 56002, Germany

Xavier Bourrain
Service Etat des Eaux Evaluation Ecologique, Agence de
l'Eau Loire-Bretagne, Ploufragan 22440, France

Elke Fries
Water, Environment and Eco-technologies Division,
Bureau de Recherches Géologiques et Minières (BRGM),
Orléans 45100, France

Cécile Grosbois
GéoHydrosystèmes Continentaux (GéHCO), Université
Francois Rabelais de Tours, Tours 37000, France

Jörg Klasmeier
Institute of Environmental Systems Research, Universität
Osnabrück, Osnabrück 49074, Germany

Teresa Marti
Investigación y Proyectos Medio Ambiente S.L. (IPROMA),
Castellón de la Plana 12005, Spain

Ralph Urbatzka
Interdisciplinary Centre of Marine and Environmental
Research (CIIMAR), Porto 4050-123, Portugal

A Dick Vethaak
Unit Marine and Coastal Systems, Deltares and Institute
for Environmental Studies, VU University Amsterdam,
Amsterdam 1081, The Netherlands

Margrethe Winther-Nielsen
Environment and Toxicology, DHI, Hørsholm 2970,
Denmark

Michael Faust and Thomas Backhaus
Faust & Backhaus Environmental Consulting,
Fahrenheitstr. 1, Bremen 28359, Germany

Carolina Vogs, Stefanie Rotter and Rolf Altenburger
Department Bioanalytical Ecotoxicology, UFZ - Helmholtz
Centre for Environmental Research, Permoser Str. 15,
Leipzig 04318, Germany

Janina Wöltjen and Andreas Höllrigl-Rosta
Federal Environment Agency, Wörlitzer Platz 1, Dessau-
Roßlau 06844, Germany

Lillemor K Gustavsson
Karlskoga Environment and Energy Company, Box 42,
Karlskoga 69121, Sweden

Sebastian Heger, Henner Hollert and Steffen H Keiter
Department of Ecosystem Analysis, Institute for
Environmental Research, RWTH Aachen University,
Aachen 52062, Germany

Jörgen Ejlertsson
Scandinavian Biogas Fuels AB, Holländaregatan 21A, Stockholm 11160, Sweden

Veronica Ribé
School of Sustainable Development of Society and Technology, Mälardalen University, Västerås 72123, Sweden

Krisztina Vincze, Katharina Graf, Volker Scheil, Heinz-R Köhler and Rita Triebskorn
Animal Physiological Ecology, Institute for Evolution and Ecology, University of Tübingen, Konrad-Adenauer-Strasse 20, Tübingen 72072, Germany

Printed in the USA
CPSIA information can be obtained
at www.ICGtesting.com
JSHW051441221024
72173JS00006B/1547

9 781632 396273